D1265541

The Structure of Urban Systems

As the world's population is increasingly concentrated in urban centres, urban systems analysis has become a critical field of research – one that overlaps the traditional academic territories of geography, economics, and regional science. John Marshall defines urban systems analysis as a study of the spatial organization of networks of urban centres at regional, national, and international scales. In this introduction to the subject he presents a framework for its study.

Marshall maintains that the study of the structure and development of urban systems should be guided by a principle-based framework in which the themes of location, economic functions, and population size are in the foreground. He outlines how urban systems analysis seeks to provide 'insight into the roles performed by urban centres as elements of the grand process of settlement and development of the earth by humankind.'

JOHN MARSHALL teaches in the Department of Geography, York University.

JOHN U. MARSHALL

THE STRUCTURE OF URBAN SYSTEMS

UNIVERSITY OF TORONTO PRESS
Toronto Buffalo London

Toronto Buffalo London
Printed in Canada

ISBN 0-8020-5756-X (cloth)
ISBN 0-8020-6735-2 (paper)

Printed on acid-free paper

Canadian Cataloguing in Publication Data

Marshall, John U. (John Urquhart)
The structure of urban systems

Includes index.
ISBN 0-8020-5756-X (bound) ISBN 0-8020-6735-2 (pbk.)

1. Cities and towns. 2. Urbanization. 3. Urban economics. I. Title.

GF125.M37 1989 307'.3 C89-094549-7

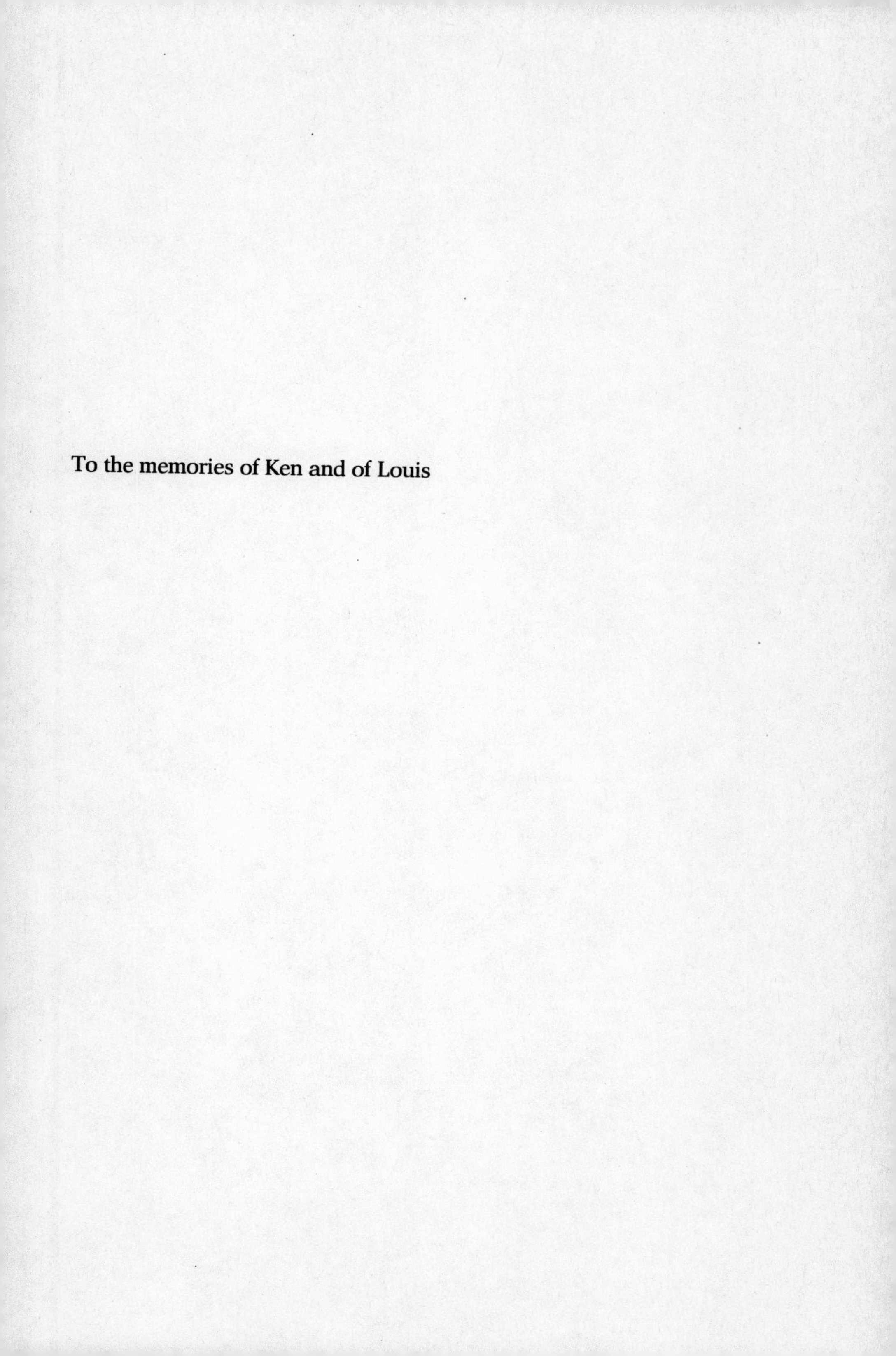

To the memories of Ken and of Louis

Contents

Preface

This book presents a survey of urban systems analysis, a field of research that overlaps the traditional academic territories of geography, economics and regional science and may be defined as the study of the spatial organization of networks of urban centers at regional, national, and international scales. The central themes of urban systems analysis are location, economic functions, and population size. The goal of the field is to provide insight into the roles performed by urban centers as elements of the grand process of the settlement and development of the earth by humankind.

Chapters devoted to aspects of urban systems analysis can be found in most general textbooks of urban geography and urban economics. These texts, however, tend to emphasize the internal characteristics of cities, and their discussions of the subject-matter of urban systems analysis are often rather perfunctory. There remains a need for a dedicated treatment that is sufficiently comprehensive to be of value not only in the classroom but also as a critical guide for those undertaking research. This book has been written to fill that need.

Although I have attempted to be comprehensive, the book is not structured as a catalog of what Thomas Kuhn would call the 'exemplary past achievements' of the field (Kuhn 1970, 175). Instead of reviewing the works of previous writers individually, I have approached each topic with the goal of providing a synthesis, blending the best and most interesting features of earlier studies in such a way that the topic appears in its most highly developed form. Included, of course, are numerous references to the works of specific authors; but the emphasis lies less on individual contributions than on the conceptual structure of the field as a whole. My aim has been to celebrate the beauty of the forest, not of the individual trees.

Throughout the book, the level of discussion is introductory. Not much use is made of statistical methods, and none at all of multivariate procedures such as multiple regression and factor analysis. These procedures do have a role to play in urban systems research but seem to me out of place in an introductory treatment. However, the book does include descriptions of certain techniques of measurement – for example, the Gini index of concentration and the Davies index of centrality. Complete descriptions of these techniques are in short supply, and a full appreciation of the meaning of 'results' – in any field of inquiry – requires that one understand the nature of the measurements on which these results are based. Our knowledge, in short, is inseparable from the manner of our knowing.

Two further preliminary points need to be mentioned. First, the book is not prescriptive. My objective is to understand, not to pass judgment or to offer normative opinions. Second, I employ both historical and ahistorical (or, as some would say, 'theoretical') approaches in the search for understanding. Geographers have tended to 'take sides' on this issue in recent years – they are not the first to have done so – and I believe that neither approach is inherently superior to the other. 'Process' and 'pattern' explanations differ not in intrinsic value but only in the kinds of questions they are well suited to address.

Parts of this book were written during 1983–4, when I held a York University Faculty of Arts Research Fellowship, and I express my gratitude to the Faculty of Arts Committee on Research, Grants and Scholarships for this generous support. I also thank Sandy Crosbie, Lyn Collins, and Ged Martin of the University of Edinburgh for inviting me to spend three months in 1985 as visiting professor in the Centre of Canadian Studies at that institution. The environments of Edinburgh – both natural and cultural – provided me with just the right setting in which to rework several troublesome passages.

The illustrations were drawn in the Cartography Office of the Department of Geography, York University. I thank Hania Guzewska, Carolyn Gondor, and Carol Randall of that office, as much for their unfailing cheerfulness as for their painstaking draftmanship. The editorial skills of John Parry are also greatly appreciated.

Finally, with the intention that it be far more than a ritual observance, I thank my wife and my children, whose support never wavered and who always knew that the finished book would be worth the effort.

THE STRUCTURE OF URBAN SYSTEMS

1

Fundamental concepts

The nature of an urban system

To put the matter simply, an urban system is the set of urban centers above some convenient threshold of population size within a specified region. This short definition, however, contains a number of vague terms to which we must immediately give attention.

THE MEANING OF 'URBAN'

In its most general sense, the term *urban* identifies any nucleated human settlement whose inhabitants are supported chiefly by non-agricultural pursuits. This definition, it should be noted, embodies considerations of both form and function. With regard to morphology, there is an implicit contrast between nucleation and dispersal. With respect to function, farming is contrasted with other kinds of economic activity. In order that a particular settlement may qualify as an urban place, the morphological and functional requirements must both be fulfilled. Mere nucleation of population is not enough. Farm villages, which are nucleated in form but whose inhabitants live mainly or exclusively by farming the surrounding lands, are therefore not normally regarded as being urban in character. Such 'agro-towns' are less common today than in the past, although they can still be found in many underdeveloped countries and even in such relatively advanced regions as eastern Europe, where they may attain a population of several thousand persons.

In North America we have grown accustomed to thinking of agricultural settlement as being dispersed, with each farm family residing on its own parcel of land at some little distance from neighboring families. Dispersion has certainly been the norm throughout the period of European settlement on this continent, but examples of nucleated farm villages are not unknown. Many of the early Puritan settlements of New England were of this type, as were the much later Mennonite villages of Manitoba. It should not be forgotten, also, that the farm village was the dominant form of settlement of the sedentary corn-growing Indian tribes of much of the wooded eastern half of North America in the days before European colonization began.

In general, then, an urban place is any nucleated settlement with a predominantly nonfarm economy. But when we come to examine the term *urban* as it is actually used in research, we find that in most cases a further element is added to the definition: namely, a lower limit of population size. Such a limit is always arbitrary and is justified by practical rather than theoretical considerations. For example, the adoption of a fixed limit in a national census makes possible valid comparison of the proportions of the total population that may be labeled 'urban' in different regions of the same country or in a single region at different points in time. Unfortunately, international comparisons are hampered by the fact that various countries use different size limits; in the United States, for instance, an urban place must have at least 2,500 inhabitants, whereas the corresponding figure in Canada is only 1,000 persons. When several countries are being studied simultaneously these differences can be quite troublesome, although they can usually be overcome (with a little effort) by adopting a fairly high figure and applying it uniformly to all countries, retabulating the population statistics in order to eliminate smaller places as required. It is sometimes appropriate, in fact, to use a rather high threshold of size for 'urban' places, even when a study is set within a single country. For example, finely disaggregated data on the composition of the labor force are not readily available in North America for cities of less than about 50,000 inhabitants, and detailed studies of employment structure are therefore compelled to accept 50,000 as the effective lower size limit of urban status. Another sort of limitation arises in certain studies concerned with the relative sizes of the largest cities in a number of different areas. Here the focus of attention is on a fixed *number* of cities – say, the largest twenty cities in each area – with the result that the effective definition of 'urban' in terms of population size may be not only quite large but also somewhat variable from area to area.

In all these cases, however, it should be understood that the adoption of a specific lower size limit in a given study is the outcome of practical considerations; it is not a repudiation of the general conception of 'urban' as a term referring to any nucleated nonfarm settlement, regardless of size. The importance of the general concept is well seen, in fact, in many studies of local market towns (termed 'central places' by the urban systems analyst), which restrict the study area enough to permit detailed fieldwork and usually include even the smallest hamlets as full members of the regional urban system. To sum up, our phrase 'above some convenient threshold of population size' means simply that we should use the smallest population figure that is reasonable in the light of availability of data, study objectives, and time and other resources at our disposal.

THE GEOGRAPHIC CITY

We must also emphasize the fact that the urban systems analyst is concerned invariably with the 'geographic city' or 'urban agglomeration' rather than the city as a legal entity. The distinction is appreciated most easily by considering the political structure of a representative metropolis – Vancouver. From the point of view of political boundaries, the legal entity known as the City of Vancouver is an area of some forty-four square miles (114 square kilometers) containing, at the time of the Canadian census of 1981, some 414,000 inhabitants. The geographic city of Vancouver, however, is considerably larger in both land area and population. It includes such additional political units as the cities of New Westminster, North Vancouver, and Port Coquitlam, together with the district municipalities of Burnaby, Richmond, and Surrey; this list, by no means complete, demonstrates that the geographic city normally extends well beyond the political limits of the dominant legal unit from which the entire agglomeration takes its common name. In essence, the geographic city is the continuous built-up area covered by the central city together with its suburbs and any formerly separate towns that have been engulfed by the expanding suburban ring. In the case of a nation's leading cities, the concept of the geographic city is nowadays officially recognized by the creation of large statistical units known generally as metropolitan areas: the official designation in the United States is 'Metropolitan Statistical Area' (abbreviated to MSA), and in Canada, 'Census Metropolitan Area' (CMA). The Vancouver CMA covers 1,076 square miles (2,787 square kilometers) and contains 1.3 million inhabitants. Thus, in Vancouver's case, the legal city accounts for only 33 per cent of the total population of the geographic city (i.e. the CMA) and for a mere 4 per cent of the latter's total surface area. These percentages, it should be noted, can vary widely from city to city.

The basic justification for using the whole urban agglomeration rather than merely the legal city lies in the fact that the former provides a much better approximation of the limits of the local labor market. Many people in a modern metropolis work outside the political unit in which they reside. In particular, many persons who work in the central city reside in the surrounding suburban municipalities. Moreover, there is a distinct tendency for the suburban zone to be the place of residence of a disproportionately large share of the professional and managerial classes in the city's labor force. It follows that exclusion of the suburban areas could lead to an incomplete and distorted picture of both the city's size and its industrial structure, and hence to a mistaken idea of the city's role as an element of the nation's or region's economic life. The concept of the geographic city is an attempt to circumvent this difficulty by identifying an area that may be regarded as a meaningful functional whole in economic terms. Throughout this book, unless explicitly stated to the contrary, it is the geographic city, not the legal city, that is the focus of attention.

Although the concept of the geographic city is undeniably valuable, there are cases in which it is difficult to decide exactly where the geographic limits of a city should be drawn. Commuters flow into the central city not only from contiguous suburban areas but also from satellite communities that are spatially separate from the main built-up area. How should these satellite centers be treated? Should they be left as cities in their own right (assuming that they exceed the lower limit of population adopted in the study in question), or should they be included as part of the nearby metropolis to which they are economically tied? For example, should Hudson, Ohio – a town with several distinctive historical and economic characteristics – be treated as a discrete entity or regarded simply as part of nearby Akron? Should La Baie, Quebec, be lumped in with Chicoutimi? In these examples the towns in each pair are very different in size, a factor that may make the 'loss' of the smaller center easier to accept. But in other cases the differences in size are not great. Should Scranton and Wilkes-Barre, Pennsylvania, be combined or left separate? How about Winston-Salem, Greensboro, and High Point in North Carolina, or Guelph, Cambridge, and Kitchener-Waterloo in Ontario? Problems of this sort are by no means confined to large cities; the question of whether to combine two or more places arises for scores of small centers. Sometimes the answer is predetermined by the nature of the available statistics. Census data for MSAS, for example, are compiled on the basis of whole counties (except in New England, where minor civil divisions are used), and hence the separation of two or more physically discrete towns lying within a single county is simply not feasible without access to unpublished and often confidential information in census bureau files. But where no such 'given' answer applies, the decision on how best to combine (or not combine) neighboring places simply has to be left to the judgment and discretion of the individual researcher. Detailed local knowledge will obviously be an asset, especially knowledge of the strength of commuting flows, of social interconnections, and of the sense of identity held by each community. In the final analysis, however, individual judgment is the crucial factor.

Related to the concept of the geographic city is the fact that urban systems analysis is concerned with cities as points rather than as areas. The contrast between these two viewpoints is the most fundamental distinction in urban geography, serving to divide the field into two complementary domains. One of these domains is the study of the internal structure of the city. Here the city is viewed as an area: that is, as a phenomenon having two-dimensional extension and exhibiting internal areal differentiation in terms of land use and population characteristics. The other domain, in which the city is treated as a point, is the subject of this book. Each study in urban systems analysis deals with a relatively large section of the earth's surface, often a whole country and sometimes an area that is international in extent. The scale of study is such that the urban network can be visualized as being essentially a punctiform distribution. The urban systems analyst 'sees' the towns and cities of his or her study area in the manner of an observer in a high-flying aircraft; he or she is interested in such things as their locations, their size, and the parts they play in the region's economy, but not in their internal layout. The student of urban internal structure, in contrast, 'sees' each city in a more literal sense of the word. He or she works at what may be termed the townscape scale, or the scale encompassing those features that would ordinarily be visible to an observer moving through the city by car or on foot. Urban systems analysis might be defined in negative terms as being concerned with those problems in urban geography that do not require for their solution a knowledge of the internal spatial differentiation of cities.

THE LIMITS OF AN URBAN SYSTEM

We have already noted the difficulty that may arise in setting limits to a particular geographic city. A similar problem is encountered in deciding where to place the boundaries of the region whose urban system is to be analyzed. Ultimately, urban systems analysis seeks to give an account of the towns and cities of the entire surface of the globe. For obvious practical reasons, however, few individual studies are more than national or at most continental in scope. The bulk of research, in fact, is subnational in coverage. With regard to the problem of establishing regional boundaries, it is widely agreed that the study area should possess some sort of unity that can be expressed in terms of general forces that may significantly affect the fortunes of the area's urban centers. A whole nation makes a reasonable study area because it can be argued that the urban network is bound to be affected (though not uniformly across space) by governmental policies that are national in scope, such as policies concerning immigration, defense, or industrial location. At the subnational scale, the use of major administrative divisions such as states or provinces can be given an analogous political justification.

Judging by the published literature, however, many writers feel that a single state or province makes a less satisfactory study area than a group of contiguous states or provinces with substantially similar economic histories and common current problems of economic development. Thus, most urban systems analysts would prefer to study the US South, reasonably defined, rather than merely Georgia or to study Canada's prairie provinces as a whole rather than merely Saskatchewan. The idea of treating several adjoining political units as a single region frequently has physiographic overtones, as both of the preceding examples suggest: the unity of the South is partly climatic in origin, and that of the western interior of Canada is largely geomorphological. Indeed, physiographic factors sometimes come to the fore as the most satisfactory basis for regional delimitation. Aquitaine and the Ganges Valley, to mention two examples, would seem to be regions of this kind.

A special type of regional unity, not necessarily unrelated to political and physiographic features, is the unity displayed by the area that forms the contiguous sphere of influence of a prominent city. A city is involved in the life of its surrounding area in many ways. For example, it is a shopping center for consumer goods and services, a place of entertainment, a source of newspapers, magazines, and radio and television broadcasts, and perhaps the destination for a permanent change of residence. The larger the city, the more varied and also the more specialized are the services and the opportunities that it offers, and the larger, normally, is its surrounding sphere of influence. By means of fieldwork and other methods it is possible to establish the boundaries of the area that is oriented toward a particular city more strongly than toward other cities of comparable size and influence. Because it is certain that the development of the smaller urban centers within this area is affected by the fortunes of the dominant city, this kind of region makes a highly satisfactory study area for urban systems research. We shall have more to say about urban spheres of influence in due course. At present the important point is that the delimitation of a study area is not something to be treated lightly or carried out in a haphazard fashion. Researchers should always explicitly seek a sense of regional unity in terms of features that are relevant to the development of the urban network.

WHOLE AND PART: THE MEANING OF 'SYSTEM'

Let us assume that a study area has been delimited, that a minimum population size for urban status has been selected, and that problems related to the geographic limits of the urban places within our area have been satisfactorily solved. We must now note that the region's urban system comprises *all* the towns and cities that meet the size criterion we have established. It is axiomatic in urban systems analysis that no town that satisfies this criterion may be excluded. By definition, an urban system with missing members is not an urban system. In other words, a study in urban systems analysis is not a study of some of an area's towns, selected at random or on a whim; it is always conceived as a study of the complete network of towns.

The reason for this requirement is that the urban systems analyst seeks to compare, explicitly or implicitly, the characteristics of individual towns with the corresponding characteristics of the regional urban system taken as an aggregate. The system as a whole serves as a benchmark, or standard of reference, against which the individual town is evaluated. Examples of this kind of comparison will be found throughout this book. Clearly, comparative statements relating individual towns to the characteristics of the whole system will be something less than accurate if the system's characteristics are determined from incomplete information. To put the matter another way, urban systems analysis is permeated by a functionalist, or 'part-and-whole' method. The whole – the entity whose essence we seek ultimately to comprehend – is the complete urban system. The parts are the individual urban centers, their characteristics being elucidated in terms of their roles in sustaining the identity of the larger systemic totality to which they belong.

This functionalist conception, moreover, accounts for the use of the term *system* in urban systems research. This word serves to remind us that a town is not an isolated entity leading a self-contained life devoid of relationships with the world beyond its own borders. Towns and cities are interdependent: any one town's size, economic character, and prospects for growth are affected by the nature and strength of its interconnections with other towns. The events that occur in the development of an individual town cannot be understood fully without consideration of trends affecting the entire urban system. As in other contexts, the term *system* draws attention to the importance of the interdependencies that link the various parts together to form a larger whole. The habit of thinking constantly on the systemic level is perhaps the most important prerequisite for work in urban systems research. Recognition of the importance of this perspective is certainly fundamental to an understanding of the literature in the field.

However, the use of the word 'system' should not be taken to imply that the set of towns in a particular study is closed to outside, or exogenous influences. Obviously the only urban network that is truly a closed system is that of the entire planet. (The words 'network' and 'system' are treated as synonyms in urban systems research.) The analyst normally assumes, however, that the towns and cities within any circumscribed region have closer ties with one another, on the average, than with towns and cities elsewhere. Generally speaking, the larger the region studied, the more self-contained its urban network is likely to be. Nevertheless, just as the individual town is viewed as an element in a regional urban system, so the regional network of towns must be conceived as an element in some larger system which, in the limit, is world-wide in scope.

Finally, and closely related to the previous point, the urban systems analyst recognizes that a particular network's systemic properties – that is, its characteristics as a whole – are dependent on the location of the boundary that separates the region of study from the rest of the world. For example, a town may exhibit a high rate of population growth in relation to other towns in the same state; but if this happens in a slow-growth state, the same town may be only an average performer by the standard of the urban system of the whole nation. Or again, the proportion of a city's labor force employed in a particular economic activity (for instance, manufacturing) may be greater than that in other cities in a limited region, but quite unremarkable within a more extensive urban system. This sort of relativity is unavoidable, but it is not a cause for alarm so long as careful attention is paid to the drawing of regional boundaries, especially when it is desired to compare the properties of two or more different urban systems.

Now that the concept of an urban system has been described in some detail, we may close this section by indicating which features of such systems are regarded as important objects of study. The literature of urban systems analysis, though outwardly diverse, shows persistent concern with three main aspects of towns: first, their population size, which leads naturally to consideration of rates and causes of growth; second, the pattern of location of towns across the earth's surface; and third, the economic functions that towns perform. These three basic concerns are so pervasive that it is possible to formulate a broad definition of the field itself in terms of them, as follows: urban systems analysis seeks to provide an understanding of the sizes, spatial arrangement, and economic activities of the towns and cities above a specified minimum size within a defined area. In pursuing these interrelated goals, urban systems analysis contributes to its parent field of human geography by enhancing our appreciation of the variable character of humanity's imprint upon the earth.

General features of spatial organization

Every discipline has a core of fundamental ideas concerning the manner in which its subject-matter behaves. These central ideas, molded over long periods of time by the interplay between theory and observation, provide a conceptual framework for the field and serve as guidelines for the furtherance of research. The remainder of this chapter introduces a group of interrelated concepts of spatial organization that lie at the heart of urban systems analysis. The concepts are not difficult to grasp, but their simplicity must not be allowed to obscure their importance. Combined with one another in various ways, they serve as the foundation for much of the literature in the field.

POINT, LINE, AND AREA FACTORS

The factors that explain the locations of cities fall naturally into categories defined by the basic geometrical concepts of point, line, and area. Factors that operate at point locations are termed site factors. Factors that are linear or areal in their influence are known as factors of situation. The term *regional setting* is sometimes used in place of *situation*.

Points

Out in the field, it is usually site factors rather than situational factors that are readily apparent to an observer. This is essentially a matter of scale. The advantages of a particular location in relation to its regional setting are not normally visible at ground level, whereas the importance of a factor that is present at the site itself is often easily perceived. Mining towns, for example, are obviously sited at points of access to economically valuable minerals. Similarly, most resort centers owe their origins to the presence of fixed natural resources such as good ski slopes, sandy beaches, or hot springs.

In the past, town sites were often selected with military considerations in mind. One cannot visit Bern, with its ancient nucleus moated on three sides by a meander of the Aare River, or Quebec, perched on its rocky promontory, without being impressed by the defensive potential of their sites. Site advantages are equally striking, though in the context of trade rather than defense, in the case of great natural harbors such as those of San Francisco, Halifax, and Rio de Janeiro. On a humbler level, a high proportion of towns of all sizes throughout the eastern United States and Canada occupy waterpower sites along the area's numerous streams, a fact that reflects the original importance of these centers as gristmill and sawmill locations during the early period of European settlement.

The site alone, however, can never provide a complete explanation for the location of a town. After all, towns are not found in every meander loop, on every hilltop, or at every site with waterpower potential. Even mining towns and resorts require reasonable proximity to their markets; sun-drenched beaches in remote and thinly populated corners of the globe have yet to spawn towns. The full explanation of urban locations involves a consideration of the linear and areal factors that relate each town to the total pattern of human settlement.

Lines

The linear element in the spatial organization of an urban system arises from the fact that towns are nodes on transportation networks. Trade and manufacturing – the two most important functions that cities perform – involve the continual movement of raw materials and finished products from place to place. Cities are not only the points at which raw materials are assembled and processed, but also the centers where most of the finished products are consumed. In addition, cities provide the maintenance and repair services that the transportation facilities themselves require for their continued efficient operation. Every city, no matter how small or remote, is connected to the global system of transport routes in some fashion, and a city's location can always be interpreted partly in terms of the linear movements of people and products along these routes.

Cities are related to transport routes in a number of ways. Before the advent of the railroad, movement was almost invariably cheaper and easier by water than by land, and so navigable rivers became the chief arteries of commerce. From Memphis on the Nile to Memphis, Tennessee, river ports have always been prominent among the world's major cities. A river port may have one or more of several important features. For example, it may lie at the so-called head of navigation, the upstream limit of easily navigable water. This is the case at Basel on the Rhine, at Minneapolis, and at the famous Fall Line towns of the Atlantic coastal plain, including Richmond, Raleigh, and Augusta. Towns situated at heads of navigation perform break-of-bulk and transshipment functions in connection with the change from water to land transportation and vice versa. These functions have also supported the growth of towns where major rivers are crossed by important overland routes. Examples include Duisburg, Hannover, Leipzig, and other cities at points where the principal north-flowing German rivers are intersected by the line of the Hellweg, the great east-west medieval trade route which followed the northern edge of the Hercynian Uplands. American examples are provided by Zanesville and Chillicothe, situated where Zane's Trace, the earliest improved road across southeastern Ohio, crossed the Muskingum and Scioto rivers, respectively. Another type of river settlement is the confluence town, situated where two navigable streams join together. An oft-cited example is Koblenz; its very name – from the Latin *confluens* – reveals the character of its location. Lastly, mention should be made of towns situated at points where rivers can conveniently be bridged. London and Paris are good examples of this type.

The coming of the railroad led to a substantial decline in the relative importance of navigable rivers for transportation. Understandably, however, railroads were generally constructed so as to connect those cities that had already risen to prominence during the pre-railroad era. Thus, although the actual lines of movement were often relocated, the principal nodes at which these routes intersected remained the same. But the relocation of traffic could lead to the stagnation of towns unlucky enough to be bypassed by the railroad lines. Many small general-cargo ports on the Ohio-Mississippi-Missouri river system suffered this fate during the second half of the nineteenth century.

Before the introduction of motor vehicles, travel by road was divided into a series of distinct segments or 'stages' (whence the term *stagecoach*). The end of each stage represented an overnight stop or at least a meal and a change of horses. Towns spaced at appropriate intervals along main roads could derive considerable income from the provision of food, lodging, and other services to travelers. In a few cases towns were founded with this source of support explicitly in mind. For example, when the Canada Land Co. built the Huron Road from Stratford to Goderich, Ontario, in order to attract settlers into the Huron Tract, the towns of Mitchell, Seaforth, and Clinton were laid out at equal intervals along the route to serve both as supply centers for incoming settlers and as staging points for through traffic. More commonly, staging points were selected from among settlements already in existence. In every case, however, traffic along the road had a beneficial economic effect on the line of towns thereby connected. Along the railroads, too, benefits accrued to towns that served as the 'division points' where trains underwent a change of crew. Because of the relatively meager development of alternative sources of urban support west of the Mississippi, the division-point function was especially important in promoting town growth across the American West. The building of the Union Pacific transcontinental line, for example, was the chief stimulus for the early growth of the towns of North Platte, Cheyenne, Ogden, and – on the branch line to Portland – Pocatello.

Areas

The nature of the areal factor in the location of cities is well expressed in the following quotation: 'A fundamental trait of both town and city, in all ages, has been that they serve as institutional centres (commercial, cultural and administrative) for a surrounding territory. It is only in recent times that industry [i.e., modern manufacturing] has become a primary cause of urban growth' (Dickinson 1961, 3). Although the details of the relationship between town and country have changed over time, the city's role as the focus of economic, social, and political life for its surrounding region long antedates the Industrial Revolution and may justifiably be viewed as the role that most nearly expresses the fundamental nature of urbanism. On the commercial level, the historic city was the marketplace where farmers from the surrounding lands could sell surplus foodstuffs and buy the products of craft industries as well as specialized commodities such as salt and spices. In modern cities the details of this commerce are highly complex, yet the essential function of exchange remains the same, with the city acting as a collection center for regional surpluses and as a distribution center for a wide variety of consumer goods and services. Throughout history the city has also been the focus of cultural activities of a largely noncommercial character. The city is the home of the great cathedrals, art galleries, museums, theaters, libraries, and universities, all of which serve areas extending far beyond the city's own boundaries. Finally, many cities are political centers with legal jurisdiction over administrative territories of varying extent. The commercial, cultural, and administrative functions performed by the city for the larger region create a bond that unites town and countryside in a relationship of mutual dependency. Without cities, rural areas would be deprived of markets and other essential services. Without a rural population to serve, cities would not be called into existence. Town and country dwell together in symbiotic harmony.

The mutual interdependence of town and country has one consequence so obvious that it is easily overlooked: at the global scale, cities are generally confined to areas capable of supporting a permanent agricultural population. Moreover, within any area possessing a broadly uniform level of agricultural productivity, there is a rough but definite association between the density of the rural population and the average spacing of cities above any chosen minimum size. For example, cities of 10,000 or more inhabitants are farther apart, on the average, in sparsely populated Kansas or Saskatchewan than in the more thickly settled rural sections of Ohio or southern Ontario. The presence of large numbers of manufacturing employees, together with their dependents, in certain cities may partly obscure this general relationship, but most of the world's major concentrations of manufacturing activity are fundamentally market-oriented and are therefore found in areas that were relatively densely settled before modern industrial growth began. Hence the occurrence of manufacturing tends to reinforce, rather than counteract, the correlation between rural population density and the spacing of towns.

A few cities can fairly be said to have no rural tributary areas to serve. Mining towns in the northern regions of Canada and the Soviet Union tend to be of this sort, but even in these harsh environments a handful of dairy and vegetable farms may spring up to help feed the urban population. The clearest examples of urban centers with no rural surroundings are isolated oases situated along trade routes in the great Old World desert zone from the Western Sahara to Central Asia. The normal pattern of human settlement, nevertheless, is one in which town growth is primarily a response to the needs of an agricultural population. As Mark Jefferson wrote, 'Cities do not grow of themselves. Countrysides set them up to do tasks that must be performed in central places' (Jefferson 1931, 453). The use of the term *central place* to refer to a town's role as a service center for rural population is frequently thought to have originated in the important work of Christaller (1933). It was Jefferson, however, who introduced this now standard term, though he himself did not pursue the topic of town/country relations to any great depth.

Corridors and central places

To recapitulate, the factors influencing the location of any town include both the characteristics of the actual site (point factors) and the nature of the regional setting (linear and areal factors). The urban systems analyst, though not blind to the significance of point factors for the individual town, is interested in the spatial organization of whole networks of towns at regional and national scales. He or she is therefore concerned primarily with the factors of situation rather than those of site. It is for this reason that the field is sometimes referred to as 'the study of the external relations of cities.'

Two broad conceptualizations of systemic spatial structure have evolved, parallel to the two geometrical categories of situational influences, those of line and those of area.

First, corresponding to the category of linear factors, there is the transportation, or corridor model, in which an urban network is viewed as being essentially a set of nodes strung along transport routes like beads on a string. The heavier the traffic, the larger the beads. This concept has received a particularly lucid exposition in the work of Whebell (1969), but it underlies many other studies (e.g. Conzen 1975; Burghardt 1979) and can be traced back to the mid-nineteenth century (Kohl 1841; see Lukermann 1966, 17–20, for details). Closely allied with the corridor model is the concept of 'gateway' cities, or 'points of entry' (Rose 1966; Burghardt 1971). Many gateway cities, naturally, are seaports, and all derive their significance from control of key locations on long-distance routes.

Second, corresponding to the areal element in city locations, there is the cellular, or central place model, in which towns are viewed as being essentially the nuclei for surrounding tracts of rural territory. The emphasis is placed on areal coverage rather than linear movement. The crucial element of the spatial structure is taken to be the fact that towns are uniformly dispersed throughout the settled area, not the fact that they are strung out in linear fashion along traffic arteries. This model, like the corridor concept, can also be traced well back in time (Galpin 1915), but its modern development is associated primarily with the names of Christaller (1933) and Berry and Garrison (1958a, 1958b). Unlike the corridor concept, that of central places has given rise to a fairly substantial body of formal theory, described in detail in chapter 5, below.

The corridor and central place models should be regarded as complementary, not mutually exclusive. Each has had its champions, but experience has shown that neither model on its own provides a satisfactory account of the complete spatial distribution of towns over a wide area. If exclusive reliance is placed on the corridor concept, little or nothing is said about towns that do not happen to lie on the principal transport routes. Conversely, undue infatuation with central place principles can distract attention from significant linearities that may lie embedded within an otherwise uniformly dispersed pattern of towns. In short, as Harris and Ullman (1945) pointed out many years ago, linear and areal influences are at work simultaneously, and both should be taken into account.

HIERARCHICAL STRUCTURING

The concept of hierarchical structuring is a refinement of the more general concept of cities as central places; it recognizes the obvious fact that, broadly speaking, the size of a city's tributary area is directly proportional to the size of the city itself. The essential idea is that urban centers fall into distinct groups, or 'orders,' according to their size and to the particular set of goods and services that they provide for their surrounding areas. The smallest centers – hamlets – provide no more than a handful of basic services: a gas station (formerly a smithy), a general store, a post office (usually housed within the store), and perhaps a diner or tavern. Because these simple services do not attract customers over long distances, hamlets have very small tributary areas, often no more than a mile or two in average radius. Centers of the second order – villages – provide a more extensive array of goods and services and draw their customers from somewhat farther afield. Because villages provide certain services that are not available in the hamlets, hamlets themselves are included within the tributary areas of villages. In other words, village tributary areas are not merely larger than those of hamlets, they include urban places (hamlets) as well as rural territory. In addition, the set of services provided by the villages includes the set of services provided by the hamlets. Each village, in effect, is a hamlet as well as being a second-order center. Thus a village has not one tributary area but two, one for each level at which it operates within the hierarchy. Its hamlet-level tributary area is completely contained within its village-level tributary area. From the point of view of areal coverage, the country is now blanketed by two cellular nets of tributary areas, each of which exhausts the total territory. The fine-meshed first-order net has hamlets (some of which, of course, are also villages) as the nuclei of its cells. The more open, second-order net represents the tributary areas of the villages.

This description – deliberately presented in simple terms at this early stage – can be extended systematically to the higher orders of the urban system. Thus, third-order centers – call them towns – provide all the services found in villages and hamlets together with an additional group of services not found on the village and hamlet levels. Town-level tributary areas are larger than those of villages and include both villages and hamlets as well as rural population. Each third-order center has three concentric tributary areas: one as a hamlet, one as a village, and one as a town. Similarly, fourth-order centers offer a still longer list of goods and services and have four concentric tributary areas. In theory the addition of further layers can continue indefinitely, but in practice the upper limit is about eight or nine orders, and fewer than this in areas containing no really large cities.

Two distinctive features of a fully developed hierarchical system should be noted. First, provided that the region served by the system has a uniform or near-uniform density of rural purchasing power, the various orders of urban centers will exhibit discrete levels of population size. Differences in the sizes of centers within a single order will be small (ideally, zero), whereas differences between adjacent orders will be readily apparent. The size differences between the orders are related to quantum jumps in the sizes of tributary areas. The constraints of geometry (see chapter 5) ensure that the tributary areas on each level are at least three times as large, on the average, as tributary areas on the level immediately below. Broadly speaking, we should therefore expect a similar threefold jump in the sizes of the urban centers themselves.

Second, the centers of any given order are less numerous and farther apart than the centers of the order immediately below. Like the discrete stratification of population size, these characteristics follow logically from the quantum jumps in the sizes of tributary areas. Because higher-order centers have larger tributary areas than lower-order centers, fewer higher-order centers are required in order to serve all parts of a given territory. Moreover, the rough uniformity of the areal extent of tributary areas on any one level implies that higher-order centers are separated by greater distances than lower-order centers. These features are readily observable in the world around us. Normally, the larger the town we start out from, the greater is the distance we have to travel to reach the nearest center of the same or larger size.

In spite of the obvious relevance of the concept of hierarchical structuring for an understanding of the size and spacing of cities, this concept was never intended to provide a comprehensive model of the spatial organization of an urban system. It is a partial concept, applying to cities only in their role as central places. It does not apply to other functions that cities perform, such as mineral extraction, the resort function, and, above all, manufacturing. The presence of these other functions may obscure the more or less regular relationships that would exist between size and spacing in a 'pure' system of central places. In particular, cities containing heavy concentrations of manufacturing activity tend to occur as distinct clusters, the members of each cluster being much closer together than would be the case with pure central places of similar size. This effect can be seen, for example, in the Pittsburgh-Youngstown-Akron-Cleveland district and in Canada's 'Golden Horseshoe' (Toronto–Hamilton–St Catharines–Niagara Falls). Arising out of this superimposition of activities, a persistent theme in urban systems research is the attempt to evaluate the relative contributions of central place activity, manufacturing, and other urban functions to the complete spatial pattern of towns.

Within the category of central place activity itself, it is customary to deal with the political function separately from the commercial function of providing consumer goods and services. The choice of locations for political capitals at local, state, and even national levels is frequently determined in large part by factors other than ease of access to a consumer market. Moreover, the 'natural' commercial tributary areas of cities may not correspond particularly well with political boundaries. As a result, a capital city may not be a member of the highest order of commercial centers within the area over which its political jurisdiction extends. Ottawa, Canberra, and Brasilia, to give three of the best-known examples at the national scale, do not belong to the highest levels of the commercial hierarchies in their respective countries. Many American state capitals fall below the highest commercial order within their states, and analogous examples can be found at the level of county seats. In short, there is often a rather low level of spatial concordance between the political and the commercial hierarchies. For convenience, therefore, the political function is usually set aside or at least dealt with separately in detailed studies of the hierarchical structuring of urban systems.

THE RANK STABILITY OF LEADING CENTERS

Once the initial influx of settlers into a region has come to an end, the largest cities in the regional urban system normally exhibit considerable inertia: their rank order tends to remain fairly stable for long periods, which may easily exceed a century. The usual path of evolution is one in which each city grows significantly in absolute size while its relative importance within the regional network remains constant or changes only slowly.

Two examples – one national and one regional in scope – are shown in Table 1.1. In the case of France, eight of the ten largest cities in 1810 were still among the top ten 165 years later, giving an overlap of 80 per cent. For the top twenty cities, the corresponding overlap is 60 per cent. The main changes over this long period were the decline of certain historic regional centers in the Paris Basin (Orléans, Amiens, Caen, Reims) and the rise of cities specializing in manufacturing (Lens, Valenciennes), tourism (Nice, Cannes), or both (Grenoble). Although these changes are not without interest, the general impression is one of stability, especially for the very largest cities in the system. Moreover, the period 1810–1975 witnessed great changes in the absolute sizes of the cities. The combined population of the top twenty centers rose from 1.6 million to 18.5 million, with Paris alone increasing from 630,000 to 9.9 million. Thus, while the total population of the top twenty French cities grew more than elevenfold, a majority of the cities maintained essentially the same relative importance within the system.

The data for southern Ontario tell a similar story. In combination, the top twenty cities contained 257,000 inhabitants in 1871, compared with 4.7 million a century later. Despite this eighteenfold absolute increase, however, the degree of overlap in the ranking is 70 per cent for both the top ten and the top twenty cities. Relative decline has affected general cargo ports along the St Lawrence–Lake Ontario water route (Port Hope, Belleville, Brockville) together with some inland regional centers (Guelph, Chatham). As in France, there is evidence that the rise of modern manufacturing has been a factor promoting advances in rank (Windsor, Kitchener, Sarnia, Oakville). On the whole, however, dramatic examples of 'rank jumping' are not numerous. So far as the relative importance of

TABLE 1.1
Rank stability of leading cities: two examples

Rank position	Cities of France		Cities of southern Ontario	
	1810	1975	1871	1971
1	Paris	Paris	Toronto	Toronto
2	Lyon	Lyon	Hamilton	Ottawa
3	Bordeaux	Marseille	Ottawa	Hamilton
4	Rouen	Lille	London	London
5	Marseille	Bordeaux	Kingston	Windsor
6	Lille	Toulouse	St Catharines	Kitchener
7	Nantes	Nice	Brantford	St Catharines
8	Strasbourg	Nantes	Belleville	Oshawa
9	Toulouse	Rouen	Guelph	Kingston
10	Orléans	Grenoble	Oshawa	Sarnia
11	Metz	Toulon	Chatham	Niagara Falls
12	Nîmes	Strasbourg	Peterborough	Brantford
13	Amiens	St-Etienne	Niagara Falls	Cambridge
14	Caen	Lens	Windsor	Oakville
15	Toulon	Nancy	Cambridge	Guelph
16	Montpellier	Le Havre	Port Hope	Peterborough
17	Reims	Cannes	Brockville	Cornwall
18	Clermont-Ferrand	Tours	Cobourg	Welland
19	Nancy	Clermont-Ferrand	Kitchener	Brampton
20	Rennes	Valenciennes	Stratford	Markham

SOURCES: Data for France in 1810 from Dupeux (1981); all other information from national censuses of France and Canada for years indicated

cities is concerned, a resident of nineteenth-century Ontario magically transported to the present would encounter essentially familiar ground.

The rank stability of large cities is a reflection of the fact that, broadly speaking, the attractiveness of a city for new investment is directly proportional to its size. Fundamentally, cities are agglomerations of jobs; remove the jobs and the city will soon wither away. By the same token, city growth depends on the addition of new jobs. (Retirement communities are a partial exception but are few in number.) New jobs, in turn, are created by new investments of funds in both the private and public sectors of the economy. In general terms, the volume of new investment in both profit-oriented and nonprofit enterprises is governed by the magnitude of the expected returns. City size, as an indicator of past success, is understandably perceived as a reasonable guide to success in the future, and therefore new job-creating investment is roughly proportional to city size. A large city offers several important advantages to the potential investor, including a large local market, a large and diversified supply of labor, a complex infrastructure of business and community services (law firms, advertising agencies, financial institutions, technical consultants, universities and colleges, medical facilities, and so forth), and an established network of commercial connections with the rest of the nation. The larger the city, the more pronounced are these advantages. In the words of the old proverb, 'nothing succeeds like success.'

The existence of a positive relationship between city size and the volume of new investment per unit of time can account for rank stability once differences in city sizes have become established. However, the question of how cities come to achieve different sizes in the first place presents quite a separate problem. The sorting process by which urban centers become differentiated in size belongs to the earliest years of settlement in each region, whereas the phenomenon of rank stability belongs explicitly to the post-frontier period. Early differentiation according to size is traditionally explained by the doctrine of initial advantage (Pred 1965, 1966; Muller 1976, 1977). The specific initial advantages of particular locations are manifestations of the point, line, and area factors already introduced. The greater the perceived advantages of a location, the larger will be the urban nucleus established there during the first years of settlement. A few false starts may be made owing to incompleteness of information, but these temporary aberrations are soon corrected and the principal features of the urban network become fixed very early in the region's history. Once differences in the sizes of towns are established, these differences themselves become the dominant (though not always the only) controlling factor in subsequent development. In short, it is useful to distinguish a system's origins from its ensuing evolution. An explanation of origins involves an appreciation of significant differences among locations in terms of initial advantages. Subsequent development is largely a matter of inertia. The evolution of urban systems is everywhere guided by the principle of resistance to sudden and dramatic change.

Summary

The first part of this introductory chapter explained the meanings of basic terms such as *urban, geographic city,* and *urban system.* The field of urban systems analysis was defined as the study of the sizes, spatial arrangement, and economic functions of the towns and cities above some specified minimum size within a particular area.

The second part of the chapter discussed three fundamental aspects of the spatial organization of urban systems: the locational significance of point, line, and area factors; the principle of hierarchical structuring; and the rank stability of leading centers. It was noted that the corridor and central place concepts of spatial organization are complementary rather than competitive and that the principle of hierarchical structuring is intended to apply only to the role of cities as central places, not to other functions that cities may perform. The rank stability of large cities was related to the strong tendency for city size to be positively correlated with the volume of new job-creating investment attracted to each city per unit of time.

2

World urbanization: a historical outline

The ancient world

The urban tradition to which the modern cities of the Western world belong originated in the Fertile Crescent, the roughly semicircular area extending from the Levant to the head of the Persian Gulf and including the alluvial lowland watered by the Tigris and Euphrates rivers.

The oldest nucleated settlement for which physical remains have been excavated appears to be Jericho, situated near the point where the River Jordan flows into the Dead Sea. Archeologists have dated the settlement at Jericho as early as 8000 BC, but the evidence suggests that urban functions on this site were extremely rudimentary and that the settlement should probably be regarded as an agricultural village rather than a true town. Moreover, there does not seem to have been a continuous history of nucleated settlements linking Jericho with later centers. Ancient Jericho was evidently quite isolated in both space and time, a precocious harbinger of things to come.

For the beginnings of a continuous and unmistakably urban tradition we must turn to Lower Mesopotamia, where the civilization of Sumer was established in the fourth millennium BC. The Sumerian cities, of which Ur and Eridu are perhaps the best known, were important primarily as administrative and religious centers, but they also housed concentrations of craftsmen and traders. They were thus multifunctional in character, bringing together political, ecclesiastical, industrial, and commercial activities in a single settlement.

Statements concerning the causal factors that lay behind early urbanization must necessarily be highly speculative, but it is generally agreed that both social and technological changes were involved. On the technological side, an indefinitely long period of trial and error by early tribes had ultimately led to the domestication of cereals, to the invention of the plow, to the use of draft animals (notably the ox), to the invention of the wheel, and, above all, to the development of techniques of irrigation. These innovations, when applied to the naturally fertile alluvial valleys of the Middle East, brought about the production of an agricultural surplus on a more or less permanent basis for the first time in human history.

Without such a surplus, and without the means of transporting it, the establishment of cities would have been inconceivable in antiquity, as indeed it would be inconceivable today; in their dependence on foodstuffs brought regularly from beyond their own borders, modern cities are no less vulnerable than those of ancient Sumer. But technological progress by itself was not sufficient to ensure the emergence of the first cities. An alteration in social structure was also necessary: specifically, the division of labor into specialized occupational classes. Once this division began to occur, it is reasonable to suppose that the desire for safety and the natural gregariousness of humans would soon produce the idea of concentrating the nonagricultural occupations – and no doubt some farmers as well – in nucleated settlements of a size distinctly greater than that of purely agricultural villages. In addition, even for this embryonic stage in the development of civilization, it is probably justifiable to assume that economies of scale and agglomeration played a significant role. Quite apart from considerations related to defense, to royal whim, or to the supposed sacred importance of certain sites, the formation of towns made good economic sense in promoting a level of efficiency in commerce, manufacturing, and administration that would have been impossible to achieve with a completely dispersed population.

Without tracing the steps by which empire succeeded empire, we may observe that some forty urban sites are currently known for Mesopotamia in the pre-Christian era, though it should be noted that not all of these cities flourished simultaneously (Hammond 1972). During this long period, also, urban centers appeared in areas beyond the Fertile Crescent. Egypt, the Indus Valley, and the North China Plain all saw the emergence of networks of urban places during the third and second millennia BC. Each of these areas is sometimes held to have been a region of independent invention of the idea of living in cities, but the possibility that they were influenced in some degree by the Mesopotamian experience cannot be definitely excluded. The Indus Valley civilization died out around 1500 BC, but cities reappeared later in peninsular India. In China the diffusion of cities from the hearth area of the Middle Hwang Ho Valley kept pace with the gradual spread of permanent agricultural settlement across eastern and southern China, with the most active city construction occurring in the four centuries straddling the birth of Christ (Chang 1963).

In the West, the city spread initially from Mesopotamia and Egypt to the coastal regions of the eastern Mediterranean. The transition from a riverine to a thalassic environment was accompanied by a notable development of seafaring skills; it is symbolic that the Battle of Salamis, arguably the event that marks the birth of 'Europe' as a historical and geographical entity, was a naval engagement. The Mediterranean itself is a relatively tranquil sea, but it is bordered for the most part by rugged, inhospitable terrain in which good farmland is highly fragmented, occurring mainly as small pockets of coastal lowland watered by short rivers flowing from the encircling hills. In the lowland pockets around the Aegean Sea, sometimes inland but chiefly on the coasts, there emerged numerous small communities of the type that came to be known in Greek as the polis, or city-state. The typical polis was a self-governing unit of territory consisting of a single small urban center together with its contiguous agricultural lands, the effective limits of the miniature state being determined by the configuration of the adjacent hill country.

From a geographical viewpoint the most significant feature of this type of settlement is the limitation on population growth imposed by the physical environment. The small amount of farmland available to each polis severely restricted the number of people who could be supported by local agricultural production, and it is not surprising that the inhabitants turned to fishing and to seaborne trade in an effort to ensure adequate supplies of food. But even fishing and commerce were not enough, and the net result was a massive exercise in colonization. Whenever the pressure of population on resources became intolerably great, the polis responded by establishing a daughter colony at a previously unclaimed coastal site, and thus a new polis was born. Miletus, an important Ionian Greek city on the Aegean coast of Asia Minor, is reported to have given birth to more than eighty daughter colonies in this way (Smailes 1966, 16). Not infrequently, daughter settlements in due course were obliged to found colonies of their own. As a consequence, urban centers sprang to life not only in the Aegean area but also around the shores of the Black Sea and along all the northern coastlands of the Mediterranean as far west as the Strait of Gibraltar.

Among the many new settlements established during this period, one was destined to play an outstanding part in world history. This was Rome, founded about mid-eighth century BC. By the beginning of the

Christian era, the remarkable successes of the Romans as conquerors had brought them an empire that extended throughout the entire Mediterranean basin and penetrated northward as far as the Danube, the Rhine, and lowland Britain. For the urban geographer, the greatest significance of classical Rome lies in the fact that the needs of imperial defense and administration led to the first widespread appearance of urban centers north of the Alps. The primitive Celtic tribes that inhabited these northern regions had constructed many defensive earthworks at strategic points, but only a few of their very largest settlements might reasonably be viewed as urban. The Romans commandeered a large number of Celtic forts and transformed them into well-appointed towns to serve as centers of administration, commerce, and craft industries. In Gaul there were approximately 115 Roman towns and in Britain a further 40 (Pounds 1969, 148–52). Among the largest were Lyon (Lugdunum), Bourges (Avaricum), Dorchester (Durnovaria), and London (Londinium). Some centers, known as coloniae (colonies), were specifically identified by the Romans as military retirement communities and were populated largely by former soldiers and their families. Most of the coloniae were located in the Italian peninsula, but Cologne (Colonia Agrippensis), York (Eburacum), and Gloucester (Glevum) were also members of this class.

Rome at the height of its influence was almost certainly the largest city the world had yet produced. Estimating the populations of ancient cities is fraught with difficulties (Chandler and Fox 1974, 2–9), but Rome at the beginning of the Christian era contained probably something like 750,000 inhabitants of all classes, including slaves. This was a remarkable achievement and depended entirely on the steady importation of massive quantities of grain and other foodstuffs from various parts of the empire under the constant protection of Roman naval and military power. The great size of classical Rome, however, should not be allowed to obscure the fact that the vast majority of the urban centers of the ancient world were tiny settlements of not more than two or three thousand persons. Large size could be attained only if a city extended its influence in such a way that it could draw on the agricultural surpluses of areas beyond its own immediate vicinity, and few cities of antiquity possessed sufficient wealth or military power to exercise such dominance over wide areas. Prior to the rise of Rome it is doubtful that more than a handful of cities could have exceeded 100,000 in population, among them Babylon, the Egyptian Thebes, Athens, Alexandria, and perhaps the very largest centers of China and India. Most 'cities' were mere villages by modern standards of size. It should be kept in mind, however, that the total population, urban and rural combined, was also very much smaller than at present, and therefore the towns would appear as important settlements in the context of a thinly peopled world.

The Middle Ages

REGIONAL ASPECTS OF URBAN DEVELOPMENT

The decline of Rome – this itself posing historical problems of the first rank – was accompanied by losses of population and importance by most of the towns that had grown up in Western Europe under Roman influence. The depopulation of the towns could have curious results: at Arles in Provence the population shrank so drastically that the entire town was rebuilt inside the old Roman arena, the walls of the latter serving as defensive ramparts for the whole settlement (Mumford 1961, 248). The major force sustaining urban life during the 'Dark Ages' (AD 400–1000) was the Christian church. Towns that were chosen as the seats of bishops and archbishops generally had the best prospects for continued growth. So far as commercial activity was concerned, however, most towns served only as market centers for their own local districts. The movements of long-distance traders, though never completely stilled, appear to have sunk to a low level. All of Western Europe suffered an economic recession that lasted for several centuries.

The city of Rome itself also lost population, especially after Constantinople emerged as the capital of the Eastern Roman Empire. By AD 1000, Rome had shrunk from its former grandeur to a town of perhaps 40,000 (still a large city by the standards of the day), while Constantinople had blossomed into a thriving metropolis of almost half a million inhabitants. Also worth noting is the rise of Islam from the seventh century onward. Although the Moslems confined their interests mainly to the Middle East and to the southern shores of the Mediterranean, they had a significant impact on the development of towns in Spain. North of the Pyrenees and the Alps, however, urban life was moribund.

The gradual revival that occurred from the eleventh century onward was partly the result of growing political stability and partly a reflection of a renewed increase in the total population. The pace of economic development began to quicken. The main changes that took place as medieval Europe emerged from the 'Dark Ages' are effectively summarized in the following passage:

> Anyone who had set out to survey the cities of the Old World about the year AD 900 would have found urban life vigorous in China, in India, in the Islamic world from central Asia and Sind to Spain, and in the Eastern Roman Empire ... Tenth-century Italy and north-western Europe would have looked like an exceptional patch of non-urban territory on the map of the Old World at this date. This derelict fragment of the Roman Empire had sunk back into a primitive economy of subsistence farming. Yet, within the next few centuries, northern Italy and Flanders each gave birth to a cluster of city-states that could challenge comparison with the Sumerian cluster at its zenith. Venice and Genoa planted colonies in the Levant and round the shores of the Black Sea that rivalled the former colonies of Greek Miletus, Megara and Chalcis in the intensity of their commercial activity. The host of Hansa towns on the rivers and coasts of the north German plain opened up the Baltic and imposed their domination on Scandinavia, while, in south Germany, chains of commercial city-states were conjured into existence by the overland trade between Italy and the Low Countries. (Toynbee 1967, 26)

Some idea of the medieval transformation of the European urban network can be gained from comparison of Figures 2.1 and 2.2, which show all towns having at least 20,000 inhabitants in AD 1000 and AD 1500, respectively. On the earlier map, all but a handful of the towns are accounted for by five regional clusters, as follows.

1. *Southern Spain.* This region, comprising the Guadalquivir Valley and the adjacent Mediterranean coastlands, was dominated by Cordoba (450,000), which had served for generations as the capital of the Islamic Ommayad dynasty. Cordoba at this date was five times as large as Seville, the region's second city.

2. *The Balkans.* In this area the leading city was Constantinople (450,000), which vied with Cordoba for the honor of being the largest city in the world. The dominance of Constantinople in the Balkan cluster was even greater than that of Cordoba in Andalusia; Constantinople outranked Salonica (Thessaloniki) by a ratio of about eleven to one. Constantinople and Cordoba were the only cities exceeding 100,000 anywhere in Europe at this time. In addition, both were first and foremost political and religious capitals rather than commercial centers, and both belonged to the Mediterranean world rather than the world of north-western Europe.

3. *Southern Italy and Sicily.* Palermo (75,000) was the largest city of this group, which also included Rome and the Naples-Amalfi-Salerno subgroup.

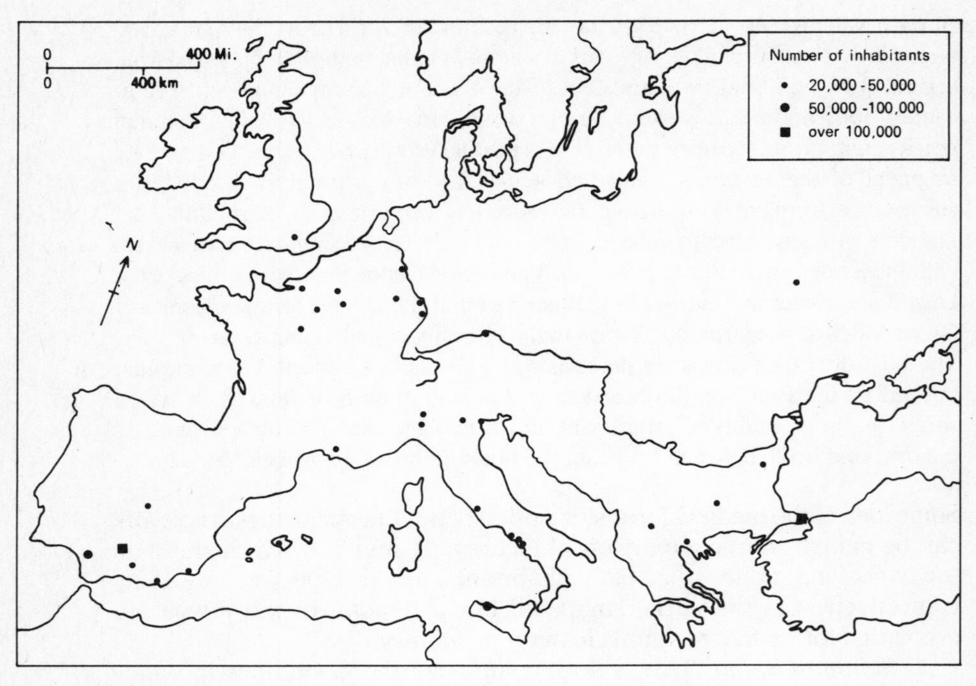

Figure 2.1
Cities with 20,000 or more inhabitants in Europe, C. AD 1000. Note the clustering of centers in southern Spain, northern and southern Italy, the Balkans, and the area extending from the Paris Basin to the Rhine. SOURCE: Chandler and Fox (1974)

4. *Northern Italy*. Here the four cities of Venice (45,000), Milan (30,000), Pavia (30,000), and Verona (20,000) in the Po Valley represented the emergence of a cluster of city-states that were to rise to world prominence in the coming years. The clusters of cities in both northern and southern Italy were composed of towns of broadly similar size; no single city dominated as in the Balkans and southern Spain. In the south, Palermo was slightly more than double the size of both Rome (35,000) and Amalfi (35,000), but the Naples-Amalfi-Salerno subgroup as a whole contained about the same total population as Palermo. In the north, Venice was not even twice the size of Milan and Pavia. These observations support the view that the Italian merchant cities were genuinely able to compete with one another on something like equal terms. No overriding political or religious authority existed to bring about the centralization of wealth and power in a single dominant metropolis.

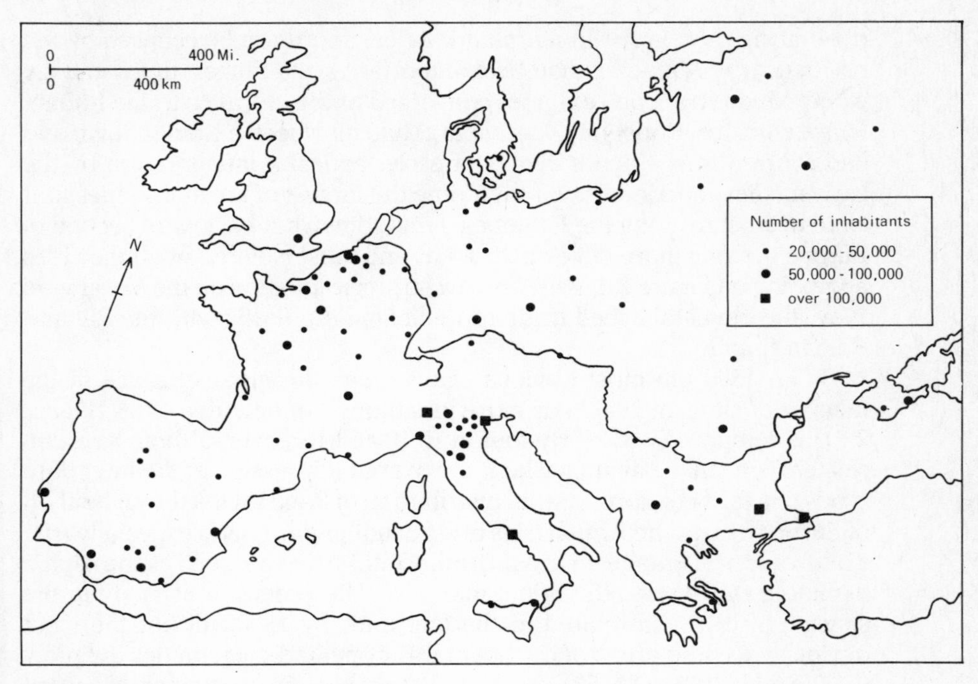

Figure 2.2
Cities with 20,000 or more inhabitants in Europe, c. AD 1500. By this date, centers of this size were present in almost all sections of the continent. Major clusters were evident in northern Italy and in the Low Countries. Compare with Figure 2.1.
SOURCES: Chandler and Fox (1974), Dickinson (1961), Mols (1972), Russell (1972), Ennen (1979), and various histories of individual cities

5. *The region from the Paris Basin to the Rhine*. The similarity in the sizes of the cities in this final cluster was even more striking than in Italy, with all nine of the centers having between 20,000 and 30,000 inhabitants. At this early date, the cities of the Low Countries proper (Bruges, Ghent, Antwerp, Brussels) had yet to break through the threshold of 20,000 population. The cluster reflects, in fact, a stage of development in these northern districts when inland transportation, including movement on major rivers such as the Rhine, the Seine, and their tributaries, was more important than coastal shipping.

Four large centers lay outside the five regional clusters. London, Britain's largest city from Roman times down to the present day, stood at the lowest convenient bridging point on the Thames, the principal English river facing the European mainland. Arles, shortly to be eclipsed by the rise of nearby Marseille, had been important since Roman times and lay where Mediterranean sea-lanes converged at the entrance to the Rhône-Saône corridor. Ratisbon (Regensburg) had also been a Roman town and had evolved into a major center of ecclesiastical administration by the early medieval period. It was for a time the largest of the many cities that were to emerge along the Danube as this somewhat backward section of Europe became more developed. Last, and most isolated of all the cities appearing on Figure 2.1, Kiev lay at a strategic location on the Varangian Way, the old-established trade route linking the Baltic with the Eastern Roman Empire.

By AD 1500 the most obvious change was the great increase in the number of cities of 20,000 or more inhabitants: ninety-five cities (Figure 2.2), as compared with thirty-four in AD 1000. Moreover, although certain clusters can still be distinguished, the overall impression at the later date is one of a fairly uniform spatial distribution of large towns throughout all of Europe except the British Isles and Scandinavia. This is especially true if allowances are made for areas of inhospitable terrain such as the Alpine mountain ranges and the Pripet marshes. The general uniformity in the spacing of towns is related to the fact that, by 1500, all but the most obviously useless areas of continental Europe had come under the plow (C.T. Smith 1978, 117–82). As each district became permanently settled and began to produce an agricultural surplus, a network of villages and towns grew up to serve as market centers. Typically, a single well-placed town in each district emerged with a dual commercial function, serving not only as the major focus of local trade but also as a center for trade over long distances, linking the local district with the wider world. These principal commercial nodes, which invariably lay on major routeways and generally also possessed both political and religious importance, served local districts of broadly similar extent and thus tended to be uniformly spaced throughout the settled regions of the continent.

Against this general background of uniform spacing, northern Italy stands out on Figure 2.2 as an area containing a densely packed cluster of large towns. Fifteen of the ninety-five cities on this map lie within the rough quadrilateral bounded by Venice (115,000), Milan (104,000), Genoa (62,000), and Florence (70,000) – the four largest cities in this Italian group. All these cities profited from their key location on 'the main axis of international trade, an axis that had Italy as its pivot and joined northwestern Europe with the Levant' (Lopez 1971, 96). The main commodities originating in the east were spices, precious stones, raw silk, and ivory. In the opposite direction the leading trade goods were woolen and linen cloth, arms and other iron wares, glassware, and timber. Virtually all the trade between northwestern Europe and the Orient passed through northern Italy or was carried in Italian ships, or both. The cities of northern Italy became synonymous with expertise in commerce. 'Genuensis ergo mercator' – a Genoese, therefore a merchant – went the medieval proverb (Ennen 1979, 138).

At the northern end of the great axis of trade a second distinct cluster of cities could be discerned. Led by Bruges (90,000) and Ghent (80,000), this group included Antwerp, Brussels, Lille, and Valenciennes in Flanders proper, with Liège, Cologne, Dieppe, Amiens, Rouen, and Paris all situated in the same general area. Collectively these cities were noted not only for trade but also for their production of manufactured goods, especially textiles. The group profited from its location in the area where the shortest channel crossings from England to the European mainland met the lowland routes connecting Iberia and Aquitaine with Germany and the Baltic. In the latter region, trading activity was controlled largely by the Hanseatic League, a group of cities federated together for mutual benefit and including Cologne, Hamburg, Lübeck, Danzig, and other places, some of them very small.

Although the Moors had been expelled from Spain by 1500, the cluster of cities that had developed in Andalusia under Moslem influence was still a prominent feature of the urban pattern. Cordoba, however, had lost most of its former glory and had shrunk to 35,000 in size, while Granada at 70,000 had become the largest city in Spain. In the central part of the Meseta a smaller cluster of four cities – Valladolid, Toledo, Segovia, and Medina del Campo – had also emerged by 1500. The rise of Madrid, however, still lay in the future.

Far to the east, a diminished Constantinople of 200,000 inhabitants was still Europe's largest city, despite having fallen to the Turks in 1453. Second place appears to have been a tie between Adrianople and Naples, at 125,000. Venice and Milan were the only other centers to exceed 100,000 (Chandler and Fox 1974). As in AD 1000, no city north of the Alps had managed to reach this figure.[1]

It would be a mistake to assume, of course, that the developments that transformed the European urban pattern of the year 1000 into that of 1500

1 There has been controversy over the size of late medieval Paris, with some scholars arguing for 200,000 people in the fourteenth century and others for around 80,000. On balance, the smaller figure seems more likely. For details see Ennen (1979, 186–7) and Russell (1972, 150).

proceeded at a uniform rate throughout the intervening five centuries. The normal state of affairs in many cities, in fact, seems to have been one of ups and downs rather than one of steady growth. Some cities, including Reims, Cartagena, and Kiev, appear on the map for the year 1000 but had fallen below the arbitrary threshold of 20,000 inhabitants by 1500; today, of course, these towns all stand far above this figure. A few other centers, such as Pisa and Louvain, appear on neither figure but exceeded 20,000 inhabitants for some portion of the five hundred years that separate the two maps. Accounting for the variable fortunes of individual cities is a task that lies well beyond the scope of this introductory survey. In general, however, the period from 1000 to 1350 was one of marked expansion; the following century and a half saw slow growth and even absolute decline.

The period to 1350 witnessed not only the revival of many towns of Roman or Celtic origin but also the creation of numerous small centers where no settlement had previously existed. Although none of these new foundations attained the status of major centers (20,000 or more population) by 1500, they collectively formed an important stratum of the complete urban network; most of them, indeed, survive today. Essentially, new towns were established wherever new lands were brought into cultivation. Notable among them were the towns in Gascony known as bastides – the name is related to the verb 'bâtir' and signifies a newly built town as opposed to one received from the past. At least 112 bastides were constructed by the English in Gascony between 1250 and 1320 (Beresford 1967). After the mid-fourteenth century, however, rural depopulation became common in all parts of Europe, the founding of new settlements came virtually to a halt, and many established towns of all sizes suffered absolute population losses. The major cause of this dramatic reversal was the Black Death, which blanketed the continent in 1348–50 and continued to flare up intermittently for the next three hundred years (Mols 1972, 8–9). Nevertheless, as Dickinson (1961, 279) has noted, 'the overwhelming majority of the settlements of today, throughout the whole of western, central and southern Europe, were in existence at the end of the Middle Ages.'

TRADE AS A FACTOR IN TOWN GROWTH

Reference has already been made to the general role of trade as a causal factor in medieval city growth. Two specific points need particular emphasis in this connection: the difference between local and long-distance trade, and the role of trade in town origins.

Local and long-distance trade

It is important to distinguish between local trade and long-distance trade. The difference, as the terms imply, is primarily one of scale, but certain other features should also be noted. Local trade, by definition, invariably serves the final market; that is, the goods received are intended for private use and are not for resale. The purchaser, or some member of the purchaser's household, is the ultimate consumer of the goods or services in question. With long-distance commerce, in contrast, the purchaser is a middleman, acting either as the paid agent of a distant consumer or as a speculative trader in his own right. Long-distance trade thus serves a nonfinal market: the buyer or agent is not the ultimate consumer. However, not all transactions serving a nonfinal market involve movement over long distances. A middleman may stand, for example, between a producer of manufactured goods and customers for those goods living in the same community. The latter case does not come under the heading of local trade as ordinarily used, and this exclusion points up the fact that local trade is retail trade rather than wholesale trade. The term *retail trade*, however, must be interpreted broadly so as to include consumer-oriented services (for example, barbers and doctors) as well as trade in material goods.

Because consumers in general are neither willing nor able to travel great distances in order to satisfy their wants – and this, while obviously true of the Middle Ages, is a potent factor even today – the concept of local trade implies a state of spatial interdependence between an urban center and a surrounding area of limited extent. In the medieval world, peasants from the countryside would assemble at a nearby town on specified market days in order to participate in the buying, selling, and perhaps simple bartering of agricultural surpluses and the products of craft industries. Once or twice in each year a somewhat longer trip might be made to a major fair in a larger center, but almost all the 'shopping' done by the typical peasant was carried out very close to home. Local trade thus generates a cellular spatial structure in which the countryside can be visualized as being divided up into a large number of relatively small districts or cells, each cell having a market town as its nucleus.

Long-distance trade produces, in contrast to this cellular pattern, the type of structure known to mathematicians as a graph; that is, a geometric figure composed of vertexes (points) and edges (lines). The vertexes, of course, are the cities, which act as the centers of articulation and control for long-distance commercial activity, and the edges represent the routes along which the goods actually move.

It is axiomatic that the vertexes of the long-distance trading graph are a subset of the nuclei in the cellular pattern of local trade. In other words, some towns serve only as local market centers, whereas others are important at both the local and the long-distance scales. These latter towns, the Fernhandelsstädte (long-distance trading cities) of the German historical literature, understandably tend to emerge as the largest cities in the urban system (Carus-Wilson 1958; Reynolds 1977, 56–60).

The distinction between local and long-distance trade corresponds to the distinction drawn in chapter 1 between the central place and the corridor concepts of spatial organization, respectively. Given this correspondence, the fact that local trade and long-distance trade coexist in all urban systems affirms the point made earlier that the concepts of central place and corridor are complementary and not competitive.

Pirenne and town origins

There has been confusion in the literature with regard to the role of long-distance trade as a creative force in medieval urbanization. The confusion stems from a certain ambiguity in the work of the great Belgian economic historian Henri Pirenne (1862–1935). On the one hand, Pirenne was well aware that nucleated settlements existed long before the eleventh-century revival of long-distance trade. He pointed out, for example, that the emerging merchant class frequently formed a small suburb outside the city walls, obviously not possible unless the city antedated the merchants (Pirenne 1925, 101–3, and 1937, 39–42). On the other hand, he seemed reluctant to accord full urban status to the fortified local market centers and ecclesiastical seats of the 'Dark Ages.' Although never unequivocal, he wrote as if he regarded significant long-distance trade as a necessary condition of town status.

This ambiguity has led to a sort of double standard. Lewis Mumford, for example, first attributes to Pirenne the view that medieval towns originated because of the revival of long-distance trade and later criticizes Pirenne on the grounds that nucleated settlements antedated this revival (Mumford 1961, 253–6). The position Mumford attributes to Pirenne is tenable only if long-distance trading activity is a sine qua non of urban status, but the subsequent criticism implies the acceptance of a much broader definition of towns. Mumford seems to want to have it both ways.

Once this kind of confusion is recognized and eliminated, three conclusions are warranted. First, given our general definition of a town as *any* nucleated nonfarm settlement, it is perfectly clear that the vast majority of medieval urban centers were not created by the revival of long-distance trade. Second, the founders of some genuinely new towns in the Middle Ages did aspire thereby to generate long-distance commerce. For example, one of the motives of Edward I of England in planting bastides in Gascony was to promote trade in wine and other agricultural products, thus increasing royal revenues by way of commercial taxes (Trabut-Cussac 1954; Beresford 1967). Third, the general revival of long-distance trade acted as a powerful stimulus to renewed city growth. Many towns that had lain dormant for centuries experienced unprecedented surges of expansion. At the same time, the growth of cities acted reciprocally to stimulate trade. As Lopez (1971, 86) has observed, 'medieval urbanization and commercialization were mutually supporting phenomena.' The fundamental causes for the growth of both cities and commerce are to be found in the steady increases of both population and productivity that appear to have characterized practically every part of Europe throughout the four centuries preceding the outbreak of the plague.

The mercantile era, 1500–1800

The European Middle Ages conventionally come to an end with Columbus's discovery of the New World. By general agreement, the history of the modern world begins about 1500. It is convenient to divide the modern period into two parts: a mercantile era, from 1500 to 1800, and an industrial era, from 1800 to the present. With regard to the development of cities, the most significant general features of the mercantile era were the gradual emergence of modern nation-states and the increasing involvement of most of these states in overseas commercial expansion. These processes, of course, did not come to a halt in 1800, but the profound impact of the Industrial Revolution after this date makes it appropriate to treat the nineteenth and twentieth centuries as a separate era in the historical geography of urbanization.

DEVELOPMENTS WITHIN EUROPE

One effective way to analyze the evolution of an urban system is to look at changes in rank. As the system develops, it is most unlikely that all of its members will grow at the same rate. During a given period of time, towns that grow faster than average will tend to rise to higher ranks within the system; towns experiencing slower growth rates will tend to get left behind, thus falling to lower ranks. Table 2.1 represents an analysis of the

TABLE 2.1
Rank mobility of European cities, 1500–1800

	Rank[a]				Rank mobility[b]		
City	1500	1600	1700	1800	1500–1600	1600–1700	1700–1800
London	24	4	2	1	0.714	0.333	0.333
Constantinople	1	1	1	2	0.0	0.0	−0.333
Paris	8	3	3	3	0.455	0.0	0.0
Naples	2.5	2	4	4	0.111	−0.333	0.0
Moscow	44	16.5	9	5	0.455	0.294	0.286
Lisbon	12.5	9.5	5	6	0.136	0.310	−0.091
Vienna	32	72.5	13	7	−0.388	0.696	0.300
St Petersburg				8			(0.853)
Amsterdam		38.5	6	9	(0.448)	0.730	−0.200
Adrianople	2.5	5	14	10	−0.333	−0.474	0.167
Berlin			99.5	11		(0.007)	0.801
Madrid		18	12	12	(0.697)	0.200	0.0
Dublin		81.5	16.5	13	(0.107)	0.663	0.119
Rome	42	11	7	14	0.585	0.222	−0.333
Palermo	28.5	12	11	15.5	0.407	0.043	−0.170
Venice	4	6	8	15.5	−0.200	−0.143	−0.319
Milan	5	8	10	17	−0.231	−0.111	−0.259
Hamburg	78	47	21	18	0.248	0.382	0.077
Barcelona	57.5	23.5	18	19	0.420	0.133	−0.027
Lyon	8	14	19	20	−0.273	−0.152	−0.026
Marseille	32	30.5	15	21	0.024	0.341	−0.167
Copenhagen		47	28.5	22	(0.365)	0.245	0.129
Bordeaux	90	72.5	43.5	23	0.108	0.250	0.308
Seville	24	7	16.5	24	0.548	−0.404	−0.185
Genoa	15	19.5	25.5	25	−0.130	−0.133	0.010
Cadiz			66.5	26		(0.206)	0.438
Glasgow				27.5			(0.572)
Rouen	10	19.5	24	27.5	−0.322	−0.103	−0.068
Edinburgh		72.5	60.5	29.5	(0.164)	0.090	0.344
Valencia	24	15	40	29.5	0.231	−0.455	0.151
Manchester				31			(0.530)
Florence	12.5	21.5	23	32	−0.265	−0.034	−0.164
Prague	12.5	13	38	33	−0.020	−0.490	0.070
Liverpool				34			(0.496)
Stockholm			38	35.5		(0.453)	0.034
Warsaw		55		35.5	(0.295)	(−0.295)	(0.480)
Birmingham				37.5			(0.458)

TABLE 2.1
(*continued*)

City	Rank[a] 1500	1600	1700	1800	Rank mobility[b] 1500–1600	1600–1700	1700–1800
Nantes			43.5	37.5		(0.398)	0.074
Granada	12.5	9.5	21	39	0.136	−0.377	−0.300
Oporto			99.5	40		(0.007)	0.427
Bologna	18	25	27	42.5	−0.163	−0.038	−0.223
Bristol			89.5	42.5		(0.060)	0.356
Brussels	52	35	21	42.5	0.195	0.250	−0.339
Turin		96.5	41.5	42.5	(0.023)	0.399	−0.012
Wrocław	82.5	60.5	48	45	0.154	0.115	0.032
Cork				46			(0.374)
Salonica	38	35	48	47	0.041	−0.157	0.011
Dresden				48			(0.356)
Rotterdam			31	49		(0.530)	−0.225
Lille	66	63	30	51	0.023	0.355	−0.259
Verona	41	33	33.5	51	0.108	−0.008	−0.207
Zaragoza	90	92.5	76	51	−0.014	0.098	0.197
Budapest	90	87		53.5	0.017	(−0.074)	(0.307)
Ghent	8	66	36	53.5	−0.784	0.294	−0.196
Antwerp	43	30.5	25.5	55.5	0.170	0.089	−0.370
Königsberg			68.5	55.5		(0.192)	0.105
Leeds				57.5			(0.274)
Livorno				57.5			(0.274)
Toulouse	38	47	48	59	−0.106	−0.011	−0.103

SOURCES: Population data on which rankings are based chiefly from Chandler and Fox (1974), with partial corroboration from Dickinson (1961), Ennen (1979), Mols (1972), Russell (1972), and a number of historical studies of individual cities

a As is customary in working with ranks, ties are accommodated by assigning each affected city the average value of the tied ranks. For example, if two cities are tied in sixth and seventh place, each is given the rank of 6.5. A blank space indicates that a city was ranked lower than number 100 in the year in question.

b Mobility indexes in parentheses are estimates. See text for full explanation.

changes in rank experienced by large European towns during the mercantile era. All towns that had 50,000 or more inhabitants in 1800 are included, and the table lists these fifty-nine places in rank order according to their sizes in the final year. So far as the available data allow, the rank of each of these towns is recorded at intervals of one hundred years back to 1500. A blank space indicates that a town was not numbered among the largest one hundred places in Europe in the year in question.

The main difficulty encountered in the study of changes in rank stems from the fact that a shift of any given number of rank positions is more easily achieved by a small town than by a large town. In absolute terms, a change from, say, tenth to sixth place in the ranking is the same as a change from fiftieth to forty-sixth: both changes involve a rise of four rank positions. However, because smaller places are closer together in actual population size than larger places, the movement from tenth to sixth position is a more remarkable achievement than that from fiftieth to forty-sixth. We can overcome this difficulty by the use of a simple index of rank mobility, M, defined as follows:

$$M = (R_0 - R_1)/(R_0 + R_1),$$

where R_0 and R_1 represent the town's rank at the beginning and the end of a given time period, respectively. In words, the index is simply the difference between the two rank positions, divided by their sum; the result is positive if the city rises and negative if it falls. The theoretical limits of the index are -1.0 and $+1.0$, with a value of zero signifying no change in rank.[2]

Mobility index values for the three centuries of the mercantile era are shown in the last three columns of Table 2.1. Values in parentheses are estimates for cases in which the rank at the beginning or end of the century is known only to be 101 or lower. Each estimate is given as the closest value to zero that could possibly occur, calculated by assuming that the unknown rank is in fact 101.

A consideration of the frequency distribution of the index values suggests that a city is doing well if its mobility index for a particular century is $+0.300$ or higher, and doing poorly if its index is -0.300 or lower. Although the performances of individual cities can be quite variable from century to century (Warsaw evidently being exceptional in this respect), virtually all the highest positive index values are accounted for by three general classes of towns.

1. *Major political capitals*. Into this category fall Copenhagen, Madrid, Moscow, and Paris in the sixteenth century; Lisbon, Stockholm, and Vienna in the seventeenth; and Berlin, Budapest, Edinburgh, St Petersburg (now Leningrad), and Vienna in the eighteenth. London, the largest

2 A negative index of rank mobility does not necessarily imply loss of population in absolute terms. A city may be growing yet fall in rank when overtaken in size by other places growing more quickly.

city in Europe in 1800, may be added to these lists in all three centuries. Every major European power of the early modern period is represented in this roster, and it is safe to conclude that the strong growth of these capitals was closely connected with the emergence of the great nation-states, notably Austria-Hungary, England, France, Portugal, Prussia, Russia, and Spain. Among these states, Russia is unique in having contributed two capitals: first Moscow, which continued to do well throughout the three centuries covered in Table 2.1, and later St Petersburg. The latter, founded by Peter the Great in 1703 as Russia's 'window on the west,' holds the record for a meteoric rise to prominence within this era.

2. *Ports*. Particularly notable are ports that were well situated to become involved in the overseas enterprises that began with Columbus. The chief examples are Amsterdam, Cadiz, Hamburg, Oporto, and Rotterdam; the French centers of Bordeaux, Marseille, and Nantes; and the British ports of Bristol, Glasgow, and Liverpool. All these cities, except Marseille, have direct access to the important Atlantic shipping lanes, and this accessibility undoubtedly promoted colonial expansion by the westernmost European powers. Several cities, including Amsterdam, Copenhagen, Lisbon, and London, were both ports and political capitals, thus benefiting from the mercantile era's two main sources of urban growth. (The Dutch still identify Amsterdam as their capital city, even though their parliament now sits at The Hague.)

3. *Manufacturing cities*. This small group is found only in the eighteenth century and only in the British Isles and consists chiefly of Birmingham, Leeds, and Manchester. Here we can detect the earliest effects of the Industrial Revolution upon the evolving system of cities. These effects were felt also before 1800 in the larger ports, such as Glasgow, Liverpool and London. The British Isles alone account for almost half of all European cities having rank mobility indexes of +0.300 or higher during the eighteenth century. Britain was already moving toward the position of world dominance that she occupied for a full century following the defeat of Napoleon at Waterloo.

At the other end of the scale, generalization is more difficult, partly because index values of −0.300 or lower are relatively uncommon. However, downward mobility was characteristic of the cities of southern Spain, northern Italy (Turin and Livorno being exceptions), and the Balkans, all highly urbanized regions during the Middle Ages (Figures 2.1 and 2.2). The changes of rank in the urban network evidently reflected the decisive northward shift in economic activity that took place after 1500.

Yet not every large city north of the Alps fared well during the mercantile era. For example, Bruges and Ghent were eclipsed in the sixteenth and seventeenth centuries by Amsterdam, Antwerp, and Hamburg. Bruges fell from sixth to eighty-first place between 1500 and 1600; Ghent fell from eighth to sixty-sixth. Bruges does not appear in Table 2.1 because it had fewer than 50,000 inhabitants in 1800, but its sixteenth-century mobility index of −0.862 is a record for negative values. Antwerp in turn fared poorly during the eighteenth century, as did Brussels. The political and commercial developments that affected the fortunes of all the towns of the Low Countries during this era make a fascinating story, but one that is too complex to be dealt with effectively in a few words.

DEVELOPMENTS OVERSEAS

The influence of Europeans on urban development in foreign areas between 1500 and 1800 was confined largely to the Americas. Except for the activities of the British and Dutch East India companies, major involvements in Asia, Africa, and Australasia did not materialize until the nineteenth century. Within the western hemisphere the earliest developments took place under Spanish auspices, during the sixteenth century. However, urban cultures did exist in the New World before the Spaniards arrived. The Aztec, Chibcha, Inca, and Maya peoples of Central and South America all built cities, and there is ample evidence that their larger centers were important trading cities as well as spiritual and ceremonial foci.

The Spaniards in America, like the ancient Romans in Gaul, took control of the native urban centers and used them as a framework for the administration and exploitation of the newly conquered lands. The conquerors also built a number of new towns, many of which were conceived as part of a general policy aimed at concentrating the native population into compact settlements in order to facilitate conversion to Christianity. Unfortunately, this policy of concentration also served to promote the rapid spread of European diseases, notably smallpox, against which the natives had little or no immunity. It has been estimated, for example, that the indigenous population of Mexico fell from about eleven million in 1520 to less than three million in 1650, owing largely to the ravages of diseases unwittingly introduced by the Spaniards (Cook and Simpson 1948).

It was the custom of the Spanish authorities in the New World to classify towns, according to size and political importance, as ciudad (city), villa (town), or pueblo (village). Toward the end of the sixteenth century, the leading category of ciudad was represented by some eighty centers. The number of villas was about the same, and the pueblos numbered several hundred (Houston 1968). Although Mexico City and Lima were the most important cities from the political standpoint, the largest center was the silver-mining town of Potosi in modern Bolivia, which attained a population of 148,000 in 1600 (Chandler and Fox 1974, 191). At this date Potosi boasted even more inhabitants than Seville, then the largest city in Spain. Some time after 1600, however, Potosi entered a long period of decline. By 1700, first place in population had been taken over by Mexico City, which has remained the largest city in Hispanic America down to the present day.

Urban development elsewhere in the Americas was later in getting started and more limited in scope. On the eastern side of South America the Portuguese laid the foundations of the urban network of Brazil, their leading centers being the ports of Bahia (now Salvador), Rio de Janeiro, and Pernambuco (Recife). Far to the north, the seventeenth and eighteenth centuries also witnessed the beginnings of city growth in the areas that were to become the United States and Canada. In these regions no native urban tradition existed, and the Dutch, French, and English traders and colonists were obliged to create, in Bridenbaugh's apt phrase, 'cities in the wilderness' (Bridenbaugh 1938). Naturally the favored sites for the earliest towns were ports, and the backbone of the urban network of colonial America thus came to consist of the five leading coastal cities: Boston, New York (initially, under the Dutch, named New Amsterdam), Philadelphia, Baltimore, and Charleston. It is of interest to note that only one of these ports lies south of Cape Hatteras. South of this point, the eastern coast of North America is emergent, offering few good natural harbors. To the north, however, submergence has progressively drowned the mouths of the rivers, providing sheltered harbors and easy access to the interior in estuaries such as Delaware and Chesapeake bays.

Aside from the estuaries of the eastern seaboard, the other natural entrance-way to North America for transatlantic shipping is the St Lawrence River. Along the banks of the St Lawrence the French established their colony of New France, the chief urban centers being the ports of Quebec and Montreal. Quebec remained the larger throughout the mercantile era but was overtaken by Montreal early in the nineteenth century when the settlement of Upper Canada (southern Ontario) was actively under way.

At the beginning of the nineteenth century North America contained only nine towns of 10,000 or more inhabitants (Table 2.2). Only one of these nine, the twin center of Albany-Troy, was not directly accessible to

TABLE 2.2
Cities of 10,000 or more inhabitants, North America, 1800

Rank	City	Population
1	Philadelphia	68,200
2	New York	63,735
3	Boston	35,248
4	Baltimore	26,519
5	Charleston	18,844
6	Quebec	12,000
7	Montreal	11,000
8	Albany-Troy	10,315
9	New Orleans	10,000

SOURCES: Chandler and Fox (1974), Nader (1975), and US Census of 1800

ocean-going ships. New Orleans was under Spanish and later French control during the mercantile era as here defined, becoming part of the United States only with the Louisiana Purchase in 1803. Greater Philadelphia, including Germantown, Southwark, and overspill in the adjacent liberties, was the largest American city at this time, though it would soon be outstripped by the New York–Brooklyn agglomeration. Had Philadelphia been in Europe in 1800 it would have ranked only fortieth in size (between Granada and Oporto – see Table 2.1), but it held third place in the Americas, behind Mexico City (128,000) and Bahia (75,000).

In the United States in 1800, agricultural settlement had barely begun to penetrate west of the Appalachians. Nevertheless, the former British colonies contained, in addition to the towns listed in Table 2.2, about thirty small urban places having between 2,500 and 10,000 inhabitants. Half of these smaller centers were located in southern New England, and most of the remainder lay in the Middle Atlantic states (Ward 1971, 22–3). Richmond, Petersburg, and Norfolk formed an isolated cluster in the large gap between Baltimore and Charleston. West of the Appalachian barrier the largest center was Pittsburgh, at the Forks of the Ohio; in 1800 this center contained a mere 1,565 inhabitants.

The industrial era

The nineteenth and twentieth centuries have witnessed a veritable explosion of city growth throughout the world. A few general statistics will indicate the magnitude of this expansion. In Europe, which was certainly the most highly urbanized part of the world at the beginning of the industrial era, there were only twenty cities with more than 100,000 inhabitants in 1800; today there are four hundred. The corresponding figures for the Soviet Union are two (St Petersburg and Moscow) and 270; for the United States, zero and 250. Urban growth on this massive scale is without precedent in human history and must be interpreted as the outcome of two distinct processes. First, there has been an extremely large increase in the world's total population since 1800. Second, there has been a substantial rise in the proportion of the total population living in cities. I will comment on these two processes in turn.

THE DEMOGRAPHIC TRANSITION

The unprecedented growth of the world's total population since 1800 is attributable to the process known as the demographic transition. The central feature of this process is the tendency for each country's rate of population growth to follow a bell-shaped curve over time, being low at first, then rising to a peak, and finally falling back to a low level. The details of this sequence can be summarized in the form of three developmental stages. First, there is a long, preindustrial stage, during which the birthrate is high and the death rate is only slightly lower. Because there is not much difference between these two rates, the net rate of population growth is small. The second stage is ushered in by the arrival of the Industrial Revolution, which produces, among numerous other changes, relatively sudden and dramatic advances in medical care, nutrition, and general sanitation. The result of these advances is a sharp drop in the rate of mortality, particularly for infants. The birthrate, however, remains high, and therefore the difference between the birthrate and the death rate becomes comparatively large. Consequently the overall rate of population growth increases. But the birthrate does not stay high indefinitely. For reasons that remain partly obscure (and are the subject of continuing debate), the increased prosperity associated with continued industrialization is accompanied by a falling birthrate. This brings about a third and final stage, which is characterized by low birth and death rates and, as in the beginning, a small net rate of population growth.

With some exceptions – for example, parts of eastern Europe, where birth and death rates appear to have fallen simultaneously rather than at separate times – the concept of the demographic transition provides a good general description of the mechanics of population growth in the more highly developed countries of the world since 1800. Most of the underdeveloped world, however, has entered the second stage of the process only within the past few decades, and it is an open question whether the newly developing countries will in due course experience the falling birthrates that have occurred in the more fortunate parts of the world (Chung 1970). It is clear, nevertheless, that worldwide diffusion of the decline of death rates since about 1800 has been the immediate cause of the marked increase in the rate of world population growth during the industrial age. In round figures, the world's total population has increased from 900 million at the beginning of the nineteenth century to 4.8 billion at the present time, a fivefold increase in 180 years. This alone, other things being equal, accounts for a significant amount of city growth.

THE DRIFT TO THE CITIES

But a second and even more powerful force has also been at work: namely, the process by which cities have become the home of an increasing proportion of each country's total population. This process reflects the steadily increasing productivity of the agricultural segment of the economy, a trend that owes much to farm mechanization and is therefore part and parcel of the 'industrial' revolution. The nature of the link between farm mechanization and the drift of rural population to urban centers is somewhat ambiguous, giving rise to a puzzle of the 'Which came first?' variety. Did rural folk migrate to the towns because farm mechanization put them out of work, or were farmers forced into mechanization because labor was abandoning the countryside in favor of the higher wages and other attractions of city life? The causal sequence may well have varied from one region to another. But the net result in either case was an increase in the proportion of the population making its home in the cities.

This trend can be further illuminated by considering the concept of distinctive 'sectors' in the labor force. The *primary* sector is concerned with the production of raw or incompletely processed materials. It includes mining, fishing, logging, and trapping activities as well as agriculture; but, in terms of numbers employed, it is dominated in most regions by farming. The *secondary* sector is concerned with transportation (of both raw materials and finished products), with fabrication, and with storage. It is overwhelmingly dominated by employment in manufacturing. Finally, the *tertiary* sector is concerned with the delivery of goods and services to the ultimate consumer. It is dominated by employment in retailing and in the personal service industries, such as entertainment, education, and the health services. Some writers identify a *quaternary* sector, covering corporate management and the civil service, but for most purposes these office jobs can be satisfactorily dealt with as part of the tertiary sector.

Since the beginning of the industrial era, the proportion of total employment in the primary sector has been declining, while the proportions in the other sectors have been increasing. The normal pattern sees an initial shift from the primary to the secondary sector – or, more simply, from agriculture to manufacturing. The rise of the tertiary proportion begins somewhat later but is even more rapid and pronounced than that of the secondary proportion (Simmons 1986). These changes are shown schematically in Figure 2.3. In today's most industrialized countries, the primary sector has typically fallen from about 90 per cent of the total labor force two centuries ago to less than 10 per cent at the present time.

For cities, the significance of these changes lies in the fact that employment in the secondary and tertiary sectors is by nature urban, whereas almost all work in the primary sector is rural (mining towns and fishing ports excepted). Therefore, as the secondary and tertiary sectors have increased their relative importance, jobs have become more and more concentrated in cities; and the spatial distribution of jobs naturally governs that of the total population.

In developed countries, the expansion of the secondary and tertiary sectors has neatly (and somewhat mysteriously) kept pace with the contraction of the primary sector, ensuring more or less full employment throughout the process. Underdeveloped countries have not been so lucky. There, the surplus of labor that accumulates in rural areas is the result of the high fertility of the population, not of farm mechanization. The labor surplus migrates to the towns in the hope of finding work but is all too often frustrated by the inadequate rates of growth of the secondary and tertiary sectors. The results are the squatter settlements and the high rates of unemployment and underemployment that have become chronic problems in the cities of the developing world (McGee 1967, 155–70; Breese 1966, 1969). In general, in rich countries the cityward drift of population has been powered by 'pull' factors in the cities more than by 'push' factors in the countryside; in poor countries the reverse has been the case.

We may also note, though without elaboration at this point, that the process of concentration outlined above has taken place not only between

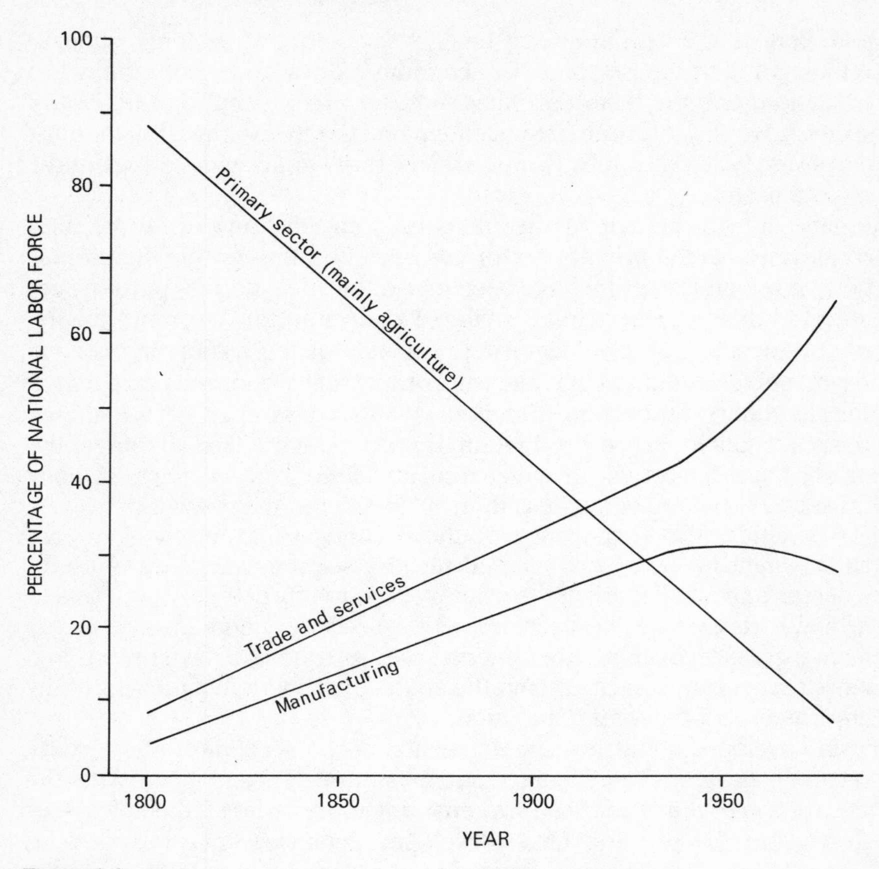

Figure 2.3
Schematic representation of the changing structure of employment in industrialized countries since 1800

the countryside and the cities but also within the urban system itself. In other words, the share of a nation's total population accounted for by its larger cities has tended to rise faster than the share accounted for by its smaller cities. In fact, as employment came to be increasingly concentrated in the very largest cities, many of the smallest urban centers of the developed countries actually experienced absolute losses of population during the late nineteenth and early twentieth centuries. The growth potential of the largest cities, however, does not appear to be unlimited. In the United States during the 1970s, for example, the very largest centers grew significantly more slowly than cities of small and intermediate size (see chapter 10).

The marked expansion of cities during the industrial era, then, has been the result of increases in the world's total population and in the proportion of the total population living in cities. It is interesting to attempt to measure, even if crudely, the relative strength of these two sources of city growth. As noted earlier, the world's population has increased fivefold since 1800. Taking this factor of five into account, let us define the minimum size of a 'large city' as 20,000 inhabitants in 1800 and as 100,000 inhabitants today. According to the estimates painstakingly compiled by Chandler and Fox (1974, 323–6), the world in 1800 contained some 535 'large cities' (i.e. with at least 20,000 inhabitants), and these cities in combination housed a total population of some thirty-four million persons. If cities grew merely at the same rate as the general population, we should expect today's large cities (100,000 and over) to contain something like 170 million inhabitants (thirty-four multiplied by five). In fact the world's 2,500 cities of 100,000 and more inhabitants currently harbor a combined population of about 962 million, or more than five times the expected number. Accordingly, the drift of rural population to the towns seems to have been a substantially more important factor than the general growth of world population in accounting for the rise of cities during the industrial age.

EUROPE AND AMERICA SINCE 1800

The complex series of technological and organizational innovations which are known collectively as the Industrial Revolution appeared first in England from about 1760 onward, spreading from there to the European mainland and ultimately to other parts of the world during the nineteenth and twentieth centuries (Ashton 1948; Deane 1969). In general terms, the two main ingredients of the process of industrialization were the development of the factory system and the invention of machines that greatly increased average productivity per worker.

Throughout Europe, the change from charcoal to coke in the vitally important iron and steel industry, together with the increasing significance of steam-powered machinery, led to the rapid growth of towns on the major coalfields. The building of canals and later of railroads reduced the real costs of transportation, thereby permitting the expansion of market areas and encouraging towns to specialize in the production of goods for which they were particularly suited. Although the new industrial centers on and near the coalfields were the most rapidly growing class of towns, industry was attracted also to existing large cities, which offered a ready market and a good supply of labor. Thus the ports and political capitals that had shown strong growth during the mercantile era tended to maintain their importance after 1800 by adding manufacturing as a further major activity. This tendency became even more pronounced when electric power began to replace the steam engine late in the nineteenth century. London and Paris, for example, became in absolute terms the leading manufacturing centers of their respective countries, though the ratio of manufacturing workers to total employment remained lower in these cities than in smaller but more specialized industrial centers such as Manchester and Lille.

In the United States, as in Europe, the rise of modern manufacturing and later the growth of the tertiary sector have strongly influenced the character of the urban network. In contrast to the situation in Europe, however, the American urban network has evolved since 1800 in a context involving radical changes in the distribution of the rural population. These two aspects – technological change and the westward advance of settlement – have been combined by Borchert to produce a descriptive model of American urban development consisting of four eras (Borchert 1967; see also the related studies by Lukermann 1966; B. Duncan and Lieberson 1970; Conzen 1977; and Meyer 1980).

1. *Sailing vessel and horsedrawn wagon (to c. 1830).* By 1830 the agricultural frontier had moved west as far as the Mississippi Valley, with the Ohio River and its tributaries serving as the main avenues of movement west of the Appalachian barrier. Because good roads were few and the railroad had yet to appear, waterborne transportation was the rule, and the new towns of the developing interior were invariably river ports. Examples include Cincinnati (the largest trans-Appalachian center in 1830), Pittsburgh, St Louis, Indianapolis, Columbus, Louisville, and many smaller places. Almost all the agricultural produce that was shipped from the Ohio country to the east coast was sent by riverboat down the Ohio-Mississippi system to New Orleans and thence by ocean-going vessel to the Atlantic seaboard, a route that reflected the great difficulty of transporting heavy loads across the Appalachians. Westward movement, however, went by way of Buffalo or Pittsburgh because upstream navigation on the Mississippi was hazardous before the introduction of steamboats. Near the end of this epoch, in 1825, New York interests scored a major coup by completing the Erie Canal, which greatly facilitated the movement of goods and passengers in either

direction along the Hudson-Mohawk corridor between Albany and Buffalo. From then on, New York's commercial supremacy over Boston, Philadelphia, and Baltimore was assured (Rubin 1961).

2. *Steamboat and iron horse (1830–70)*. The mid-nineteenth century saw the frontier of continuous settlement pushed northwestward into southern Minnesota, westward to include eastern Nebraska and Kansas, and southwestward into eastern Texas. Isolated pockets of settlement appeared on the Colorado piedmont, in Utah, and in the far west, especially around San Francisco. The period was marked by the widespread adoption of the steamboat and by the advent of the railroad. The steamboat made two-way traffic on the navigable rivers a reality and thus stimulated the growth of many river towns of all sizes throughout the continental interior. Steamboats also plied the Great Lakes, making Lake Erie in particular the *mare nostrum* of American westward expansion. The railroads, most of which were financed by eastern merchants intent on enlarging the commercial hinterlands of the major Atlantic ports, ran chiefly east–west, and a certain amount of conflict developed between this lateral orientation and the predominantly north–south movement of waterborne traffic on the Mississippi river system. One outcome of this conflict was the relative decline of New Orleans. Initially New Orleans controlled the trade of the entire Mississippi Valley, but the building of the railroads resulted in New York and other eastern centers capturing the northern portion of this commercial empire.

The railroads could stimulate phenomenal growth as well as decline. Chicago, the undisputed focal point of the evolving continental railroad network, became the fastest growing city in the United States, rising from obscurity in the 1830s to attain a population of 299,000 in 1870 (Still 1941). During this epoch, too, the effects of the Industrial Revolution began to be felt, although manufacturing remained largely concentrated in the older cities of the Atlantic seaboard (except for Charleston). The North's possession of the bulk of the country's industrial strength was an important factor in the defeat of the South in the Civil War, and the war powerfully stimulated the further development of manufacturing in many cities. The war also delayed the construction of the first transcontinental railroad, but in 1869, just as this epoch ended, the line was completed, linking Omaha and San Francisco.

3. *Steel rail and electric power (1870–1920)*. The introduction of the steel rail was significant for several reasons. First, it greatly reduced the costs of maintenance and replacement that were associated with the softer iron rails of the earlier years. Second, it could support much heavier

loads. Third, replacement of iron by steel provided an opportunity to standardize the railroad gauge throughout the country. This opportunity was not wasted, and the standardized gauge was an important element in the creation of an integrated transport network at the national scale. Together with the introduction of more powerful locomotives, the steel rail made possible the carrying of heavier loads over longer distances at higher speeds, which in turn tended to favor the concentration of industry in the largest cities. The major cities were also the first to benefit from the large-scale thermal-electric generating stations that began to appear during the 1880s. Industry dependent on small waterpower sites was increasingly at a competitive disadvantage. The epoch was marked by the rapid expansion of manufacturing not only in large eastern centers but also in the leading cities of the interior. Thus, in addition to seeing the continued strong growth of Chicago, Cincinnati, and St Louis, the period witnessed the rise to industrial prominence of Buffalo, Pittsburgh, Cleveland, Detroit, Milwaukee, and Minneapolis–St Paul. As long-haul rail transportation came to the fore, short-haul movements and river traffic were de-emphasized, leading to the stagnation of many smaller towns which had flourished as general cargo ports during less centralized times.

4. *Automobiles, air travel, oil, and amenities (from 1920).* The American frontier, defined as the moving edge of the area of agricultural settlement, ceased to exist about 1900, but this did not mean that no further changes took place in the spatial distribution of the population. The technical innovations of the automobile age have not only promoted the continued movement of people from the farm to the city but have also stimulated a gradual redistribution of population in favor of the southern and western regions of the continent. This 'sunbelt effect,' involving the rapid growth of southern and western cities and retarded growth in northeastern cities, reflects several interrelated trends, including the rise of oil (found mainly in western and southwestern states) together with the relative decline of coal; the growth of tertiary employment; the increasing integration of the national economy through the medium of air travel; the growth of both real incomes and leisure time; and the desire to abandon the old cities of the snowbelt for the warmth, cleanliness, and recreational opportunities of beach, mountain, and desert.

Since 1920, and especially since the end of the Second World War, the pace-setting American cities have not been those of the east coast or the midwest but rather such places as Los Angeles, Houston, Miami, and, more recently, Phoenix, Denver, and Atlanta. Since 1970, in fact, a few of

TABLE 2.3
North American cities that overtook New Orleans in size in each decade from 1860 to 1980

1860–70	1870–80	1880–90	1890–1900
Chicago Cincinnati St Louis	Pittsburgh San Francisco	Buffalo Cleveland Minneapolis–St Paul	Detroit Milwaukee Washington
1900–10	**1910–20**	**1920–30**	**1930–40**
Los Angeles Montreal Toronto	Kansas City	–	–
1940–50	**1950–60**	**1960–70**	**1970–80**
Houston	Dallas–Fort Worth Miami Seattle	Atlanta Denver Phoenix San Diego Tampa–St Petersburg Vancouver	San Jose

SOURCES: Censuses of the United States and Canada
NOTE: New Orleans never experienced an absolute decline in size. It grew from 169,000 in 1860 to just over one million inhabitants in 1980. Over the same period, it fell in rank from fifth to thirtieth. Buffalo, which overtook New Orleans during the 1880s, was itself overtaken by New Orleans during the 1970s.

the northeastern metropolises have actually lost population in absolute terms, including Cleveland, New York, and Pittsburgh, and this has led one pair of authors to suggest that a new epoch of 'slow growth' should be added to the four identified by Borchert (Phillips and Brunn 1978). Such losses, however, even assuming that they are not temporary, are perhaps better viewed as a natural, though not inevitable, concomitant of the trends that define Borchert's fourth epoch.

An interesting perspective on the evolution of North America's largest urban centers is provided by Table 2.3, which shows every city that overtook New Orleans in size during each decade between 1860 and 1980. New Orleans stood fifth in rank in 1860, behind New York, Philadelphia, Baltimore, and Boston; it had held that position since about 1810, when it first surpassed Charleston. Today New Orleans stands thirtieth. Late in Borchert's second epoch (1830–70), New Orleans was overtaken by Chicago, Cincinnati, and St Louis, the great trio of northern cities whose development depended so heavily on the steamboat and the railroad and whose growth epitomized so clearly the general mid-nineteenth-century increase of population density in the Midwest. During the next epoch, from 1870 to 1920, New Orleans was overtaken by a further twelve cities, two-thirds of which were located in the rapidly industrializing northern interior. The exceptions were Montreal (arguably an 'interior' city at the continental scale), the two leading cities of California, and Washington, DC. A definite break is evident in Table 2.3 between 1920 and 1940. Since the latter date, New Orleans has yielded to eleven more centers. Without exception, members of this most recent group are restricted to the great arc of southern, southwestern, and western territory extending from Florida to British Columbia. The effects of the factors that characterize Borchert's auto-air-oil-amenity epoch could hardly be more strikingly demonstrated. All in all, the timing of rapid growth in the cities that now exceed New Orleans in size reveals very effectively the sequential importance of eastern seaboard colonization, inland river and railroad transportation, Midwestern industrialization, and the twentieth-century sunbelt effect.

The urban world today

The world as a whole now contains some 2,500 cities of at least 100,000 inhabitants, including about 220 centers that exceed one million in size (Table 2.4). These totals are obviously subject to frequent upward revision.

The Soviet Union is the country with the greatest number of cities over 100,000, but the United States holds the lead for cities of a million or more. Six countries stand out with over one hundred cities of at least 100,000: the Soviet Union, the United States, India, Japan, China, and Brazil. These six countries account for 51 per cent of the world's cities of 100,000 or more, and for 54 per cent of the world's total population. Five of the six – Japan is the exception – are among the world's largest countries in terms of territorial extent.

The population living in cities of 100,000 or more inhabitants currently represents 22 per cent of the world's population. This figure may be taken as a general index of the level of world urbanization. A somewhat higher percentage would be obtained if cities of less than 100,000 inhabitants were included in the urban portion of the total population. Unfortunately, the lack of detailed population data for many countries makes it impossible to use with any confidence a figure smaller than 100,000. An

TABLE 2.4
Cities of 100,000 inhabitants and over, c. 1980

	Number of cities		
Country	100,000–999,999	1 million and over	Total
Soviet Union	255	22	277
United States	238	35	273
India	205	12	217
Japan	187	8	195
China	148	33	181
Brazil	119	10	129
West Germany	62	4	66
France	56	3	59
United Kingdom	55	2	57
Mexico	50	4	54
Italy	45	4	49
Spain	47	2	49
Yugoslavia	37	2	39
Poland	38	1	39
Nigeria	33	1	34
South Korea	29	4	33
Indonesia	25	5	30
Philippines	29	1	30
Turkey	22	3	25
Pakistan	21	3	24
Canada	21	3	24
Rest of world	544	59	603
Grand total	2,266	221	2,487

SOURCES: United Nations *Demographic Yearbook* and Philip's *New Geographical Digest*, 1986

abstract calculation based on the rate at which city size declines with falling rank in the global urban system suggests that a threshold of 10,000 inhabitants would raise the urban percentage to a value between 45 and 50 per cent.[3] However, because many places of less than 10,000 inhabitants are definitely urban in character, it is probably safe to say that ours is now an urban world in the sense that residents of towns outnumber residents of rural areas.

As might be expected, there is considerable variation from country to

3 The calculation is an application of the rank-size principle, which is explained in chapter 11.

TABLE 2.5
Relationship between level of urbanization and general standard of living

STANDARD OF LIVING	Least urbanized	LEVEL OF URBANIZATION		Most urbanized
Richest		Czechoslovakia Irish Republic	Austria Belgium East Germany Finland Greece Hungary Italy Libya Norway Poland Saudi Arabia Switzerland	Australia Bahamas Canada Denmark France Israel Japan Kuwait Netherlands New Zealand Puerto Rico Qatar Soviet Union Sweden United Arab Emirates United Kingdom United States Venezuela West Germany
	Malaysia	Costa Rica Dominican Republic Fiji Mauritius Nicaragua Réunion Turkey Yugoslavia	Bahrain Bulgaria Cuba Cyprus Iran Jamaica Lebanon Mongolia Panama Peru Portugal Romania South Africa Trinidad and Tobago Tunisia	Argentina Brazil Chile Gabon Iraq Mexico Netherlands Antilles Spain Surinam Uruguay

GE	Angola Cameroon Liberia Mozambique Papua–New Guinea Sierra Leone	Ecuador El Salvador Ghana Guatemala Guinea–Bissau Guyana Honduras Ivory Coast North Korea Paraguay Philippines Senegal Thailand Zambia Zimbabwe	Bolivia Congo Egypt Morocco Syria	Colombia Jordan South Korea Taiwan
Poorest	Afghanistan Bangladesh Benin Burkina Faso Burma Burundi Central African Republic Chad Ethiopia India Indonesia Kenya Laos Madagascar Malawi Mali Mauritania Nepal Niger Somali Republic Sri Lanka Sudan Tanzania Togo Uganda	Guinea Haiti North Yemen Pakistan South Yemen Vietnam Zaire	Cambodia	

SOURCES: *Geographical Digest 1981* (Willett et al 1981); and the map, *A Developing World*, prepared by the Canadian International Development Agency (Ottawa: Ministry of Supply and Services 1982).

NOTE: The level of urbanization is measured by the percentage of the national population living in cities of 100,000 inhabitants or more. The standard of living is measured by the average per capita annual income. Both axes of the table are divided into quartiles.

country in the percentage of the population living in cities. Retaining the threshold of 100,000 inhabitants, the urban component is found to vary from more than 50 per cent in the United States, Canada, Japan, and Chile to less than 5 per cent in Tanzania, Burkina Faso, Nepal, and Bangladesh. In general there is a positive correlation between levels of urbanization and national standards of living: the poorer the country, the less urbanized it is likely to be, and vice versa. This relationship is shown in Table 2.5, in which countries are arranged in a contingency table according to their average income per capita and the percentage of their population living in large cities. This contingency table contains 134 countries and is worldwide in coverage; the only excluded areas are the city-states of Hong Kong, Macau, and Singapore (all of which are almost 100 per cent urbanized) together with countries having no city of 100,000 or more inhabitants. All told, the excluded territories contain less than 1 per cent of the world's population.

The cells of Table 2.5 are defined by the median and quartile values of the income and urbanization data. In other words, each row of four cells contains one-quarter of the total number of countries, and the same is true for each column. If urbanization and standard of living were completely unrelated, we would expect to find approximately eight countries in each cell, this being one-sixteenth of the total of 134. Clearly this is not the case. There is a strong tendency for countries to be concentrated along the diagonal row of cells from bottom left (poor and agrarian) to top right (rich and highly urbanized) in the table. These four diagonal cells account for seventy-six countries, more than twice the number they would contain if the two variables were unrelated. The strength of the association can be evaluated statistically by means of the chi-squared test. In this instance, a chi-squared value of 101.9 is obtained with nine degrees of freedom, a highly significant result. The positive correlation between standards of living and levels of urbanization is too strong to be attributed to chance.

The most 'favorable' cells in Table 2.5 – those combining high levels of urbanization with high per capita incomes – are dominated by the countries of the Western world, though the most favorable cell of all contains also the Soviet Union and three small oil-exporting countries of the Persian Gulf (Kuwait, Qatar, and the United Arab Emirates). The least favorable cells are occupied almost entirely by countries in sub-Saharan Africa and in southern and eastern Asia. Latin American countries generally occupy an intermediate position. Several of the most populous Latin American countries, including Brazil, Mexico, and Colombia, lie to the right of the main diagonal row of cells: they are more

TABLE 2.6
The world's fifty largest cities, c. 1980

Rank	City	Country	Approximate metropolitan population (millions)
1	New York	United States	17.8
2	Tokyo-Yokohama	Japan	17.4
3	Mexico City	Mexico	14.8
4	Los Angeles	United States	12.4
5	Shanghai	China	11.9
6	Buenos Aires	Argentina	9.9
7	Beijing	China	9.3
8	Calcutta	India	9.2
9	São Paulo	Brazil	8.5
10	Paris	France	8.5
11	Moscow	Soviet Union	8.5
12	Seoul	South Korea	8.4
13	Bombay	India	8.2
14	Chicago	United States	8.0
15	Tientsin	China	7.9
16	Cairo	Egypt	6.8
17	London	United Kingdom	6.8
18	Jakarta	Indonesia	6.5
19	Philadelphia	United States	5.8
20	Lima-Callao	Peru	5.7
21	Delhi	India	5.7
22	San Francisco	United States	5.7
23	Krung Thep	Thailand	5.5
24	Rio de Janeiro	Brazil	5.1
25	Karachi	Pakistan	5.1
26	Osaka-Kobe	Japan	4.8
27	Leningrad	Soviet Union	4.8
28	Teheran	Iran	4.6
29	Detroit	United States	4.6
30	Bogota	Colombia	4.5
31	Madras	India	4.3
32	Shenyang	China	4.1
33	Santiago	Chile	4.1
34	Boston	United States	4.0
35	Ruhr complex	West Germany	3.7
36	Houston	United States	3.6

TABLE 2.6 (*continued*)

Rank	City	Country	Approximate metropolitan population (millions)
37	Dacca	Bangladesh	3.5
38	Manila-Quezón	Philippines	3.5
39	Ho Chi Minh City	Vietnam	3.4
40	Washington, DC	United States	3.4
41	Wuhan	China	3.3
42	Dallas	United States	3.3
43	Sydney	Australia	3.3
44	Kwangchow	China	3.2
45	Madrid	Spain	3.2
46	Toronto	Canada	3.1
47	Baghdad	Iraq	3.0
48	Berlin (whole)	East and West Germany	3.0
49	Athens	Greece	3.0
50	Istanbul	Turkey	2.9

SOURCES: United Nations *Demographic Yearbook* and Philip's *New Geographical Digest*, 1986

highly urbanized than their general standard of living would suggest. No entirely satisfactory explanation for this tendency comes to mind, but it may be significant that these countries were strongly affected by Spanish and Portuguese urbanization during the colonial period.

To conclude this global survey, Table 2.6 provides a ranked list of the world's fifty largest metropolitan areas, c. 1980. Every continent is represented, although Cairo (rank 16) is the only African center and Australasia is represented only by Sydney (rank 43). Sub-Saharan Africa is the most extensive region having no city among the top fifty. The largest urban centers of this region are Kinshasa (the capital of Zaire) and the Johannesburg-Witwatersrand gold-mining conurbation, each with approximately 2.4 million inhabitants. The United States, with ten of the top fifty cities, appears in the list more often than any other country.

New York and Tokyo-Yokohama, which head the list of world cities, are very similar in size. Mexico City, currently in third place, is growing much faster than either of the two leaders and has the potential to become the world's largest metropolis by 1990, if not earlier. For North Americans, however, the most interesting question concerns New York and Los

Angeles. New York has evidently ceased growing, at least for the moment. Los Angeles continues to expand rapidly, though one cannot help wondering how long it will be before California's problem of fresh water supply – or perhaps a major earthquake – will retard the city's growth. Our children may live to see Los Angeles replace New York as the most populous urban agglomeration in the United States. Fifty years from now, Angelenos vacationing in Mexico City may well be traveling between the two largest cities in the world.

Summary

This chapter presented an overview of the main features of urbanization during four major historical periods: the ancient world, the Middle Ages, the mercantile era, and the era of modern industrial growth.

The distinction between local trade and long-distance trade was stressed, and it was indicated that this distinction parallels the contrast between the central-place and corridor concepts of spatial organization. In medieval Europe, the revival of long-distance trade was an important stimulus to urban growth. However, the main force behind the creation of wholly new towns was the need for central places in newly settled regions.

A method of analyzing urban system development by means of indexes of rank mobility was described. The application of this method to European cities during the mercantile era showed that political capitals, ports, and some manufacturing towns experienced the strongest growth.

For the United States since 1800, city development was described in terms of Borchert's four epochs. The chapter concluded with a consideration of the state of world urbanization at the present time.

3

Urban functions and the economic classification of cities

Introduction

THE NATURE OF THE DATA

It is a matter of common observation that cities differ from one another in terms of the products or services for which they are chiefly noted: Pittsburgh for steel, Detroit for motor vehicle manufacturing, Las Vegas for its casinos, Sudbury for mining, and so forth. The detailed study of these differences is one of the major traditions of the field of urban systems analysis.

Research of this kind relies heavily on national census figures that describe the composition of the urban labor force according to various types of economic activity. And here we must immediately distinguish between two separate species of labor force statistics: namely, those dealing with occupations and those dealing with industries. In the 'occupation' tables of the census, each person in the labor force is classified according to the nature of the specific job which he or she performs. In the 'industry' tables, by contrast, the basis of classification is the type of good or service produced by the corporation or agency in which the worker is employed. The difference between these two kinds of classifications may be illustrated by the following hypothetical example involving three persons employed by a manufacturer of aircraft. The first worker is employed on an assembly line in the actual manufacturing plant; the second is a sales representative who spends much of the time on the road negotiating contracts; and the third is a highly trained engineer who helps to design the company's products. The 'occupation' classification would place these persons in three different categories: namely, those of production workers, sales staff, and professional and technical personnel, respectively. The 'industry' classification, however, would report all three under the single heading of the manufacture of transportation equipment.

In the study of urban economic differences, our attention is focused on intercity variation in product mix, and therefore it is the 'industry' data rather than the 'occupation' data that are required. The 'occupation' data are of no value in this context because they do not explicitly identify the actual goods and services that each city produces. It is quite possible for two cities with similar occupational profiles to differ widely in product mix, and such differences will obviously be concealed if reliance is placed on occupational statistics. Throughout this book, in conformity with standard practice, only 'industry' data are considered.[1]

Two additional points concerning labor force statistics must now be emphasized. First, the term *industry* is not synonymous with the term *manufacturing*. The 'industry' tables in the census cover all members of the labor force, not just those who happen to work in factories. Accordingly, retailing, wholesaling, education, public administration, and so forth are all regarded as types of 'industry.' Even agriculture is identified in the census as an 'industry,' although this category, for obvious reasons, is normally excluded from studies of urban systems. In short, an 'industry' in the eyes of the census bureau is simply any distinct type of economic activity. This comprehensive usage of the term is potentially confusing because journalists and other popular writers often treat the words 'industry' and 'manufacturing' as if they were more or less synonymous – as in the phrase 'industrial land use,' for example. Fortunately, the context usually makes it clear which meaning is intended.

Second, industry statistics can be assembled at different levels of aggregation. If the number of categories in the classification is small – as a rule of thumb, fewer than ten – we say that the information is 'highly aggregated' or that we have a 'coarse breakdown' of the data. If the number of categories is large – meaning more than about thirty – we say that the statistics are 'highly *dis*aggregated,' or that we have a 'fine breakdown.' Studies that make use of highly disaggregated data are not

1 Occupational data are of value to sociologists interested in such topics as intergenerational social mobility, but geographers rarely use them. One example is provided by Britton (1973).

necessarily superior; in fact, the finer the breakdown, the greater the danger that important general features will become lost in a mass of details. Most researchers prefer to work with intermediate levels of aggregation: that is, with classifications containing between ten and thirty categories. Depending upon the objectives of the study, the categories of the initial breakdown can later be placed in groups in order to facilitate analysis of the urban system's major characteristics. For example, the different categories of manufacturing – such as textiles, metal fabricating, and chemicals – may be combined into a single class if the focus of attention is on manufacturing in general and not on particular kinds of manufacturing industries. However, statements based on one particular breakdown of the data will not necessarily remain true if the level of aggregation is changed.

The usual first step in any analysis of urban employment structure is to express the numbers of workers engaged in the various activities as percentages of each city's total labor force. The use of percentages, rather than absolute numbers of workers, permits cities of different sizes to be compared with one another directly. The set of percentages for an individual city – summing, of course, to 100 per cent – is known as the 'profile' of industry (or of employment) for that city. The terms *structure* and *mix* are also frequently used in this context.

Employment profiles vary considerably from city to city. For example, if we add together the numbers employed in all the different types of manufacturing in order to obtain a total for the manufacturing sector as a whole, we find that the percentage of the metropolitan labor force accounted for by this sector in Dayton, Ohio, is four times as high as the percentage in Billings, Montana, and six times as high as the percentages in Washington, DC, and Columbia, Missouri (Table 3.1). Columbia, however, far exceeds Dayton, Billings, and Washington with respect to the percentage of total employment in education. In Washington, as might be anticipated, the industrial profile is dominated by employment in public administration (that is, in the civil service); in Sudbury, Ontario, mineral extraction plays the leading role (Table 3.1). Specific comparisons of this kind naturally give rise to a series of broader questions about the urban system as a whole. What does the profile of the *average* city look like? Do some industries vary from city to city more than others? If so, which activities are the most variable, and why? Can we identify distinctive types of cities in such a way that the variability of employment profiles is strongly marked between the different types but relatively insignificant within each type? Finally, if such distinctive types can be

TABLE 3.1
Employment profiles of selected cities (percentages of total labor force)

Industry	Dayton, Ohio	Billings, Montana	Columbia, Missouri	Washington, DC	Sudbury, Ontario
Manufacturing sector					
Foods and beverages	1.2	3.0	0.8	0.4	1.2
Textiles and clothing	0.3	0.1	0.0	0.1	0.1
Printing and publishing	3.6	1.5	1.8	2.5	0.5
Primary metals	1.2	0.0	0.0	0.1	10.2
Metal fabricating and machinery	12.2	1.0	0.3	0.8	0.9
Transportation equipment	3.8	0.1	0.1	0.2	0.1
Electrical products	6.7	0.1	1.7	0.8	0.1
Chemical products	0.7	0.3	0.2	0.2	0.5
All other manufacturing	8.2	3.0	1.2	1.4	1.0
Subtotal	(37.9)	(9.1)	(6.1)	(6.5)	(14.6)
Other industries					
Construction	4.4	7.1	6.0	5.7	8.9
Transportation, communication, and utilities	4.5	10.9	3.9	6.5	5.8
Wholesale trade	3.0	8.6	2.4	2.8	3.1
Retail trade	13.9	18.8	13.3	12.9	11.1
Finance, insurance, and real estate	3.4	5.6	6.7	5.8	3.0
Health services	6.2	7.9	11.4	7.9	5.2
Education	7.3	9.4	32.7	8.8	7.6
Personal services	7.4	12.3	8.6	8.4	6.6
Business services	4.0	4.8	4.4	8.7	3.0
Mineral extraction	0.2	0.8	0.5	0.1	26.5
Public administration	7.8	4.8	4.1	25.8	4.6
Subtotal	(62.1)	(91.0)	(94.0)	(93.4)	(85.4)

SOURCES: Calculated from industry tables of national censuses (United States 1970; Canada 1971)
NOTE: Workers in agriculture, forestry, fishing, trapping, and 'industry not specified' are excluded. Values are correct to one decimal place. Owing to rounding, the columns may not sum to exactly 100 per cent.

recognized, what locational patterns do they reveal when placed on maps?

Finding answers to these questions is the principal task to which the remainder of this chapter is addressed.

THE AVERAGE CITY

A moment's reflection will show that the employment profile of the statistically defined 'average city' cannot be expected to remain constant from country to country or from one historical period to another. If the necessary data were available, we would no doubt be able to show that the average profile for English cities during the fourteenth century was quite different from their average profile today. Similarly, the average profile of American cities at the present time is unlikely to be the same as the average profile of, say, Peruvian cities. For illustrative purposes, however, it will be convenient to develop a single example in detail. Accordingly, the set of cities analyzed in this chapter and in chapter 4 consists of all centers in the conterminous United States and in Canada having 50,000 or more inhabitants at the time of the 1970–1 censuses of industrial employment (1970 in the United States, 1971 in Canada). In conformity with the concept of the geographic city, the data refer to metropolitan areas, not to central cities. In all, 268 cities are included: 234 in the United States and 34 in Canada.

Treating American and Canadian cities as members of a single urban system does not imply that Canadian cities and American cities are alike in all respects. Indeed, as Goldberg and Mercer have recently demonstrated, Canadian cities differ significantly from their American counterparts along a number of important social and political dimensions (Goldberg and Mercer 1986). However, viewed from the perspective of economic history, American and Canadian cities have generally responded to the same kinds of forces in the same kinds of ways. With respect to employment structure as such, cities on both sides of the border are cut very largely from the same cloth. Reflecting this similarity in urban economic experience, the Canadian and American censuses use almost identical sets of categories in their respective classifications of industries.[2]

The average industrial profile of the 268 cities is shown in Table 3.2. For the sake of clarity, the activities that comprise the manufacturing sector are shown separately from all other industries. Each entry in the

2 Two specific differences should be noted, however. In the United States, the postal service is classified under 'public administration,' whereas in Canada it is classified under 'transportation, communication, and utilities.' In addition, the Canadian census includes members of the armed forces as part of the labor force, while the American census does not, although civilian employees in military installations are included in 'public administration.' In order to achieve comparability, I have discarded the armed forces from the Canadian data and transferred the Canadian postal service from 'transportation, communication, and utilities' to 'public administration' (where, in fact, it was assigned in Canadian censuses prior to 1961).

TABLE 3.2
Average composition of the urban labor force

Industry	Average percentage[a]	Standard deviation	Spatial distribution[b]
Manufacturing sector			
Foods and beverages	2.1	1.7	S
Textiles and clothing	2.1	3.3	S
Printing and publishing	1.4	0.6	U
Primary metals	2.1	3.9	S
Metal fabricating and machinery	4.6	4.4	S
Transportation equipment	2.7	4.4	S
Electrical products	2.5	3.4	S
Chemical products	1.3	2.0	S
All other manufacturing	6.6	4.3	S
Subtotal	(25.4)		
Other industries			
Construction	6.3	1.4	U
Transportation, communication, and utilities	6.9	2.2	U
Wholesale trade	4.4	1.5	U
Retail trade	15.0	2.0	U
Finance, insurance, and real estate	4.9	1.5	U
Health services	7.8	2.0	U
Education	9.0	4.1	U
Personal services	9.0	2.7	U
Business services	4.1	1.4	U
Mineral extraction	1.0	2.8	S
Public administration	6.1	3.8	S
Subtotal	(74.5)		

SOURCES: As for Table 3.1
NOTE: This table is based on the 268 cities and metropolitan areas in the United States and Canada (excluding Alaska and Hawaii) with populations of 50,000 or more in 1970–1. Workers in agriculture, forestry, fishing, trapping, and 'industry not specified' are excluded.
a The percentages are not weighted by city size. Weighting has very little effect on the values and suffers from the disadvantage that no standard deviations can be given.
b An industry is classified as sporadic (S) if its standard deviation is greater than one-half of the value of its average percentage; U = ubiquitous.

column of average percentages was obtained by summing the appropriate percentage values from the 268 individual cities and dividing the result by 268.

Many people are surprised to learn that the manufacturing sector, on average, accounts for only one-fourth of the total urban labor force. The knowledge that we live in a highly 'industrialized' society seems to create the impression that jobs in factories represent a majority of all employment. However, three out of every four jobs lie outside manufacturing, with service industries such as retailing, personal services (e.g. hotels, restaurants, beauty parlors, dry cleaners, movie theaters), education, and health care being the largest employers. In the past few decades, moreover, employment has been growing much more rapidly in the service industries than in manufacturing, and we have no reason to suppose that this trend will not continue in the foreseeable future. We should therefore expect that the manufacturing sector's share of total employment will become even smaller as the future unfolds.

THE GEOGRAPHIC VARIABILITY OF INDUSTRIES

Table 3.2 also shows the value of the standard deviation associated with the average percentage for each industry. A simple but effective measure of the variability of an industry from city to city can be obtained by dividing the standard deviation by the corresponding average, producing a statistic known as the coefficient of variation. Division by the average compensates for the fact that the average values differ from one industry to another. For example, retail trade and the manufacture of chemical products have identical standard deviations, but the former has by far the larger average value and therefore a much smaller degree of relative variability from city to city. Arbitrarily, any industry with a coefficient of variation below 0.5 is here classed as ubiquitous, while the remaining industries are regarded as having locational patterns that may be termed sporadic.

All the manufacturing activities except printing and publishing are sporadic in their geographic pattern of occurrence, whereas all other industries except mineral extraction and public administration are ubiquitous. This means that, if two cities are selected at random, there is a high probability that their manufacturing sectors (expressed, of course, in percentage terms) will be markedly different, whereas their profiles in the nonmanufacturing activities are likely to be quite similar. This contrast reflects the fact that manufacturing involves the creation of a tangible product while most of the nonmanufacturing industries do not. The industries in the manufacturing sector must take account of the uneven spatial distribution of raw materials, but virtually all other industries – with mineral extraction the only major exception – are free to locate near their markets. Indeed, except for mineral extraction and, to a lesser extent, public administration, the geographic distribution of nonmanufacturing activities closely parallels the distribution of total population. Manufacturers, of course, must also be aware of the locations of their markets, but within this general constraint they tend to select specific sites where the necessary raw materials can be assembled at low cost. In many types of manufacturing, a relatively small number of large plants occupying favorable locations in terms of raw material assembly can profitably supply a product to the entire national territory. The same obviously cannot be said for the industries in the nonmanufacturing sector. The services provided by a retail store, a school, or a hospital cannot be shipped across the country like a refrigerator or a box of breakfast cereal. This limitation also applies to the construction industry, which produces a tangible product but one that generally remains fixed at the location where it is built (prefabricated homes and cottages being a partial exception). Thus nonmanufacturing activities are found wherever there is a body of customers for them to serve.

Mineral extraction and public administration violate this general pattern, but these anomalies can readily be explained. Economically valuable mineral deposits (aside from local quarries and gravel pits) are comparatively rare occurrences. Very few, in fact, are important enough to be the dominant source of employment in cities of the size considered here. Most cities have negligibly small employment in this category, but a few cities have massive amounts. Thus it is not surprising that mineral extraction has a higher coefficient of variation than any other industry. Public administration, in turn, is dominated by civil service employment at the state or provincial and federal levels. These workers, for reasons of administrative efficiency, tend to be locationally concentrated. They are highly centralized, in particular, in state and provincial capital cities and in the two national capitals, Washington, DC, and Ottawa. However, the category of public administration includes also local civic officials and the attendant municipal bureaucracies found in every city. Fittingly, public administration has the lowest coefficient of variation among all the activities classed as sporadic in Table 3.2.

Finally, we noted that printing and publishing is the only activity in the manufacturing sector to be classed as ubiquitous. This activity is dominated numerically by workers employed in the production of daily newspapers. Since every city of 50,000 or more inhabitants in North

America produces at least one daily paper, it is not surprising that the coefficient of variation for the printing and publishing industry is relatively low.

Functional classification: general aspects

When cities are grouped into distinctive types according to the overall characteristics of their economic profiles, the result is known as a functional classification of towns. Functional classifications have three purposes. First, by summarizing large bodies of data in the form of a small number of basic categories, they simplify the task of making oneself familiar with an urban system's economic character. Second, when presented in the form of maps, they draw attention to the fact that significant relationships often exist between functional types and locational patterns. The resulting appreciation of the causal connection between location and function enhances our understanding of the spatial distribution of towns. Third, they make possible the formulation and testing of specific hypotheses relating the functional differentiation of cities to other variables, such as city size or the degree of industrial diversity of the urban economy. One example of such a hypothesis is the proposition that cities that function primarily as central places (that is, as regional trade centers) are, taken as a class, more diversified than cities in which manufacturing is the major economic activity. (This hypothesis is tested in chapter 4.)

It has occasionally been pointed out that the purposes of functional classification have rarely been stated explicitly in the published literature (O.D. Duncan et al 1960, 35; R.H.T. Smith 1965a, 539). Although this comment is valid, it should be interpreted as being concerned more with style than with substance. Internal evidence indicates that most writers have had at least one of the above three purposes in mind when creating their classifications. The second objective in particular – that of elucidating the link between function and location – may fairly be said to be inherent in all taxonomic work undertaken by geographers.

The topic of functional classification received its initial impetus early in the present century from Hettner (1902) and Aurousseau (1921), and the number of taxonomic studies that have appeared over the past sixty years is formidably large. Highlights include studies by Betjeman (1943), Harris (1943), Pownall (1953), Nelson (1955), Steigenga (1955), Maxwell (1965), R.H.T. Smith (1965b), Harris (1970), and Stanback and Knight (1970). Superficially this literature is very diverse, especially in terms of analytical methods, but several general features give the field a certain unity. These general features will now be described. No attempt is made here, however, to reproduce the details of individual studies, since an overview of general characteristics is more useful than a summary of particular works.

Perhaps the most basic theme in functional classification – though not often expressed – is the idea that cities are essentially multifunctional. Even where one particular activity is popularly said to 'dominate' the local economy, as with the civil service in Washington, DC, or motor vehicle manufacturing in Detroit, it is taken for granted that other kinds of urban functions are normally also present. In most cities no obviously dominant type of activity exists, and it is often difficult even for local residents to identify confidently the functions that are comparatively well developed. On the whole, classification does not separate cities into radically different types but simply shows that each economic activity is relatively more prominent in some cities than in others. Thus, for example, labeling a town as a manufacturing center does not imply that other industries, such as wholesaling, retailing, and education, are absent, but only means that manufacturing comprises a conspicuously large proportion of the town's labor force. (The meaning of 'conspicuously large' is explored below.) Similarly, classifying a city as a resort, or as a university town, does not imply that it has no employment in manufacturing. In short, the labels attached to the categories of a functional classification must be understood to indicate no more than the relative importance of the various urban activities, most of which are present in every city to a greater or lesser degree.

Another general point is that functional classification should be based, whenever possible, on quantitative data. The earliest works in this field, such as that of Aurousseau (1921), were purely qualitative, identifying each type of city by means of verbal description alone. But this approach means that anyone who undertakes a comprehensive study of the cities of a particular urban system (a task not attempted by Aurousseau) must rely exclusively on intuitive judgment in assigning individual towns to different classes. Under these conditions, consistency in the assignment of towns to classes cannot be ensured. Later works, beginning with a justly famous study of American cities by Harris (1943), have invariably made use of quantitative data and have employed numerical rules to govern the allocation of cities to the categories of the classification. The quantitative approach eliminates the risk of misclassifying individual towns and clearly specifies taxonomic criteria for other workers to adopt or to criticize and modify.

It must be strongly emphasized that the use of numerical rules does not remove the intuitive or subjective element from taxonomic work. Subjectivity is always present in the investigator's choice of quantitative criteria. Harris, for example, used employment percentages chosen arbitrarily 'on the basis of an analysis of cities of well-recognized types' (Harris 1943, 87). Nelson (1955) and Steigenga (1955), in contrast, applied the conventional statistical concepts of the mean and the standard deviation to the percentages of the urban labor force in each activity. Other workers have employed numerical procedures of considerable complexity (R.H.T. Smith 1965b; Maxwell 1965). Each procedure, once selected, can of course be objectively applied to the data – that is, applied rigorously, without allowing any exceptions or 'special cases.' But the actual choice of technique is fundamentally subjective. Failure to recognize this ever-present element of subjectivity can lead to exaggerated claims for the 'objectivity' of the resulting classification.

A further general point is that the level of employment (in percentage terms) at which an activity may be deemed 'conspicuous' or 'significant' in any city must be allowed to vary from function to function. The force of this point is best illustrated by means of an example. Suppose that a particular town has exactly 20 per cent of its labor force in manufacturing (that is, all kinds of manufacturing combined) and a further 20 per cent in retail trade. In absolute terms, manufacturing and retailing are equally important in this town's employment structure. The equality vanishes, however, when the two percentages are compared with their respective system-wide averages. As Table 3.2 shows, 20 per cent lies below the North American urban average for manufacturing but above the average for retail trade. We might therefore feel justified in concluding that retail trade is 'significant' and that manufacturing is 'not significant' in the city in question. The 'expected' level of employment, as revealed by the systemic average, varies from function to function. Failure to allow for this natural variation leads to confusing results. Thus, if '20 per cent' were applied as a fixed criterion for both manufacturing and retail trade, manufacturing centers would be deceptively numerous and retailing cities would be misleadingly rare. In fact, out of the 268 cities on which Table 3.2 is based, 166 (62 per cent) have at least 20 per cent of their labor force in manufacturing; only 4 (less than two per cent) attain this level in retail trade. (To satisfy the curious, these four cities are Lawton, Oklahoma, and Laredo, McAllen, and Brownsville, Texas.) No one who is familiar with North American cities would regard such a result as providing a reasonable description of the relative importance of retailing and manufacturing within the urban system.

From the viewpoint of the statistician, the systemic average percentage for a given function is the proper indicator of the expected level of employment in that activity. Indeed, the average or mean value of a set of numbers is still occasionally referred to as the 'expectation' or the 'expected value' in statistics texts. However, writers on functional classification have tended to define 'significance' in terms of critical values that lie *above* the average or expected percentage. Nelson (1955), for example, used the mean percentage plus one standard deviation as the critical value for each activity. The use of criteria lying above the average values has never been convincingly justified, but it appears to be favored on the grounds that the straightforward use of the mean percentages would give each city 'too many' significant functions, thus leading to a classification too laden with detail to be easily digested. The adoption of criteria that exceed the average percentages, however, introduces the possibility that some cities may turn out to have *no* significant functions. In fact, this is precisely what occurred in 27 per cent of the cities analyzed by Nelson (1955). Nelson responded to this problem by calling these places 'diversified' cities, noting that their employment structures were more similar to the national average structure than were the profiles of the cities in any other group.

The final point to be noted concerning the taxonomic literature in general is the high degree of agreement among different workers on the issue of the actual list of functional types into which cities are to be classified. The basis of this agreement is the fact that a majority of the world's cities, especially in economically advanced countries such as the United States and Canada, are supported primarily either by trade (both local and long distance), by manufacturing, or by a blend of both. These commercial-industrial centers can be separated into classes according to the relative importance of trade and manufacturing. Stanback and Knight, for example, divided commercial-industrial cities into three groups identified as 'manufacturing,' 'mixed,' and 'nodal' centers, these terms representing segments of the functional continuum that runs from specialized manufacturing cities at one extreme to pure central places at the other (Stanback and Knight 1970, 116–56). There also exists, however, a second general group of towns, although it is a small group in comparison to the number of cities that belong to the main commercial-industrial continuum. In the cities of this second group, neither trade nor manufacturing is strongly developed, and some relatively uncommon type of activity comes to the fore. The most frequently recognized classes within this second group are mining towns, resorts, university towns, political capitals, military towns (with naval bases sometimes a separate subtype), and, particularly in the Old World, pilgrimage centers.

In effect there is a pervasive distinction in the literature between two super-classes of towns, one containing the numerically dominant commercial-industrial centers and the other containing the towns of all the minor types taken collectively. It is useful to give these two super-classes names of their own. In this book we shall refer to the towns of the main commercial-industrial continuum as normal, and to the towns of all other types as exceptional. This basic distinction is significant for two reasons. First, it serves to remind us that trade and manufacturing have been the overwhelmingly dominant urban activities, in terms of the numbers of people supported, throughout the historical evolution of the global urban network. Second, the distinction turns out to be important for our understanding of the spatial distribution of cities. Specifically, normal cities exhibit a spatial pattern embodying strong systematic (that is, nonrandom) tendencies, with clustering of cities where manufacturing is prominent and wide dispersal where trade receives the greatest emphasis. Exceptional cities, in contrast, although their locations are individually explicable, form a subsystem that consists, geometrically speaking, of points scattered at random across the territory occupied by the system's normal towns. These fundamental elements of spatial structure are illustrated in the analysis of North American cities presented below.

The fact that different researchers have produced urban taxonomies containing essentially the same set of functional classes strongly suggests that these classes –manufacturing towns, mining towns, and so forth – can be accepted as valid and useful subdivisions of the complex totality of the urban economy. Functional classes, in short, are 'natural kinds' (Quine 1969). These classes, moreover, should not be regarded as being dependent on the particular employment categories recognized by the census bureau and other statistical agencies. Rather, the categories in the census are themselves 'natural kinds,' reflecting distinctions that permeate our economic life and are accordingly used as familiar labels in everyday speech. Ultimately, a functional classification of towns is a particular way of expressing the fundamental economic concept of the division of labor, a concept that clearly has geographical as well as sociological implications.

A functional classification of North American cities

We now present, in order to illustrate the general points made above, a functional classification of the 268 American and Canadian cities having 1970–1 populations of 50,000 or more.[3] The overall taxonomic strategy is as follows. First, exceptional cities (as defined above) are separated from the full set of towns by means of a series of diagnostic criteria specially designed for this purpose. Second, cities with normal profiles are divided into three groups, according to the relative importance of manufacturing and of the trade and service activities in their industrial composition.

THE EXCEPTIONAL CENTERS

The following criteria are used to identify the four types of exceptional cities.

Mining towns. Employment in the census category of extraction is greater than 12 per cent of the city's total labor force. A city's 'total labor force' in this analysis excludes the census categories of agriculture, forestry, fishing, trapping, and 'industry not specified.' The total labor force, in other words, is the sum of the twenty industrial categories listed in Tables 3.1 and 3.2.

Government centers. Employment in public administration exceeds 12 per cent of the city's total labor force. Some government centers are actually military towns – all military bases, of course, being owned and operated by the federal government. As noted earlier, members of the armed forces are not included in the US census and have been deleted from the Canadian data in order to ensure comparability. However, some military towns have a relatively high proportion of their employment in public administration owing to the presence of significant numbers of civilian workers in military installations.

Resorts. Employment in personal services exceeds 15 per cent of the city's total labor force. The category of personal services includes workers in hotels and motels, in restaurants and drinking places, and in various

3 Since this chapter was written, employment data for the censuses of 1980 (United States) and 1981 (Canada) have been released. Archer and White (1985) have analyzed the 1980 American data. A comparison of studies completed for different dates shows that the functional structure of the North American urban system has been very stable over time. In fact, the principal features described in this chapter and in Archer and White (1985) have remained essentially unchanged since at least as far back as the 1940s (Harris 1943).

TABLE 3.3
Exceptional cities in the North American urban system

Class	Cities	Percentage of labor force in relevant diagnostic industry[a]
Mining towns	Sudbury, Ontario	26.5
	Midland, Texas	25.2
	Odessa, Texas	15.1
	Sydney, Nova Scotia	13.3
Government cities	Ottawa, Ontario	29.7
	Washington, DC	25.8
	Macon, Georgia	20.2
	Sacramento, California	19.0
	Quebec, Quebec	17.2
	Tallahassee, Florida[b]	16.9
	Huntsville, Alabama	16.1
	San Antonio, Texas	15.3
	Lawton, Oklahoma	14.8
	Texarkana, Arkansas	14.7
	Newport News, Virginia	14.7
	Regina, Saskatchewan	14.5
	Salt Lake City, Utah	14.3
	Harrisburg, Pennsylvania	14.0
	Oklahoma City, Oklahoma	13.8
	Springfield, Illinois	13.6
	Norfolk-Portsmouth, Virginia	13.3
	Victoria, British Columbia	12.8
	Austin, Texas	12.6
	Pensacola, Florida	12.3
	St John's, Newfoundland	12.1
Resorts	Las Vegas, Nevada	33.7
	Reno, Nevada	25.5
	Atlantic City, New Jersey	16.8
	Fort Lauderdale, Florida	15.6
	West Palm Beach, Florida	15.5
University towns	Columbia, Missouri	32.7
	Bryan–College Station, Texas	32.6
	Champaign-Urbana, Illinois	32.0
	Gainesville, Florida	27.6
	Lafayette, Indiana	26.2
	Ann Arbor, Michigan	24.4
	Provo-Orem, Utah	22.9

TABLE 3.3 (*continued*)

Class	Cities	Percentage of labor force in relevant diagnostic industry[a]
	Tallahassee, Florida[b]	22.8
	Madison, Wisonsin	19.2
	Bloomington, Illinois	17.8
	Kingston, Ontario	16.7
	Lansing, Michigan	16.6

a The relevant diagnostic industries are as follows: for mining towns, mineral extraction; for government cities, public administration; for resorts, personal services; and for university towns, education. See text for detailed discussion of criteria.

b Tallahassee is both a government center and a university town. No other city falls into more than one class.

branches of the entertainment industry (though not in the making of motion pictures, which is classified as a type of manufacturing). A strong showing by these amenities in the local employment profile is a characteristic feature of tourist resorts.

University towns. Employment in education is greater than 16 per cent of the city's total labor force. It should be noted that employment in education does not include students, but only persons on the payroll of educational institutions.

Each of the numerical criteria just described was selected by considering the images evoked by individual cities and by marrying these images to a list in which all cities of the system were ranked in descending order according to the percentages of their labor forces employed in the activity in question. Each cutoff point was chosen so as to identify, at the top of the ranked list, a group of cities in which the activity is commonly regarded as being dominant. Inevitably, this approach leaves room for differences of opinion as to the exact placement of each cutoff point. However, no taxonomy is exempt from subjective judgment. The results obtained with the criteria listed above are shown in Table 3.3.

It is theoretically possible for a single town to satisfy two or more of the above criteria simultaneously. Such overlap is not necessarily a weakness, although the results can be somewhat bewildering if multiple labeling is common, as in the studies by Nelson (1955) and Maxwell (1965). In the present case, only one city attains the critical threshold in more than one of the four exceptional categories: Tallahassee, Florida, is both a governmental and an educational center (Table 3.3).

Exceptional cities in general, and mining and resort towns in particular, are comparatively rare. Only 41, or 15 per cent, of our 268 cities fall into the exceptional categories. Of course, we would not expect cities labeled 'exceptional' to form anything but a small minority within the urban system. However, the relative frequency of exceptional cities would probably be higher than 15 per cent if the analysis could be extended to include small towns (which would require access to unpublished statistics). Mining towns, for instance, rarely exceed 50,000 inhabitants. Even major mining centers such as Butte, Montana, and Thetford Mines, Quebec, are well below this level. Similarly, many well-known resorts, such as Aspen, Colorado, and Boothbay Harbor, Maine, are small, often having fewer than 10,000 permanent residents. The same applies to the famous resort centers of the French and Swiss Alps, such as Chamonix, St Moritz, and Gstaad. Indeed, for both resorts and mining towns, their fame is usually out of proportion to their size. The ranks of small towns also include some government centers (e.g. Montpelier, Vermont) and university towns (e.g. Buckhannon, West Virginia). The proportion of exceptional cities might well prove to be higher among small towns than among larger centers. Nevertheless, taking all sizes together, exceptional centers would probably not exceed 20 per cent of all towns. Most small towns, in fact, function primarily as shopping and service centers for a surrounding rural region. As central places, they therefore belong to the commercial-industrial continuum of normal towns.

Mining centers differ from one another on the basis of the particular minerals they produce. Of the four mining towns appearing in Table 3.3, Midland and Odessa, Texas, are centers of oil production; Sydney, Nova Scotia, is noted for coal; and Sudbury, Ontario, is one of the world's leading producers of nickel and copper ores. The refining of mineral products is classified in the census under the heading 'manufacturing'; 'extraction' covers only the production of ores, concentrates, crude petroleum, natural gas, and nonmetallic minerals, such as sand, gravel, and building stone. Refineries, moreover, are not necessarily located at the sources of their raw materials. In Sudbury, for example, the principal mining company (International Nickel) refines copper on site but ships nickel concentrate to be refined at Port Colborne, Ontario, a small town that now forms part of the St Catharines–Niagara CMA.

Government centers may be subdivided into political capitals and others. Capitals account for thirteen of the twenty-one government cities recognized in this analysis. There are seven state capitals (Sacramento, Tallahassee, Salt Lake City, Harrisburg, Okalahoma City, Springfield, and Austin) and four provincial capitals (St John's, Quebec, Regina, and Victoria), with Ottawa and Washington, DC, heading the list (Table 3.3). The remaining eight government centers are noted for major military bases, except Huntsville, Alabama, which is the site of a major research center for rocketry and space exploration – a federal facility not devoid of military significance.

Military towns pose a problem in this study because of the exclusion of members of the armed forces from the data. As indicated earlier, civilian employment in military installations is occasionally great enough to give a town an exceptional profile even where non-military governmental activity falls within normal limits (e.g. in San Antonio, Texas, and Macon, Georgia). However, certain cities with major military installations are not exceptional on the basis of our data, most notably, San Diego and Halifax, both of which contain naval bases. (The installations at Halifax are not large by American standards, but the military presence there is stronger than in any other Canadian city included in this analysis.) Because the inclusion of San Diego and Halifax as exceptional towns would violate our chosen taxonomic criteria, I am leaving these two centers as normal cities. Their military importance, nevertheless, is duly acknowledged.

The remaining exceptional centers consist of five resorts and twelve university towns (Table 3.3). Three of the resorts – Las Vegas, Reno, and Atlantic City – owe their development more to liberal gambling laws than to the presence of any noteworthy environmental amenities. In Florida, however, climate is the basic attraction, and probably all urban centers in this part of the country derive part of their support from the tourist trade. Fort Lauderdale and West Palm Beach are merely the most favored among many beneficiaries.

The twelve university towns reflect the fact that some institutions of higher learning have been founded in small cities, possibly to create a tranquil environment removed from metropolitan noise and congestion. In a major city, even a large university may have little impact on the structure of the labor force. By contrast, a large or even a medium-sized university located in a smaller city not only concentrates employment in the educational sector but also shapes the general cultural ambiance of the town. This unique urban character justifies the inclusion of university towns as a distinct functional category (Gilbert 1961).

As Figure 3.1 shows, the university towns identified in this study are not uniformly distributed geographically: seven are clustered in a relatively small area in the Midwest. In fact, university towns have

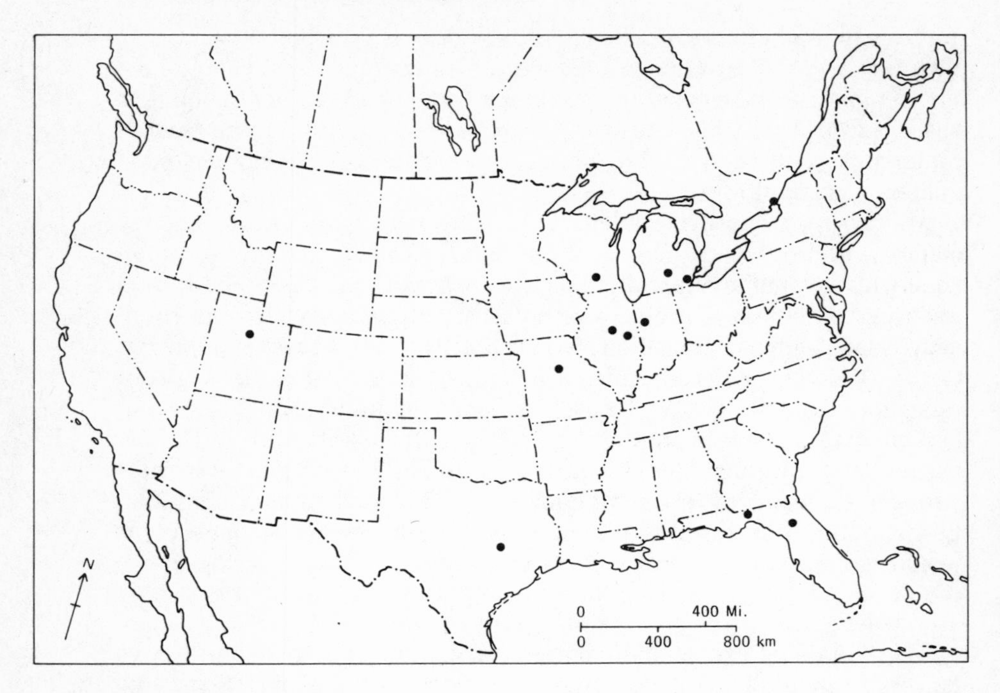

Figure 3.1
University towns (50,000 or more population, 1970). Note the cluster of centers in the Midwest. For the names of individual centers, see Table 3.3.

traditionally formed a significant element in the fabric of Midwestern urbanization, an element that is lacking or only weakly represented in other regions of North America.

TWO UNUSED CATEGORIES

So far as exceptional industrial profiles are concerned, the four types discussed above (mining towns, government centers, resorts, and university towns) are the only functional categories that are indispensable in the North American context. However, two further exceptional categories – transportation cities and religious centers – were considered for inclusion and subsequently discarded.

The category 'transportation city' has been used frequently in previous studies but is potentially misleading: as indicated in chapter 1, *every* city can legitimately be viewed as a nodal point on a transportation network. Among cities of the size considered here – 50,000 inhabitants or more – the most common reason for an unusually high percentage of employment in transportation is the presence of extensive railroad repair shops and marshaling yards, as at Roanoke, Virginia, and, most notably, Moncton, New Brunswick. However, the transportation function is not the essential economic characteristic of these towns. Most observers would classify Moncton as the leading central place of eastern New Brunswick and Roanoke as a diversified city, the principal service center for a section of the Great Valley, with a moderate amount of manufacturing. For large cities, transportation is best regarded as a ubiquitous general framework or, as Lukermann (1966) has put it, a 'circulation manifold' in which all urban settlements are embedded. Here and there, particularly in major railroad centers and in ports, the percentage of employment in transportation is slightly inflated, but a separate category of 'transportation cities' does not appear to be a useful element in a functional classification of large centers.

Even among small places, few towns would qualify as being primarily transportation centers. One contender is Gander, Newfoundland, a town sustained by its role as a stepping-stone and navigational control point for transatlantic air traffic. Another is Sept-Iles, Quebec, a port created for the transshipment of Labrador iron ore bound for the steel mills of the lower Great Lakes. (Recent mine closures have placed its future in jeopardy.) Although rare, these examples suggest that cautious use of the category of transportation centers may help in the study of the functions of small towns. But no North American city with more than 50,000 people is dominated by transportation. The nearest approaches are provided by the Lake Superior ports of Duluth-Superior (shipping iron ore and concentrate) and Thunder Bay (some iron ore, but chiefly prairie grain). Both of these cities, however, perform important central-place functions in addition to their roles as transshipment points.

As for 'religious centers,' North America is not devoid of sacred locations. As examples, the shrine at Sainte-Anne-de-Beaupré in Quebec and the Hill Cumorah site in upstate New York each attract thousands of religiously motivated visitors every year (Roman Catholics and Mormons, respectively). However, sacred sites in North America have had a negligible effect on the growth of cities. The United States and Canada have no cities in which activities connected with religious observances are vital to the local economy, as in such Old World pilgrimage centers as Lourdes, Fatima, Rome (or, more specifically, Vatican City), Varanasi, and, above all, Mecca and Jerusalem. North American cities are the products of a secular society.

There is one partial exception. As the headquarters of the Church of Jesus Christ of Latter-Day Saints (the Mormons), Salt Lake City might be regarded as an ecclesiastical capital, a faint but nevertheless recognizable analog of Rome, Mecca, or Constantinople. But despite Salt Lake City's undoubted significance as a spiritual center, the economic basis of its growth lies in its dual role as the state capital of Utah and as a major regional service center. It is therefore not classified as a religious center. In fact, it satisfies the criterion laid down above for governmental centers and is so classified in this analysis.

The great cathedral cities of Europe, like Salt Lake City, also derive their principal economic support from temporal rather than spiritual sources. The category 'cathedral towns' has a degree of explanatory value for historical studies of the development of European cities before the Industrial Revolution (H.T. Johnson 1967), but the great cathedrals have little economic significance today beyond their role as tourist attractions. Most cathedral towns are now sustained by the typical modern blend of commerce and manufacturing; they are normal rather than exceptional cities. Even at Chartres, where the contrast between the small size of the town and the massive bulk of the cathedral remains especially sharp, survival depends less on the cathedral's drawing power than on the town's role as the leading central place of the districts of Beauce and Thimerais.

NORMAL CENTERS

Normal centers can be defined in negative terms as cities that fail to meet any of the criteria given above for the identification of exceptional places. More positively, we may say that normal centers are cities supported primarily by one or both of the two historically pre-eminent urban functions: namely, trade and manufacturing. As noted earlier, normal cities form a continuum, with regional service centers (e.g. Billings, Montana) at one end and cities heavily involved in manufacturing (e.g. Dayton, Ohio) at the other. (The employment profiles of Billings and Dayton appear in Table 3.1.) On this continuum the strength of the manufacturing sector – measured in the usual way, as a percentage of each city's total labor force – tends to be inversely proportional to the strength of the trade and service activities. In other words, where manufacturing is strongly represented, trade and service industries are relatively weakly developed, and vice versa. Consequently, the position of any particular city along the commercial-industrial continuum can be judged simply by looking at the percentage of the labor force in manufacturing. If this percentage is high, we are dealing with a 'manufacturing city'; if it is low, we are dealing with a central place or 'nodal city.'[4]

Following the lead (although not the method) of Stanback and Knight (1970, 116–56), I have divided the continuum of normal cities into three segments, as follows.

Manufacturing cities. Employment in the manufacturing sector is greater than 34 per cent of the city's total labor force.

Mixed cities. Employment in manufacturing is greater than 21 per cent of the city's labor force, but not greater than 34 per cent.

Nodal cities. Employment in manufacturing does not exceed 21 per cent of the city's total labor force.

Unlike the criteria selected earlier for exceptional centers, these percentages make the three classes of normal cities mutually exclusive by definition. Choosing the critical values, of course, is a matter of personal judgment; but, once selected, these criteria can be applied consistently to allocate normal centers into three nonoverlapping functional categories. In the present case we find sixty-nine cities in the manufacturing class, eighty in the mixed class, and seventy-eight in the nodal class. All told, the normal centers account for 85 per cent of the 268 cities included in the analysis.

The spatial distributions of the manufacturing, mixed, and nodal cities are shown in Figures 3.2, 3.3, and 3.4, respectively. In contrast to the exceptional cities, the normal cities are sufficiently numerous to permit general statements to be made concerning their locational patterns. The most striking feature is the very marked spatial concentration exhibited by the manufacturing cities (Figure 3.2). Apart from two Canadian outliers (Shawinigan and Sault Ste Marie) and an arc of five southern cities reaching from Lynchburg, Virginia, to Gadsden, Alabama, the manufacturing cities are confined to a narrow belt of territory extending from southern New England and the adjacent Middle Atlantic seaboard to the Midwest. No city in the manufacturing class lies west of Cedar Rapids,

4 The term *trade and service industries* can be given a number of different operational definitions. At a minimum, it includes wholesale and retail trade. It may also be defined broadly to encompass all activities except manufacturing and mineral extraction (Table 3.2). Experiments show that the inverse relationship with manufacturing holds true regardless of whether the definition adopted is narrow or broad.

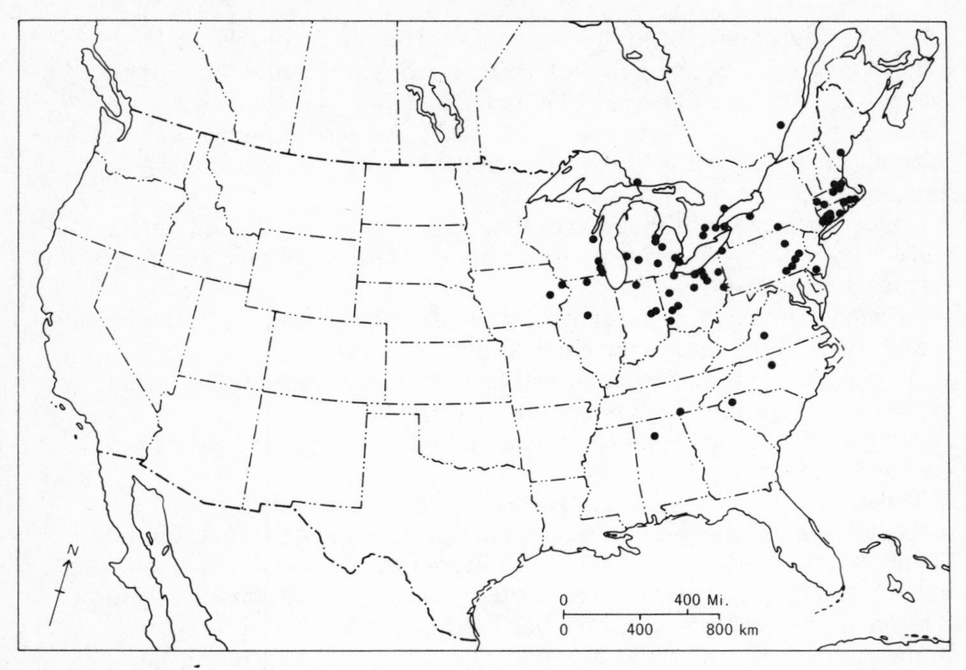

Figure 3.2
Manufacturing cities (50,000 or more population, 1970). The concentration of manufacturing centers east of the Mississippi and north of the Ohio-Potomac line is one of the most persistent features of the spatial structure of the North American urban system.

Iowa, and only the five southern cities already mentioned lie south of Hamilton, Ohio.

Now it might be objected that we should hardly be surprised to find the manufacturing cities situated within the traditional North American manufacturing belt. As Macaulay would have put it, 'every schoolchild knows' that American factory employment is concentrated in the northeastern quadrant of the country. This objection, however, puts the cart before the horse. In actual fact, it is only through studies of the sort we are now describing that a region known as the 'manufacturing belt' can be identified in the first place. Such studies might make use of data for counties or even whole states instead of cities, but without some sort of empirical analysis the idea of a manufacturing belt remains an untested hypothesis. Schoolchildren 'know' about the northeastern manufacturing

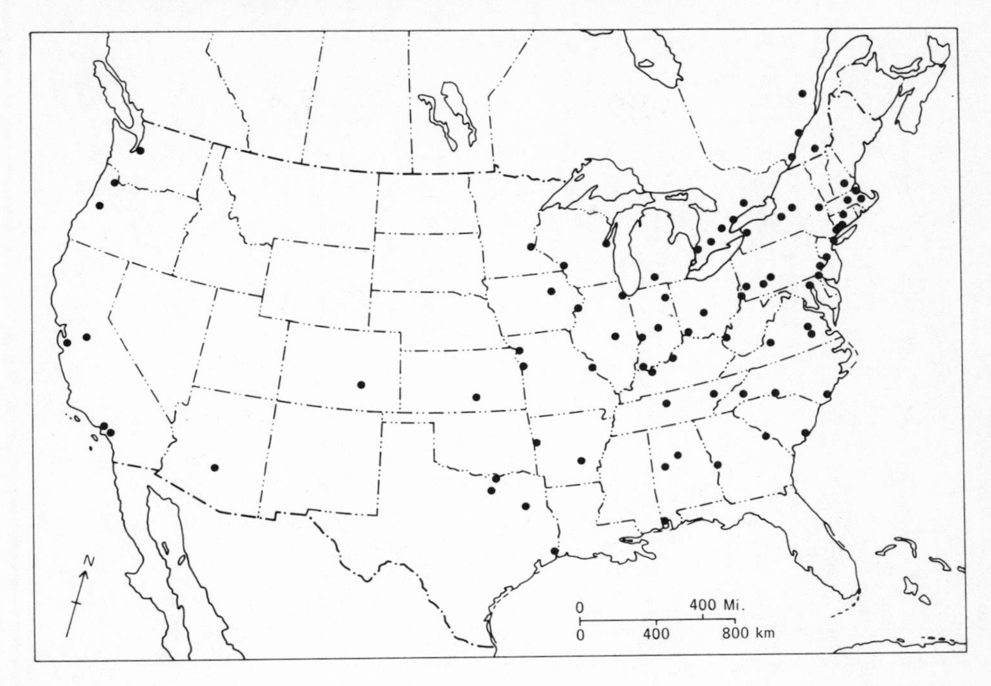

Figure 3.3
Mixed cities (50,000 or more population, 1970). In these centers, manufacturing activity and regional service activity are both well represented. Compare with Figures 3.2 and 3.4.

region only because the location of North American manufacturing activity has been studied by several writers over the years (Strong 1937; Jones, 1938; A.J. Wright 1938; Alexander 1958), and the results have spread to all levels of the educational system and thus into the layperson's image of the national space-economy. Spatial patterns, however, may change. Thus the present analysis does not merely lead to the circular statement that 'manufacturing cities are concentrated in the manufacturing belt.' Instead, we should regard Figure 3.2 as a basis for affirming that the manufacturing belt remains, as of 1970–1, a valid geographical generalization.

In contrast to manufacturing cities, mixed cities are well represented in the south and west as well as the northeast (Figure 3.3). Indeed, their spatial distribution resembles the general pattern of population density in North America more closely than do the distributions of manufacturing or

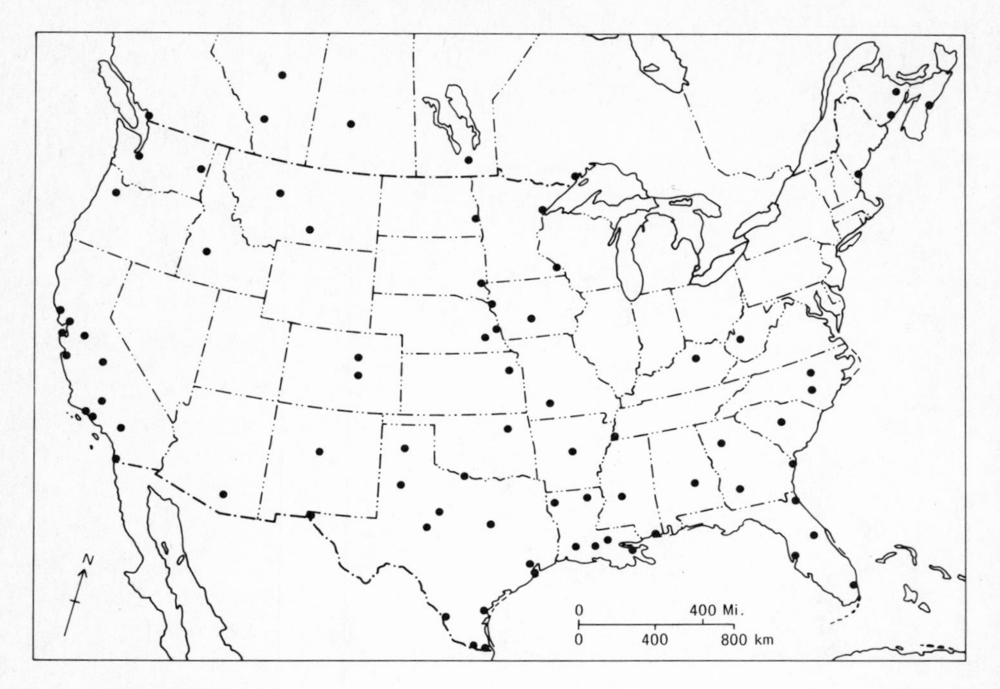

Figure 3.4
Nodal cities (50,000 or more population, 1970). In these centers, the manufacturing sector forms only a minor component of the local economy. Note the absence of nodal cities from the traditional manufacturing belt of the northeast.

nodal cities. Mixed cities are perhaps the most 'typical' or 'representative' of North American urban centers: they lie in the central section of the continuum of normal cities. They might fairly be described as the most normal of normal centers. They are 'diversified' cities, a technical term defined precisely in chapter 4. Mixed cities, particularly those of medium size, offer the most suitable social environment for market research into the acceptability of new consumer products. Peterborough, Ontario, has been used as such a 'laboratory' for many years.

Nodal cities, in turn, are widely distributed in the south and west but conspicuously absent from the traditional manufacturing belt (Figure 3.4). Nodal cities and manufacturing cities not only possess mutually exclusive employment profiles but also occupy mutually exclusive geographical spaces. To generalize still further, the northeastern manufacturing belt, including its extension into southern Ontario, contains only manufacturing and mixed cities, whereas the remainder of the United States contains only mixed cities and nodal cities, except for the strip of territory bordering the southern Appalachians, where all three classes of normal cities are equally well represented.

These generalizations, of course, are contingent on the fact that the present analysis is limited to cities of at least 50,000 inhabitants. It is likely that the spatial separation of nodal and manufacturing cities would be less clearly marked if the study were extended to include smaller towns. Even then, however, the general contrast between the northeast, with its emphasis on manufacturing, and the south and west, with their emphasis on central place activities, would still be evident. This contrast, in fact, is one of the most fundamental features of the human geography of North America.

The concentration of manufacturing in the northeast can be explained by a combination of six causal factors. The first factor is related to two fortunate natural circumstances: the northeastern seaboard of North America lies closer to western Europe than the coast south of Cape Hatteras, and it also offers a greater selection of convenient natural harbors. In consequence, it has always been the site of the principal gateways into North America for transatlantic traffic, and this, in turn, ensured that the region was the first part of North America to receive the Industrial Revolution from Europe. It is generally accepted that the northeast's early start in industrialization gave this area a significant advantage over other regions (C.M. Green 1965; Pred 1966).

Second, thanks to its gateway role and to the fact that its climate and soils are favorable for general farming, the northeast has always had a higher population density than other parts of the country. There has therefore always been a ready supply of factory labor – continuously augmented by the arrival of new immigrants – and a relatively large local market for manufactured goods. Other things being equal, manufacturers will naturally tend to concentrate where labor and markets are readily available.

Third, funds were available for investment in the construction of factory buildings and for the purchase of equipment. Some of these funds originated overseas, especially in Britain, but an additional important source was the capital accumulated by successful merchants in the leading American seaport cities during the eighteenth and early nineteenth centuries.

The fourth significant element was the presence of coal and iron, the two most fundamental raw materials for modern industrial growth. Coal was relatively plentiful, and the main Appalachian coalfields of Pennsylvania and West Virginia have remained important down to the present day. Iron was less common, but workable deposits were found in many parts of the northeast, especially in southern New England and in eastern and central Pennsylvania, and these scattered deposits were instrumental in the establishment of a fledgling iron and steel industry. These early sources of iron were small in size and quickly exhausted, but the industry's future was soon assured by the discovery of massive iron ore resources in northern Michigan and Wisconsin and, above all, in the Mesabi Range of northern Minnesota.

Mention of the Lake Superior ores brings us to the fifth factor promoting industrialization, namely the Great Lakes themselves. These inland seas have greatly facilitated the movement of both passengers and freight throughout the interior of eastern North America, and their value has been significantly enhanced by manmade improvements such as the Soo and Welland canals. The Great Lakes are perhaps best known for the transportation of iron ore and grain, but the ships that ply these waters carry cargoes of many kinds, including finished products as well as raw materials. Since the opening of the St Lawrence Seaway between Montreal and Lake Ontario in 1959, Toronto, Cleveland, Detroit, and even Chicago have become 'Atlantic' ports, directly accessible by ocean-going ships.

The sixth and final factor was the role played by the US Civil War. Significant industrialization had occurred in the northeast before war broke out, and a case can be made that the North's industrial advantage was an important factor in the outcome of the conflict. But wars in general tend to encourage rapid development where manufacturing is already under way, and the Civil War seems to have run true to form in this respect. Moreover, the war coincided with the onset of the period of most pronounced industrial growth in North America's history. The defeated South was unable to derive much benefit from the decades of rapid growth between 1870 and the First World War, while the North – including both the seaboard states and the emerging Midwest – confidently forged ahead. By 1920 the areal pattern of manufacturing had attained essentially its present form.

The above explanation of the spatial distribution of manufacturing cities (Figure 3.2) also, by inverting the argument, explains the distribution of mixed cities and nodal cities (Figures 3.3 and 3.4). In effect, mixed and nodal cities are defined by the relatively feeble showing in the manufacturing sector. Hence, areas where the factors promoting the growth of manufacturing are absent or poorly represented should contain nodal cities and mixed cities but not manufacturing cities, and this is just what does occur. To put the matter another way, the spatial structure of the continuum of normal centers takes the form of a core-and-periphery pattern, characterized by 'distance decay'; the farther one moves from the core, in which the manufacturing cities are concentrated, the more likely it becomes that nodal cities will be the dominant type encountered.

CENTROGRAPHY

One useful method of summarizing the spatial distribution of a set of points is to identify the median center of the set (Neft 1966, 29–33). The median center is located as follows. First, find the line of longitude that divides the set of points into two equal groups, with half the points lying to the east of the line and half lying to the west. If the number of points is odd, the dividing line will pass through the point having the median value of longitude in the set. If the number of points is even, the dividing line is placed halfway between the meridians of the two points occupying the midmost locations in longitude. Second, do exactly the same for latitude as for longitude; that is, find the parallel of latitude that divides the array of points equally into a northern half and a southern half. The spot where the two dividing lines intersect is the median center of the set of points.[5]

This procedure was applied to each of the three types of normal cities in our classification. The median centers lie as follows: for *manufacturing cities*, in the waters of Lake Erie, twenty miles west of Erie, Pennsylvania; for *mixed cities*, twenty miles northeast of Springfield, Ohio (a manufacturing city, not a mixed city); for *nodal cities*, fifty miles due east of Oklahoma City, Oklahoma.

Two comments on these results are in order. First, the three median centers are located approximately in a straight line running from northeast to southwest, with the extremes of the commercial-industrial continuum (that is, the manufacturing and nodal types) occupying the extreme positions along this line (Figure 3.5). Given the core-and-

5 The exact location of the median center can be made to vary by changing the orientation of the axes used to identify it. However, except where the points are few in number and widely scattered, this variability is slight and can safely be ignored. Moreover, the most natural way to proceed seems to be to align the axes by the cardinal points of the compass. For technical details see Neft (1966) and the classic paper by Sviatlovsky and Eells (1937). For an interesting commentary on the history of centrographic measures in general, see Poulsen (1959).

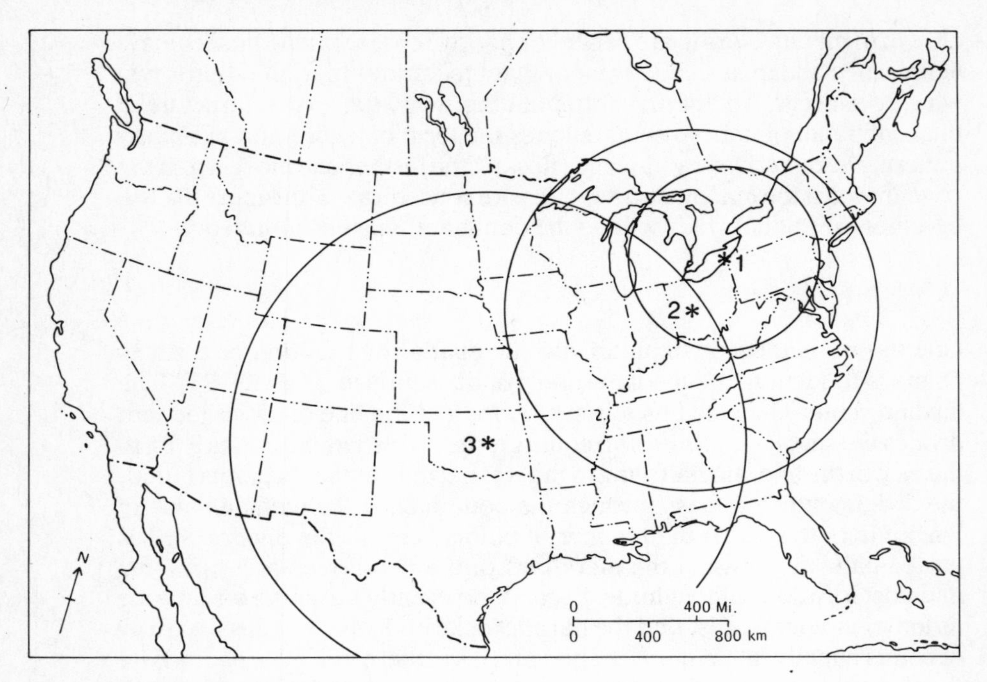

Figure 3.5
Locations of median centers. The numbers 1, 2, and 3 refer to manufacturing, mixed, and nodal cities, respectively. Each circle has the corresponding mean distance deviation as its radius.

periphery pattern described above, this arrangement is just what we should expect. Second, Springfield, Ohio, is much closer to Erie, Pennsylvania, than to Oklahoma City. This indicates that the spatial distribution of the mixed cities resembles that of the manufacturing cities more closely than that of the nodal cities.

This brief excursion into centrographic analysis may be carried one step further by means of the concept of the mean distance deviation, or MDD. The MDD, as its name implies, is the average distance of the members of a set of points from some specified origin. The origin need not be the median center of the set of points, but the value of the MDD is minimized when the origin is defined as lying at the median center. For this reason it is customary to use the median center as the point of origin for MDD calculations. Moreover, because the MDD is at a minimum when the median center is the origin, the median center is the point of minimum aggregate travel for the set of points in question. It is the location to which all the points in the set could be transported with the lowest possible total travel distance. Thus, the MDD provides a measure of how closely the members of a set of points are clustered around their median center.[6]

For the cities in the present study, the values of the MDD are as follows (see also Figure 3.5): *manufacturing cities*, 307 miles (494 kilometers); *mixed cities*, 603 miles (970 kilometers); and *nodal cities*, 840 miles (1,352 kilometers).

These results clearly reveal the tendency for manufacturing cities to be clustered close together and for nodal cities to be spread relatively far apart. To some extent this contrast is an indirect result of regional differences in population density. Even without manufacturing, cities of any given size would be closer together in the densely populated northeast than in the more thinly settled south and west. But manufacturing cities stand to gain from mutual proximity. Typically, the components that go into the manufacture of a single finished product are fabricated in several different cities and then shipped to a common center for final assembly. The total costs of production are obviously held down if the cities involved in this process are located reasonably close together. Nodal cities, in contrast, stand to benefit from being spaced well apart, so that each of them can effectively monopolize the retail and service trade of a large area without serious competition from other centers. Since practically all cities, in the final analysis, are partly nodal centers and partly manufacturing centers, a geographical tension is present in all urban systems between the desire to be close together and the desire to be far apart. The balance of power between these opposing forces determines whether the cities of any particular region are closely or widely spaced.

Within-group and between-group differences

The question naturally arises as to how different our seven functional classes of cities are from one another. If the differences between the groups are in some sense 'not significant,' we might conclude that the

6 A word of caution concerning MDD values. If a set of points consists of two or more distinct clusters, and if the distances between the clusters are large, the MDD for the full set of points will have a high value even though visual inspection would suggest 'clustered' rather than 'dispersed.' This difficulty can be addressed by first inventing a reasonable method for delimiting distinct clusters and then reporting median center locations and MDD values for each cluster separately. In the present analysis, as indicated by Figures 3.2, 3.3, and 3.4, this problem does not arise.

taxonomic exercise has been a waste of time, or, more precisely, that the similarities among urban employment profiles are so great that the identification of distinct functional types is misleading and unnecessary. This question can be addressed most fruitfully with the aid of a simple but powerful tool known as the index of dissimilarity.

THE INDEX OF DISSIMILARITY

The index of dissimilarity is most easily introduced by means of a concrete example. Let us suppose, therefore, that we wish to measure the degree of dissimilarity between Washington, DC, and Flint, Michigan. As a first step, we write down the employment profiles of the two cities side by side, as shown in Table 3.4. Next, we calculate the differences between corresponding percentages. It does not matter whether we subtract Flint from Washington or Washington from Flint, but it is vital to select one of these two alternatives and to keep a careful record of the sign (positive or negative) of each result. In the present example, we choose to subtract Flint from Washington. The final step is to add up either the positive differences or the negative differences, but not both. The absolute value of the sum must be the same in either case. This sum, taken as a positive number, is the required index of dissimilarity.

Since the totals of the positive and negative differences are necessarily equal to each other – except, in some cases, for tiny discrepancies that result from rounding values off – an alternative method of calculation is to add up all the individual differences, regardless of sign, and then to take one-half of the result. In other words, the index of dissimilarity may be defined as half the sum of the absolute differences between corresponding employment percentages in the two cities.

The index of dissimilarity between Washington and Flint turns out to be 42.11 (Table 3.4). A substantial proportion of this figure is traceable to the strong involvement of Washington in public administration and to the fact that Flint has an unusually high percentage in the manufacturing of transportation equipment (motor vehicles). Within the manufacturing sector, Flint has a higher percentage than Washington in every category but one: printing and publishing, the most ubiquitous of the manufacturing activities (see Table 3.2) and one generally well represented in both political capitals and university towns.

The index of dissimilarity can be interpreted as the minimum percentage of the first city's total labor force that would have to change jobs (i.e. switch over to work in a different industrial category) in order to give the

TABLE 3.4
Calculation of the index of dissimilarity between Washington, DC, and Flint, Michigan

Industry	Profile of Washington, DC	Profile of Flint	Positive differences	Negative differences
Foods and beverages	0.41	0.56		−0.15
Textiles and clothing	0.07	0.32		−0.25
Printing and publishing	2.49	0.61	1.88	
Primary metals	0.06	0.34		−0.28
Metal fabricating and machinery	0.83	2.69		−1.86
Transportation equipment	0.24	36.98		−36.74
Electrical products	0.83	0.94		−0.11
Chemical products	0.17	0.43		−0.26
All other manufacturing	1.42	3.47		−2.05
Construction	5.75	3.97	1.78	
Transportation, communication, and utilities	6.49	3.68	2.81	
Wholesale trade	2.82	3.18		−0.36
Retail trade	12.92	12.97		−0.05
Finance, insurance, and real estate	5.80	3.05	2.75	
Health services	7.88	7.48	0.40	
Education	8.75	7.59	1.16	
Personal services	8.43	6.45	1.98	
Business services	8.74	2.43	6.31	
Mineral extraction	0.10	0.10	0.00	
Public administration	25.81	2.77	23.04	
Total	100.01	100.01	42.11	−42.11

NOTE: The required index of dissimilarity between Washington, DC, and Flint is 42.11.

first city an employment profile identical to that of the second city. This interpretation is perfectly symmetrical: 42.11 per cent of Flint's workforce could change categories, in appropriate numbers, to make Flint into a replica of Washington, or 42.11 per cent of Washington's labor force could change categories to make Washington into a replica of Flint. Either of these entirely hypothetical transformations could be achieved, of course, by the redistribution of more than 42.11 per cent of one city's labor force, but neither can be accomplished with anything less.

The index of dissimilarity must always lie between zero and 100. These values are the theoretical limits of the index. Zero corresponds to a situation in which two cities possess identical employment profiles. At the other extreme, a value of 100 would be obtained if the entire labor force of

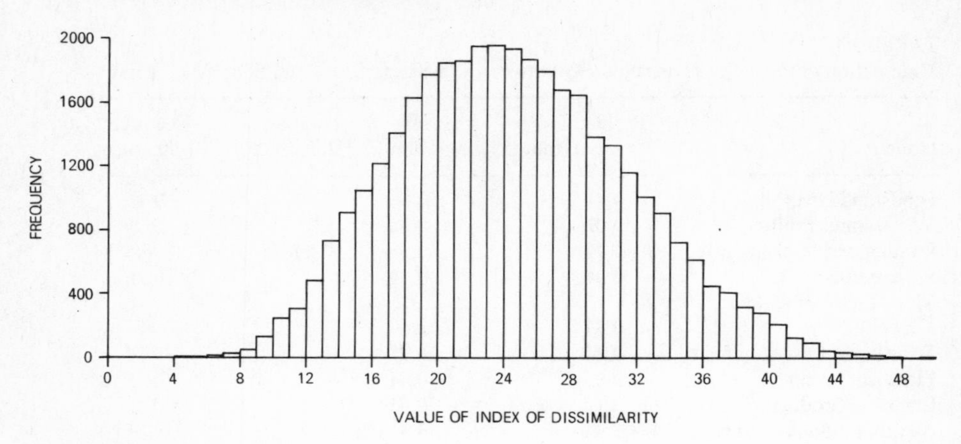

Figure 3.6
Frequency distribution of the indexes of intercity dissimilarity for the 35,778 different pairs of cities that occur in the set of 268 cities included in the functional classification. The mean value of this distribution is 24.3 and the standard deviation is 6.9.

the first city were employed in one or more categories of activity that contained no workers at all in the second city. Thus, the higher the value of the index, the greater the dissimilarity between the cities in question.

The index of dissimilarity is used in many situations where it is important to have a quantitative expression of the inequality between two matched sets of percentages. For example, in studies of the internal social structure of cities this measure is widely used to examine dissimilarities in the spatial distributions of different ethnic, racial, and occupational groups across census tracts (O.D. Duncan and B. Duncan 1955; Taeuber and Taeuber 1965; Marston 1969). In this context it is usually known as the index of residential segregation. The index of dissimilarity is also related, as described in chapter 4, to the Gini coefficient of concentration. In a word, the versatility of the index of dissimilarity is exceeded only by its mathematical simplicity.

Since the present study includes 268 cities, the total number of indexes of dissimilarity generated by the whole urban system is equal to the number of combinations of 268 objects taken two at a time:

$$C_{268,2} = 268!/266!2! = 35{,}778.$$

The frequency distribution of these 35,778 index values is shown in Figure 3.6. The distribution has a mean value of 24.3 and a standard deviation of

6.9. There is very slight positive skewness (the third moment has the value 0.35), but tests indicate that the distribution as a whole is not significantly different from a normal curve. Thus, for the urban system composed of the 268 largest North American cities, the index of dissimilarity between paired employment profiles can be regarded as a normally distributed variable.

The extremes of the curve in Figure 3.6 represent the most similar and the least similar pairs of cities within the system. The two most similar cities are Sioux City, Iowa, and Sioux Falls, South Dakota, two nodal centers that have similar names and also exhibit the property that each is the other's nearest neighbor among all the cities analyzed. The index of dissimilarity between Sioux City and Sioux Falls is a mere 5.0. The two least similar cities are the manufacturing center of Bristol, Connecticut, and the resort city of Las Vegas: their index of dissimilarity is 49.6. Within the Canadian subsystem – consisting of the country's thirty-four largest centers – the two most similar cities are St John's, Newfoundland, and Halifax, Nova Scotia (index value 8.2); the two least similar are Sudbury and Brantford, both in Ontario (index value 43.8).

DISTINCTIVENESS OF FUNCTIONAL GROUPS

The index of dissimilarity can be used as the basis of a criterion of distinctiveness for evaluating the adequacy of the various classes in our functional classification of towns. Specifically, we may regard two functional classes as being mutually distinct if the average values of the index of dissimilarity within each class are smaller than the average value of the index of dissimilarity between the classes. To make the meaning of this criterion clear, consider once such comparison in detail. The class of manufacturing centers contains sixty-nine places, and hence the number of within-group measures of intercity dissimilarity is equal to the number of combinations of sixty-nine objects taken two at a time:

$$C_{69,2} = 69!/67!2! = 2{,}346.$$

The average value of the index of dissimilarity for these 2,346 different pairs of cities within the manufacturing class turns out to be 23.4. Similarly, there are seventy-eight nodal cities, and hence the number of within-group index values for this class is

$$C_{78,2} = 78!/76!2! = 3{,}003.$$

For these 3,003 cases, the average index value is 17.1. Now, in order to

TABLE 3.5
Average values of within-group and between-group dissimilarity

	Manufacturing	Mixed	Nodal	Government	University	Mining	Resort
Manufacturing[a]	23.4	24.2	29.5	31.2	33.9	36.9	33.9
Mixed	24.2	20.3	21.4	24.4	29.0	32.1	26.8
Nodal	29.5	21.4	17.1	20.3	26.8	29.3	21.8
Government	31.2	24.4	20.3	16.8	27.6	32.3	24.9
University	33.9	29.0	26.8	27.6	19.6	36.0	31.3
Mining	36.9	32.1	29.3	32.3	36.0	20.9	31.9
Resort	33.9	26.8	21.8	24.9	31.3	31.9	17.9

a Within-group values appear along the principal diagonal of the table. For any two groups of cities, the off-diagonal between-group value is larger than either of the relevant within-group values.

calculate the between-group average value of the index, we note that each city in the first class must be compared successively with every city in the second class. Therefore, the total number, N, of between-group cases is the product of the number of cities in the two separate groups:

$$N = (69)(78) = 5{,}382.$$

For these between-group cases, the average value of the dissimilarity index is 29.5.

We may regard the manufacturing cities and the nodal cities as being mutually distinct if the average between-group dissimilarity value is larger than each of the two within-group values previously calculated. In this particular example the criterion of distinctiveness is satisfied, since the value 29.5 is larger than both of the values 23.4 and 17.1.

This method of evaluation partitions the whole urban system's grand total of 35,778 indexes of dissimilarity into within-group and between-group bundles. The average index values of these bundles are then calculated individually in order to permit comparisons.

Table 3.5 shows that the suggested criterion of distinctiveness is satisfied in every possible comparison of two functional classes. The seven within-group average dissimilarity values appear along the main diagonal of the table. For any two functional classes the two within-group averages are invariably smaller than the relevant between-group value. Thus the classification procedure has identified groups of cities possessing distinctively different employment profiles.

If the cells in each column (or, since the table is symmetrical, each row) of Table 3.5 are summed, the resulting totals vary from a low of 166.2 to a high of 219.4. Each total represents the overall similarity of one type of city to the other types. A large total indicates marked distinctiveness; a small total means general similarity. The smallest sum belongs to the column for the nodal cities; this is also the only column with no cell value greater than 30. Nodal cities, therefore, are the type from which the other types, on average, differ least. This bears out an important point made earlier: that central place activity is the function that most truly expresses the essential character of urban life. A city's most basic task is to serve as a market town for its surrounding region.

AVERAGE EMPLOYMENT PROFILES OF THE SEVEN FUNCTIONAL TYPES

Given that the seven types of cities in our functional classification are satisfactorily distinct, it is instructive to examine the average employment

TABLE 3.6 Average employment profiles for functional classes of cities

Industry	Manufacturing	Mixed	Nodal	Government	University	Mining	Resort	All cities
Manufacturing sector								
Foods and beverages	1.8	2.4	2.4	1.7	1.4	1.1	0.7	2.1
Textiles and clothing	3.5	2.4	1.1	1.0	0.8	0.1	1.3	2.1
Printing and publishing	1.5	1.5	1.3	1.2	1.4	0.8	1.0	1.4
Primary metals	4.1	2.3	0.6	0.4	1.8	5.3	0.3	2.1
Metal fabricating and machinery	8.4	4.8	2.3	2.5	1.8	1.7	1.7	4.6
Transportation equipment	5.4	2.3	1.3	1.9	2.5	0.3	1.5	2.7
Electrical products	4.4	3.0	1.0	1.2	1.7	0.7	1.2	2.5
Chemical products	1.2	2.0	1.0	0.8	0.4	1.4	0.4	1.3
All other manufacturing	10.1	7.6	4.5	3.6	2.5	1.6	3.1	6.6
Subtotal	(40.4)	(28.3)	(15.5)	(14.3)	(14.3)	(13.0)	(11.2)	(25.4)
Other industries								
Construction	5.3	6.1	7.2	6.8	6.0	7.7	9.3	6.3
Transportation, communication, and utilities	5.2	7.2	8.4	7.1	4.6	6.4	7.3	6.9
Wholesale trade	3.4	4.5	5.5	4.1	2.6	4.3	3.5	4.4
Retail trade	13.6	14.7	16.7	15.2	14.1	14.7	17.1	15.0
Finance, insurance, and real estate	3.7	5.0	5.7	5.1	5.0	3.8	5.8	4.9
Health services	6.8	8.0	8.6	7.9	8.9	5.9	6.3	7.8
Education[a]	7.3	8.4	9.2	8.7	24.3	7.2	6.4	9.0
Personal services	7.1	8.4	10.5	9.5	9.1	8.9	21.4	9.0
Business services	3.1	4.2	4.6	4.7	4.5	4.0	5.4	4.1
Mineral extraction	0.2	0.6	1.4	0.5	0.3	20.0	0.3	1.0
Public administration	3.7	5.0	6.9	16.0	6.3	4.1	5.8	6.1
Subtotal	(59.4)	(72.1)	(84.7)	(85.6)	(85.7)	(87.0)	(88.6)	(74.5)

NOTE: Values are correct to one decimal place. Owing to rounding, the columns may not sum to exactly 100 per cent.

a Tallahassee, Florida, meets the criteria for both government centers and university towns (see Table 3.3). However, for the purposes of this table only, Tallahassee is included as a university town but not as a government city.

profile of each functional type. The seven average profiles are shown in Table 3.6, which also shows, for the sake of comparison, the average employment profile of the urban system as a whole (taken from Table 3.2).

When the profiles of the various functional classes are compared with the profile of the entire urban system, several interesting features emerge. In manufacturing cities, for example, we might reasonably anticipate that the percentages would exceed their corresponding systemic averages in all activities within the manufacturing sector. However, the manufacturing centers – taken as a class – are below the system-wide averages in foods and beverages and in chemical products. These low percentages reflect the proximity of significant portions of the food-processing and chemicals industries to sources of raw materials that lie outside the northeastern manufacturing region. In particular, grain-milling and meat-packing are prominent in the otherwise weakly developed manufacturing sector in cities of the Great Plains. (Approximately two-thirds of all American cattle are raised west of the Mississippi River; two-thirds of the human consumers of this resource lie to the east.) Similarly, the processing of fruits and vegetables employs many workers in the cities of the sunbelt states, notably Florida and California. The chemicals industry, in turn, is attracted to areas rich in petroleum, natural gas, sulfur, and salt and is therefore strong in California and in the southwest, especially along the Gulf Coast in centers such as Baton Rouge, Lake Charles, and Beaumont–Port Arthur. Though the food-processing and chemicals industries are also well represented in the northeast, enough workers in these two categories lie outside the northeast to bring the corresponding percentages in the manufacturing cities down to below-average levels.

In the nonmanufacturing sector of the urban economy, we find that the mean percentages for the manufacturing cities lie below the system-wide averages in every activity (Table 3.6). The profile for the nodal cities shows just the opposite pattern, with values greater than the systemic averages in every nonmanufacturing category, even including mineral extraction. In manufacturing cities, the eleven nonmanufacturing categories account for 59 per cent of the average center's total labor force; for nodal cities, the corresponding figure is a remarkable 85 per cent. Nevertheless, nodal cities do achieve an above-average level of employment in one manufacturing category: namely, foods and beverages.

The mixed cities, somewhat surprisingly, exhibit percentages above the system-wide averages in eight of the nine categories of manufacturing, including both food-processing and chemicals. In general, however, they exceed the overall averages by a much smaller margin than do the manufacturing cities. In fact, the sum of the nine manufacturing activities is 40 per cent in the average manufacturing city, but only 28 per cent in the average mixed city – and a paltry 15 per cent in the average nodal city. Mixed cities also occupy an intermediate position in nonmanufacturing activities. Indeed, in every single nonmanufacturing category, the average percentage for mixed cities is higher than the corresponding value for manufacturing cities and lower than the value for nodal cities.

The average employment profiles for the four types of exceptional cities contain no surprises. In each case, the relevant diagnostic activity – that is, public administration in government centers, personal services in resorts, and so forth –not only stands far above its values in the other functional classes but also emerges as the largest individual value within its own profile (Table 3.6). Moreover, with one exception, all four exceptional profiles are below the system-wide averages in every kind of manufacturing. The lone exception is the case of primary metal production in the mining towns. This high value of 5.3 per cent results from the presence of smelting and refining activities – classified as 'manufacturing' in the census – in the two Canadian mining centers. Sudbury, Ontario, contains a major copper refinery for processing local ore. Sydney, Nova Scotia, has an old iron and steel complex built to exploit the coal of Cape Breton and the iron ore of nearby Newfoundland. But the coal is now depleted, the Newfoundland iron is completely gone, and markets are increasingly remote. The Sydney plant is heavily subsidized by the Canadian government and may have to be shut down in the not-too-distant future.

Overview of the taxonomic procedure

The principal steps in the creation of a functional classification of cities may be summarized as follows.

1. Bearing in mind the purpose of the classification (e.g. analysis of spatial distributions; testing of specific hypotheses about certain types of towns), decide on a list of functional categories. This is logically the *first* step. Functional categories will not arise spontaneously from the data, no matter how ingenious the analytical procedures.

2. Assemble an appropriate body of quantitative data – normally, census data on the numbers employed in the various industries – and define a set of numerical criteria that will govern the allocation of individual cities to the functional categories.

3. Apply the chosen criteria consistently to the set of cities under investigation.

4. Test the resulting functional groups of cities for mutual distinctiveness by using the index of dissimilarity to measure within-group and between-group differences. The results may indicate that certain categories should be abandoned or combined or that changes should be made in the diagnostic criteria for membership in each functional class.

The specific procedure adopted in the present study is shown schematically in Figure 3.7. This flow chart not only identifies the numerical criteria used to define each functional class but also indicates the order in which to apply them to any particular city.

One important point remains to be discussed. Although the classification is founded on the pervasive importance of trade and manufacturing and is built around the resulting concept of the continuum of normal cities, the taxonomic procedure begins with the identification of exceptional centers. There is a seeming inconsistency between the conceptual basis of the classification and the associated operational method, but this is explained by the desire for elegance on the operational level. An alternative method would be to ignore temporarily the exceptional types of towns and begin by assigning every city to one of the three normal categories – i.e. manufacturing, mixed, and nodal. Subsequently, cities that satisfy the criteria for exceptional centers could be extracted from the provisional lists of normal cities in order to complete the classification. But in this approach, which obviously produces exactly the same final results as the procedure shown in Figure 3.7, each city would have to be 'processed' twice. In short, commencing with the exceptional cases ensures algorithmic simplicity. This approach not only saves computing time but also enhances the esthetic appeal of the entire taxonomic procedure. As G.H. Hardy declared, 'There is no permanent place in the world for ugly mathematics' (Hardy 1967, 85).

Special problems

Before leaving city classification we must briefly look at two problems that cause a fair amount of anguish among researchers. Technical details are omitted and no final answers are proposed, but even an introductory treatment of urban taxonomy must at least recognize the existence of these issues.

BASIC AND NONBASIC ACTIVITIES: A CONCEPTUAL DILEMMA

We have taken it for granted that a functional classification of cities should be based on the whole of the labor force of each city – or, more precisely,

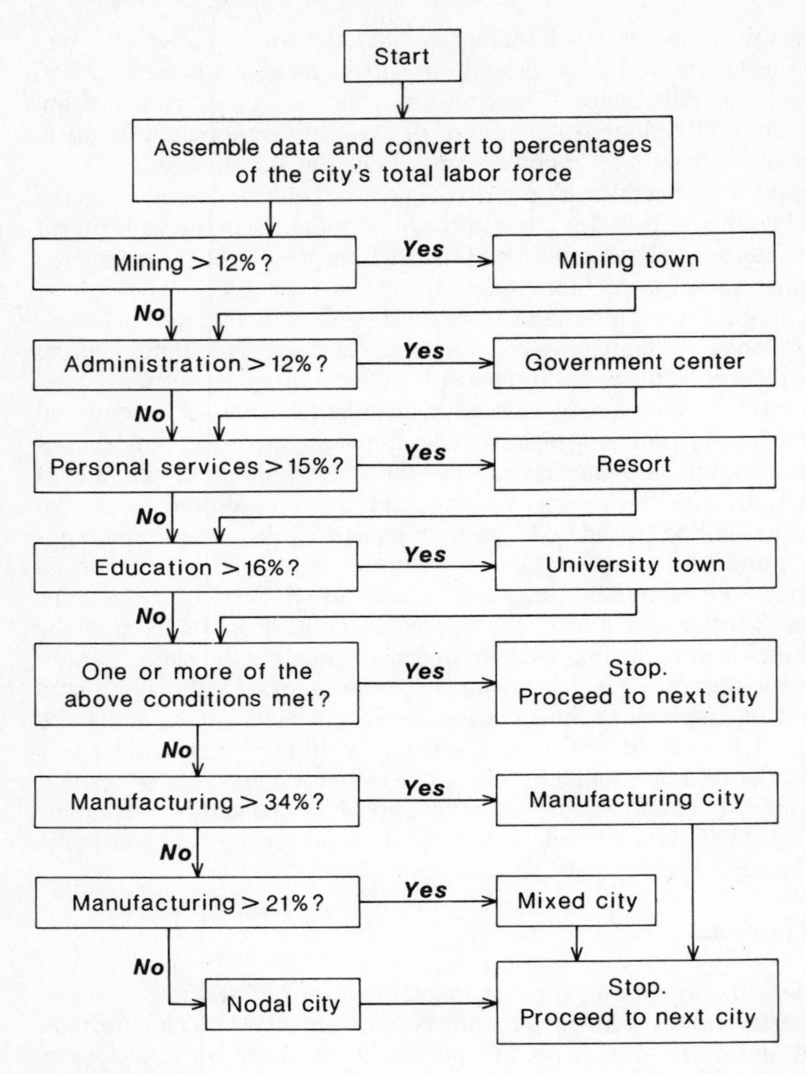

Figure 3.7
This flow chart shows the sequential procedure followed in generating the functional classification of North American cities described in the text.

the whole of the labor force recorded in the census, except for such nonurban industrial categories as agriculture and forestry (see the notes to Table 3.1). An alternative view, arising out of the extensive literature on the so-called 'economic base' concept, is that a classification might reasonably be based on only *part* of the labor force of each city, and not on the whole.

The essence of the economic base concept is as follows. Some workers within a city produce goods and services that are consumed by the city's own inhabitants; others produce goods and services that are consumed by nonresidents (who may or may not visit the city to purchase or consume the good or service). Thus the consumer's place of residence – the location of the market – can be used as a basis for dividing the city's labor force into two parts. Workers – and, if need be, fractions of workers – producing goods and services for the city's own residents are termed 'nonbasic.' Workers producing goods and services consumed by outsiders are labeled 'basic.'

The concept of the urban economic base was not initially regarded as an approach to city classification. It was developed during the 1920s and 1930s as part of a technique for estimating a growing city's future requirements for land. Given that the ratio between basic and nonbasic employment could be established by some acceptable method of measurement, it was noted that a certain (assumed) increase in a city's basic employment would lead to a predictable increase in the nonbasic workforce as well, with concomitant implications for future total population and hence for the amounts of land required for new housing, stores, schools, parks, and so forth (Weimer and Hoyt 1939; A.L. Wright 1956; Pfouts 1960; Tiebout 1962). Somewhat later it was realized that the economic base concept might also have implications for the functional classification of towns (Alexander 1954; Morrissett 1958; Ullman and Dacey 1960; Maxwell 1965).

Taxonomists who have found the concept attractive have contended, in effect, that nonbasic employment is irrelevant to an understanding of urban functions because the goods and services produced by this segment of the labor force remain within the city and do not become part of the set of external transactions that define the role of the city within the national economy. Moreover, the nonbasic segment is believed to have essentially the same industrial composition (in percentage terms) in every city. To the extent that this is the case, nonbasic employment can contribute nothing to the differentiation of cities. In the nonbasic segments of their labor forces, all cities are essentially alike. The basic portion of the labor force, in contrast, varies considerably in composition from city to city and, more importantly, represents the real source of the city's economic well-being. Only basic workers, after all, earn income for the city through external sales. Thus it may be argued that a classification of cities should ignore nonbasic workers and should be concerned only with differences in employment structure within the basic portion of the workforce.

The counter-argument maintains that the basic/nonbasic distinction fails to take account of the fundamental interconnectedness of all productive elements in the urban economy. Consider, for example, a firm that manufactures automobile transmissions and sells its entire output to a car manufacturer located in the same city. The completed cars, in turn, may be assumed to be sold to outsiders. Clearly the firm that assembles the cars must be classified as basic. But what about the transmission manufacturer? In one sense this firm is nonbasic because its output is sold to a customer within the same city. But in another and equally reasonable sense the transmission firm is basic because its products are ultimately exported from the city, and the workers producing these indirect exports are just as essential to the export trade as workers at the firm that assembles the finished cars.

But this line of reasoning rapidly leads to the conclusion that every worker in the city is either directly or indirectly involved in supporting the city's export trade. Neither the automobile manufacturer nor the transmission manufacturer, nor anyone else in the city, could function effectively if the city lacked stores, schools, fire and police departments, hospitals, public transportation, water and sewage facilities, garbage collectors, municipal government, and other components of the urban infrastructure. All these elements, from the obviously export-oriented car manufacturer down to the local corner store, are inextricably intertwined, and all are necessary to the continued operation of the city. As Blumenfeld (1955, 121) concluded in one of the classic papers on this topic: 'The economy of an area is an integrated whole of mutually interdependent activities; the distinction between "basic" and "nonbasic" seems to dissolve into thin air.' Or, to borrow an appropriate phrase from the ecological literature of the 1970s, 'everything is connected to everything else' (Commoner 1971, 33).

In practice, the question of whether or not to incorporate the basic/nonbasic distinction into a taxonomic study is perhaps less troublesome than the above opposing views may suggest. Both sides implicitly recognize that there is a positive correlation between an industry's share of a city's total labor force and its 'importance,' whether to the whole of the city's labor force or, more narrowly, to the industry's possible involvement in export or 'basic' activity. Thus, for example, if Billings, Montana, has 8.6 per cent of its labor force employed in wholesale trade, while Dayton, Ohio, has only 3.0 per cent (Table 3.1), both sides would agree that wholesale trade is more important to Billings than to Dayton. Supporters of the economic base concept would want, in addition, a quantitative assessment of the extent to which wholesale trade in each city earns income from outside; and most of the controversy surrounding the concept is found in this realm of measurement (Pratt 1968; Ullman 1968; Leigh 1970; Isserman 1980; Gibson and Worden 1981). In contrast, those who are skeptical about separating basic from nonbasic employment would rest content with a qualitative statement: the higher the percentage, the more significant the activity must be in the economic life of the city as a whole.

In effect, then, the kind of functional classification described in this chapter should be acceptable to all parties. It does not explicitly separate basic from nonbasic activity, but the numerical criteria used to define functional classes do embody the principle that an industry's importance is proportional to its share of the urban labor force. Those who endorse the basic/nonbasic distinction would almost certainly apply the same descriptive labels to the various functional classes – manufacturing cities, government centers, and so forth – as would those who find this concept uncongenial.

MULTIDIMENSIONAL CLASSIFICATIONS

Functional classifications of towns may be described as being unidimensional in that they are concerned only with one aspect of intercity differences: namely, differences in economic character, as revealed by data on employment. An alternative approach is to consider several aspects of urban character simultaneously: for example, the population's age structure, income levels, ethnic origins, and educational attainments, as well as industrial composition. Studies of this type may be termed multidimensional. The classic example is an analysis of British cities by Moser and Scott (1961). Other important research includes a study of Canadian urban municipalities by King (1966) and a collection of papers edited by Berry and Smith (1972).

Multidimensional classifications invariably use one or another member of the family of quantitative techniques known as factor analysis. The details of these techniques lie beyond the scope of this book, but the basic idea is that a large number of variables can be mathematically reduced to a small number of 'factors' by exploiting the fact that the original variables are, to varying degrees, correlated with one another. Since each resulting factor normally embodies contributions from variables of different kinds, the factors are sometimes difficult to name and to interpret, although the results often reveal insights that might otherwise pass unnoticed concerning the interrelatedness of quite different aspects of urban character.

Three related problems associated with multidimensional classifications need to be noted. First, the literature shows a tendency for certain variables to be included merely because the relevant data happen to be available. Instead of starting out with a series of questions or hypotheses about specific kinds of data, some researchers appear to have combed the census indiscriminately in an effort to compile the longest possible list of variables. The guiding principle seems to be that if everything available is included in the analysis something of value will emerge. But this approach – 'induction without hypothesis' – is at best a fortuitous way of arriving at interesting results. According to modern standards of scientific inquiry, a more acceptable procedure is to begin by formulating specific questions or propositions that seem, a priori, worthy of investigation. Only germane data are then brought into the analysis (Popper 1962; Medawar 1967; Marshall 1985).

Second, by creating composite factors that encompass portions of variables of different kinds, multidimensional classifications tend to divert attention from studies of the behavior of the individual variables. There is much to be said for dealing with only one dimension at a time. Thus, age structure, ethnicity, industrial composition, and other variables can form, each in its own right, the objects of separate inquiries. Without doubt, the traditional emphasis on functional classification reflects a widely held view that intercity differences in labor force composition are much more relevant for understanding the spatial distribution of cities than are differences in other dimensions, such as ethnic origin or educational attainment. Nonetheless, separate studies of these other dimensions are feasible, and, given enough interesting questions to be answered, they should someday be undertaken.

Finally, the value of a piece of research may not be proportional to the complexity of the analytical procedures employed. It is tempting to believe that greater complexity of technique implies greater objectivity or depth of understanding, but there is no logical reason why this should be the case. Not infrequently, in fact – and multidimensional classifications are a case in point – the results obtained by means of complicated techniques such as factor analysis are so far removed from the categories of everyday experience that it is nearly impossible to attach meaning to them. Although certain kinds of research require techniques of some intricacy, generally speaking, simplicity is a virtue.

Summary

This chapter has focused on the use of labor force statistics as a basis for understanding the various economic roles played by cities within the urban system to which they belong. The early part of the chapter introduced the concept of the employment profile of a city and stressed the importance of working with percentage values in order to permit direct comparisons among cities of different sizes. It was noted that the percentages for manufacturing activities vary from city to city much more than the percentages for the trade and service industries.

A functional classification of North America's 268 largest cities (those with 50,000 or more inhabitants) into seven types was presented. Three of these types – manufacturing, mixed, and nodal cities – account for 85 per cent of all centers and form a graded series known as the continuum of normal cities. The other four types, termed exceptional cities, are comparatively rare.

The spatial distributions of the three types of normal cities reveal a distinct core-and-periphery pattern. The core, defined by its concentration of manufacturing centers, extends from the northeastern seaboard to the US Midwest, including a small portion of southern Ontario. The periphery – the remainder of the settled area of the United States and Canada – is dominated by nodal cities. Mixed cities are found in both the core and the periphery and are thus intermediate between manufacturing and nodal cities both in employment structure and in geographical distribution.

Attention was drawn to the value of centrographic measures in describing point patterns such as those representing the normal types of cities. It was shown that manufacturing centers tend to be clustered together in order to reduce production costs while nodal cities tend to be spread far apart in order to make their service areas as large as possible.

The discreteness of the seven functional classes was tested by means of the index of dissimilarity. The classes were found to be satisfactorily distinct: between-group differences are larger than within-group differences. The average employment profiles of the seven functional classes were presented and compared.

Following a summary of the principal steps involved in carrying out a functional taxonomy, brief comments were made on the urban economic base concept and on the construction of multidimensional classifications.

4

Industrial diversification

In everyday speech a city is said to be 'specialized' if it has an unusually high concentration of employment in a single activity. Thus, Atlantic City, Washington, DC, and Detroit are generally regarded as being 'specialized,' though for a different reason in each case. The common factor is that the economy of each of these places is heavily dependent on one specific type of work. Cities that are not economically specialized in this way are said to be 'diversified.' They exhibit no unusual degree of commitment to any particular activity.

This chapter has two concerns. The first is measurement: expressing industrial diversity in quantitative terms. The second is possible relationships between levels of diversity and other factors: specifically, location, city size, and the relative prominence of particular urban functions.

The scale of specialization

First and foremost, 'specialization' and 'diversification' are not two entirely different phenomena, but opposite sides of the same coin. We should think in terms of a continuous linear scale of diversity, with 'complete diversification' at one end and 'complete specialization' at the other. Every city lies at some point along this continuum. Thus the difference between a specialized city and a diversified city is one of degree, not one of kind.

Next, in order to proceed unambiguously, we need to understand clearly what the two extremes of the continuum of diversity represent in terms of urban employment structure. It is convenient to begin with the 'completely specialized' end of the scale. The most specialized situation we can imagine is one in which the entire labor force of a city is employed in a single category of activity: the whole of its labor force would be classified in the census under a single category of industry. The more closely an actual city approaches this theoretical extreme, the closer it lies to the 'completely specialized' end of the scale.

Moving now to the opposite extreme, we encounter two possible ways of defining the 'completely diversified' end of the continuum. Early studies, such as those by Tress (1938), Rodgers (1957), and Conkling (1963), saw complete diversification in a labor force equally divided among the categories of industry used in the analysis. For example, if twenty categories were used, a city would be completely diversified if it had exactly 5 per cent of its labor force employed in each category. It is not at all clear why this 'equal shares' definition was thought to be appropriate. No rationale was ever provided, and each writer apparently assumed that this was the only way to proceed. It can be argued, however, that the idea of equal shares is unrealistic and therefore potentially misleading as a definition of complete diversification. After all, if a city actually did have its labor force equally divided among all the industries, we would regard it as a very unusual place indeed: the industrial categories used by census bureaus and other statistical agencies are not designed to contain equal amounts of the urban system's total employment. They are intended, rather, to capture commonsense distinctions that are meaningful to everyone concerned. The actual system-wide levels of employment in these familiar industrial categories, though far from being equal in size, provide a more believable definition of the 'expected' employment profile of an individual city than does the needlessly simplistic concept of equal shares. Accordingly, recent work has defined a completely diversified city as one having an employment profile that exactly matches the average profile of all the cities covered by the study (Marshall 1975b, 1981). This systemic average profile is therefore the one I shall use in this chapter to define the 'completely diversified' end of the scale of specialization.

The systemic average profile can be calculated in either an unweighted or a weighted form. For a given industry, the unweighted average percentage is obtained by calculating the percentages attributable to that industry in each city separately, summing these percentages, and dividing by the total number of cities. This procedure gives equal weight to every city, regardless of size. The weighted average percentage for an industry, in contrast, is obtained by adding together the absolute numbers of workers in that industry across the full set of cities and expressing the resulting total as a percentage of the grand total of employment in all industries in the whole system. In this procedure, each city's contribution is proportional to its size.

There is no strong theoretical reason to favor one set of averages over the other. The weighted averages give a more accurate picture of the overall structure of the urban economy viewed as an undivided aggregate, but comparisons are misleading unless the effects of differences in city size are first eliminated. Fortunately, the differences between weighted and unweighted averages in large urban systems are usually very small, and hence using one type of average rather than the other has little influence on subsequent results. This decision is truly a matter of taste.

To recapitulate, a city is diversified if it resembles the systemic average employment profile, and specialized if it does not. The systemic average profile, in short, is the standard of reference with which each city must be compared. Clearly the next step is to identify some means of measuring the extent to which a city differs from the systemic norm.

Measuring specialization: two indexes

Two standard techniques are available for measuring the overall level of diversity in a city's employment profile: the coefficient of specialization and the Gini index of concentration. The Gini index is the more complicated but it is also more sensitive to small differences in employment structure and therefore holds a slight theoretical advantage. The two indexes, for reasons that will become clear as their methods of construction are described, give very similar results, although they employ different scales of measurement and are not related by any simple algebraic formula.

THE COEFFICIENT OF SPECIALIZATION

The coefficient of specialization can be dealt with almost in a single sentence: it is identical in form to the index of dissimilarity described in the preceding chapter! The index of dissimilarity was used to compare the employment profiles of two different cities. Rechristened the coefficient of specialization, the same index is also used to compare the profile of one city to that of the urban system as a whole. As before, the index ranges (in theory) from zero at one extreme to 100 at the other. The value of zero is obtained if the profile of a city exactly matches the systemic average

TABLE 4.1 Calculation of the coefficient of specialization for Pittsburgh

Industry (1)	Percentage in Pittsburgh (2)	Percentage in systemic profile (3)	Differences (4)	Location quotient (5)	Rank by location quotient (6)
Manufacturing sector					
Foods and beverages	1.56	2.09	−0.53	0.75	17
Textiles and clothing	0.41	2.06	−1.65	0.20	20
Printing and publishing	1.22	1.40	−0.18	0.87	11
Primary metals	11.71	2.09	9.62	5.60	1
Metal fabricating and machinery	6.02	4.58	1.44	1.31	2
Transportation equipment	0.94	2.74	−1.80	0.34	19
Electrical products	3.19	2.49	0.70	1.28	3
Chemical products	1.01	1.30	−0.29	0.78	16
All other manufacturing	5.87	6.63	−0.76	0.89	10
Other industries					
Construction	5.30	6.33	−1.03	0.84	13
Transportation, communication, and utilities	7.16	6.90	0.26	1.04	5
Wholesale trade	3.99	4.36	−0.37	0.92	9
Retail trade	14.92	15.05	−0.13	0.99	7
Finance, insurance, and real estate	4.55	4.88	−0.33	0.93	8
Health services	7.93	7.84	0.09	1.01	6
Education	7.47	9.02	−1.55	0.83	14
Personal services	7.37	9.02	−1.65	0.82	15
Business services	4.44	4.10	0.34	1.08	4
Mineral extraction	0.86	1.00	−0.14	0.86	12
Public administration	4.08	6.12	−2.04	0.67	18
Total	100.00	100.00	±12.45		

NOTE: The required coefficient of specialization is the sum of either the positive or the negative differences between the percentages in columns 2 and 3. See text for detailed explanation.

profile. In other words, zero signifies complete diversification. The higher the value, the more specialized (i.e. the less diversified) is a city's economy.

As an illustration, the calculation of the coefficient of specialization for Pittsburgh is set out in the first four columns of Table 4.1. (The two right-hand columns are concerned with the Gini index.) The employment profiles for Pittsburgh and for the average city are shown in columns 2 and 3, respectively. The differences obtained by subtracting the systemic percentages from Pittsburgh's profile are listed in column 4. The sum of either the positive or the negative differences (but not both) is the required coefficient. The value for Pittsburgh is 12.45.

The coefficient of specialization was computed in this way for each of the 268 American and Canadian cities included in chapter 3. The average value of the 268 resulting coefficients is 18.3, and the standard deviation is 5.5. Thus Pittsburgh, in spite of its popular image as a specialized steel-producing city, is actually well below the average level of specialization attained by North American cities of 50,000 or more inhabitants. Pittsburgh's quite high diversification is almost certainly explained by its large size. (The metropolitan area ranked ninth by population in the United States in 1970.) The relationship between industrial diversification and city size is examined more closely later in this chapter.

Using the coefficient of specialization, we find that the most specialized center in our 268-city urban system is the mining town of Sudbury, Ontario, with a value of 37.5. At the other extreme, the most diversified center – that is, the city that most closely resembles the systemic average profile – is Columbus, Ohio, with a coefficient of only 5.6: only 5.6 per cent of Columbus's total labor force would have to be shifted into different industrial categories in order to give Columbus a profile identical to that of the average North American city.

Further discussion of empirical results is postponed until after the description of our second method of measurement, the Gini index.

THE GINI INDEX OF CONCENTRATION

This alternative approach to the problem of evaluating the degree of dissimilarity or inequality between two matched sets of percentages is generally credited to the Italian statistician Corrado Gini, though the same basic idea had probably occurred earlier to the American statistical economist M.O. Lorenz (Gini 1914, 1921; Lorenz 1905). The technique has been applied in a wide variety of contexts, and the underlying theory

TABLE 4.2
Calculation of the Gini index of concentration for Pittsburgh

Rank[a] (1)	Industry (2)	Percentage in Pittsburgh (3)	Percentage in systemic profile (4)	Cumulative percentage in Pittsburgh (5)	Cumulative percentage in system (6)	Cross-product terms (7)
1	Primary metals	11.71	2.09	11.71	2.09	–
2	Metal fabricating and machinery	6.02	4.58	17.73	6.67	41.05
3	Electrical products	3.19	2.49	20.92	9.16	22.87
4	Business services	4.44	4.10	25.36	13.26	45.10
5	Transportation, communication, and utilities	7.16	6.90	32.52	20.16	80.04
6	Health services	7.93	7.84	40.45	28.00	95.09
7	Retail trade	14.92	15.05	55.37	43.05	191.01
8	Finance, insurance, and real estate	4.55	4.88	59.92	47.93	74.33
9	Wholesale trade	3.99	4.36	63.91	52.29	70.01
10	All other manufacturing	5.87	6.63	69.78	58.92	116.78
11	Printing and publishing	1.22	1.40	71.00	60.32	25.81
12	Mineral extraction	0.86	1.00	71.86	61.32	19.12
13	Construction	5.30	6.33	77.16	67.65	129.88
14	Education	7.47	9.02	84.63	76.67	190.64
15	Personal services	7.37	9.02	92.00	85.69	198.30
16	Chemical products	1.01	1.30	93.01	86.99	33.05
17	Foods and beverages	1.56	2.09	94.57	89.08	58.69
18	Public administration	4.08	6.12	98.65	95.20	215.32
19	Transportation equipment	0.94	2.74	99.59	97.94	180.81
20	Textiles and clothing	0.41	2.06	100.00	100.00	165.00
	Total	100.00	100.00			1,952.90

a The twenty industries are ranked according to their location quotients (column 5 of Table 4.1). The required Gini coefficient of concentration for Pittsburgh is 0.195. See text for detailed explanation.

has attracted the attention of professional mathematicians. The interested reader may follow the history of the relevant ideas through the works of Hoover (1936), O.D. Duncan and B. Duncan (1955), Hainsworth (1964), Alker and Russett (1964), and P.E. Hart (1971).

As with other techniques, a worked example provides the clearest introduction to the topic. We shall again use Pittsburgh as our illustration. The starting-point is provided by the first three columns of Table 4.1, where the employment profiles of Pittsburgh and of the average city (the standard of reference) are set down side by side. These paired percentages constitute the raw material for our calculations.

At this point we must introduce an entirely new concept, namely the location quotient, which is simply a ratio between two percentages. Specifically, it is the ratio between a given city's (Pittsburgh's) percentage in a particular industry and the system-wide average percentage in the same industry. For example, Pittsburgh's percentage in primary metal manufacturing is 11.71, and the corresponding average percentage for the urban system as a whole is 2.09 (Table 4.1). We divide 11.71 by 2.09 to obtain the value of 5.60 as the location quotient for primary metals in Pittsburgh. Note that a location quotient is a dimensionless quantity – that is, it has no units of measurement – and that every location quotient refers to the state of some particular activity in some specific place (city or region, depending on the scale of the analysis). The location quotients for the various industries in Pittsburgh are listed in column 5 of Table 4.1.

Quite apart from its significance in relation to the Gini index of concentration, the location quotient is an interesting measure in its own right. Clearly, a location quotient of exactly 1 signifies that the city has precisely the average or 'expected' percentage in the industry in question. (This does not happen in Pittsburgh, although retail trade and health services come very close.) If a location quotient is greater than 1, the city has what might be called 'more than its share' in that industry – more than would be expected on the basis of the system-wide average. Conversely, a location quotient of less than 1 indicates a relative deficiency. In the case of Pittsburgh, the city is obviously well endowed in primary metal manufacturing and notably deficient in the textile industry (Table 4.1). The location quotient thus serves as a handy indicator of local excesses and deficiences of employment in particular activities – the purpose for which it was designed (Florence 1929, 327–8; 1937, 622). In the present context, however, the location quotient is used merely as a basis for ranking industries. The significance of this ranking will become clear as our discussion proceeds.

The final column in Table 4.1 shows the rank of each industry in Pittsburgh by location quotient. Two points are worth mentioning. First, for calculation of the Gini index, location quotients can be ranked in ascending or descending order; exactly the same result will be obtained. Here, descending order is arbitrarily adopted. Second, the rank of an activity by location quotient is not necessarily the same as its rank by percentage within the city. (If these two rankings were always the same, there would be no need to calculate location quotients.) In Pittsburgh the smallest industrial category – textile manufacturing – also has the smallest location quotient, but this is a mere coincidence.

It is now necessary to rewrite Table 4.1 with the industries ranked by location quotient value. This is done in Table 4.2. Now we can calculate two sets of values known as cumulative percentages; one for Pittsburgh and the other for the systemic average profile. These two sets of cumulative percentages appear in columns 5 and 6 of Table 4.2. The entries in column 5 are generated from those in column 3 as follows. The first number in column 5 is simply the first number in column 3. The second number in column 5 is the sum of the first two numbers in column 3 – that is, 17.73 is equal to 11.71 plus 6.02. The third number in column 5 is the sum of the first three numbers in column 3, and so forth. The general rule is that the nth number in column 5 is the sum of the first n numbers in column 3. Exactly the same procedure derives column 6 from column 4. Except for minor discrepancies from rounding off values in columns 3 and 4, the final entries in columns 5 and 6 will be 100.0 – the cumulative total of all the elements in the entire labor force.

It is possible to proceed directly from this point to the calculation of the Gini index by means of an algebraic formula, but the nature of the index will be more clearly revealed if it is first described in geometrical terms. For the moment, then, let us ignore column 7 of Table 4.2 and focus our attention on Figure 4.1. The framework of this figure is a square box defined by horizontal and vertical axes scaled from zero to 100. The axes represent cumulative percentage values: the vertical, for Pittsburgh; the horizontal, for the urban system as a whole. In other words, the vertical and horizontal axes represent the values in columns 5 and 6, respectively, of Table 4.2. (Interchanging the axes would not alter the final result.) Two lines cross the box from bottom left to top right. The straight line ABC is simply a diagonal line that divides the box in half; its meaning will become apparent very shortly. The other line, joining A to C by way of D, is obtained by plotting the twenty pairs of cumulative percentages from Table 4.2: that is, the plotted points have the grid coordinates (2.09,

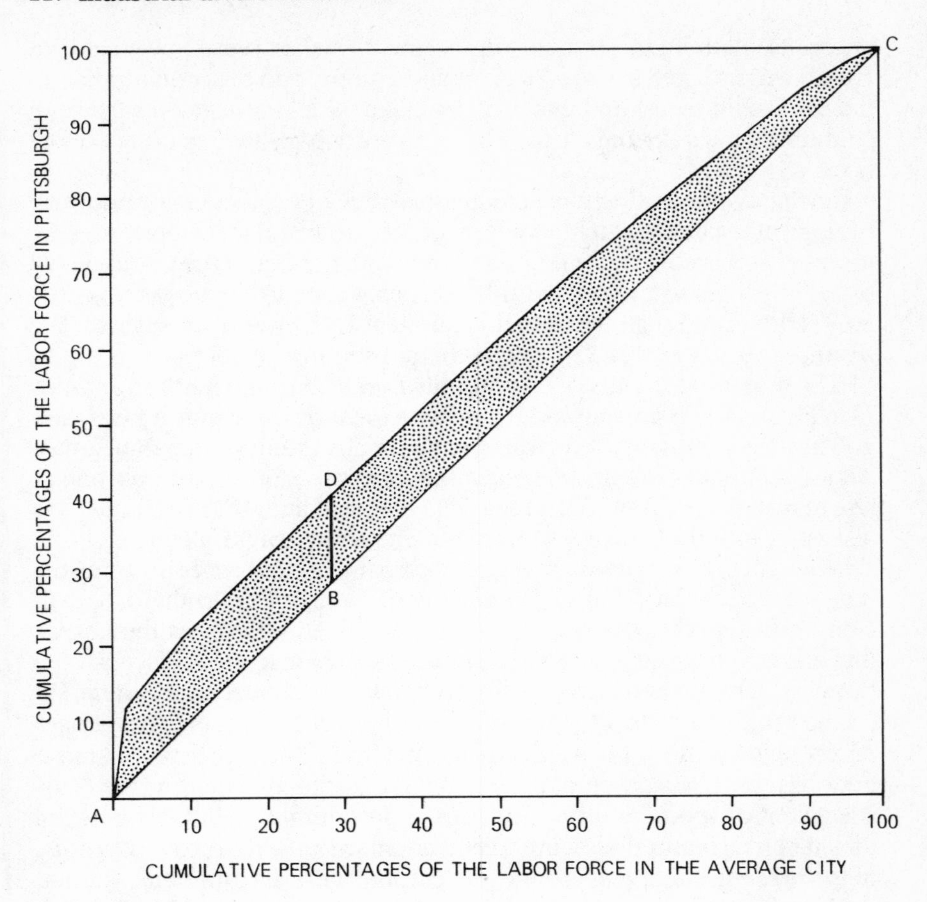

Figure 4.1
Lorenz diagram for employment in Pittsburgh

11.71), (6.67, 17.73), (9.16, 20.92), and so forth. The final point in this series has the coordinates (100.00, 100.00) and coincides with point *C* at the top right-hand corner of the box.

The line *ADC* is known as a Lorenz line (Lorenz 1905). Here, it consists of twenty straight segments, one for each of Pittsburgh's twenty industrial activities. The line is generally smooth because the segments of which it is composed are placed on the graph sequentially, according to their angles of slope, beginning with the steepest. It should now be evident why, in preparing Table 4.2, the industries had to be ranked by their location quotients. Without this ranking, the segments of the Lorenz line would not have been arranged in order of decreasing slope, and the configuration of the line would therefore have been irregular. Ranking segments by slope guarantees the smoothest possible configuration for the resulting Lorenz line.

Each location quotient is actually the value of the tangent of the angle of slope of the line segment to which it corresponds. For example, the first segment – representing primary metal manufacturing – rises at an angle of 79.9 degrees to the horizontal, this being the angle whose tangent is 5.60. The second segment, location quotient 1.31, rises at an angle of 52.7 degrees (arctan 1.31 = 52.7 degrees), and so forth.

The first six segments of Pittsburgh's Lorenz line, extending from *A* to *D* in Figure 4.1, represent industries with location quotients greater than 1. These six segments therefore rise at angles greater than 45 degrees. All subsequent segments, representing location quotients of less than 1, rise at angles less than 45 degrees. The diagonal line *ABC*, of course, has a slope of exactly 45 degrees throughout its entire length. Between *A* and *D*, therefore, the Lorenz line gets increasingly *farther away* from the diagonal; between *D* and *C*, it gets increasingly *close* to the diagonal. Consequently, *D* is the point at which the distance between the Lorenz line and the diagonal is at its maximum possible value. This fact is more than just a mathematical curiosity. When measured vertically along *DB* (or horizontally from *D* – the result is the same), and expressed as a percentage of the total height of the diagram, the maximum distance between the Lorenz line and the diagonal is exactly equal to the city's coefficient of specialization. The formal proof of this relationship is not difficult but is omitted as being somewhat marginal to our main concerns. In the present example it is easy to see from Figure 4.1 and Table 4.2 that the length of *DB* is equal to the difference between 40.45 and 28.00, which agrees perfectly with the value of 12.45 obtained earlier for Pittsburgh's coefficient of specialization.

The Lorenz line would coincide with the diagonal *ABC* if all location quotients were exactly 1.0 – if the employment profile of the city exactly matched the systemic average profile. For this reason the diagonal *ABC* is known as 'the line of complete diversification.' Moreover, the more a city's employment structure deviates from the systemic norm, the more the corresponding Lorenz line will arch upward away from the diagonal. Therefore – and here lies the essence of Gini's insight – the area between the Lorenz line and the diagonal measures the city's degree of specialization. By general agreement, since the Lorenz line must always lie wholly on one side of the diagonal, the area in question is expressed as a proportion of the triangular region above the line *ABC*, this triangle being equivalent to one-half of the complete box. The resulting index lies between zero, representing complete diversification, and 1, representing complete specialization.

Measurement of the required area can be accomplished by means of a formula. Let X_i represent the cumulative percentages for the city under investigation and let Y_i represent the cumulative percentages for the system-wide profile. Further, let N denote the number of categories of industry in the study (in our case, $N = 20$). The Gini index, G, is given by

$$G = 10^{-4} \sum_{i=1}^{N-1} \left| X_i Y_{i+1} - X_{i+1} Y_i \right| .$$

The application of this formula to Pittsburgh is summarized in column 7 of Table 4.2, which lists the successive absolute values which are the terms in the above summation. Dividing the column total by 10^4 yields the value of 0.195 as the Gini index for Pittsburgh.

Thus described in detail, calculation of a single Gini index may seem complicated and tedious – and of many, positively masochistic. However, the labor of calculation can be assigned to the nearest electronic computer. The sense of satisfaction is greatest if one does one's own programming, but personal creative involvement is not essential, provided that a competent professional programmer is available. The Gini indexes and the coefficients of specialization discussed in this chapter were calculated for our 268 cities by means of a single custom-designed program that used less than sixty seconds of computing time. While perhaps not the urban systems analyst's most valuable resource, the computer is certainly an indispensable tool.

Further consideration of Figure 4.1 in the light of the foregoing discussion should make it clear why the Gini index and the coefficient of specialization yield very similar results. In general, the more the Lorenz line is arched upward, the greater the maximum distance between this line and the diagonal of complete diversification. Thus the two measures of specialization should show a strong positive correlation with each other. For our 268 cities, the coefficient of rank correlation (Spearman's coefficient) between these two measures is an impressive +0.98. The Lorenz lines for different cities sometimes intersect each other, so that cities' ranks according to the Gini index may differ slightly from their ranks according to the coefficient of specialization. For all practical purposes, however, the two measures are interchangeable. Nevertheless, the Gini index, being a measure of area rather than length, is more sensitive to small differences between cities. For this reason the Gini index should be preferred whenever appropriate computing facilities are readily available.

Some general empirical results

DIVERSIFICATION AND CITY SIZE

The average value of the Gini index for our 268 cities is 0.267, with a standard deviation of 0.079. These values may be compared with those reported above for the coefficient of specialization: namely, an average of 18.3 and a standard deviation of 5.5. The two sets of results, as previously indicated, are based on different scales of measurement and therefore 'look different.' However, their latent similarity may be revealed by calculating the coefficient of variation – the standard deviation divided by the average value. For the Gini index the coefficient of variation is 0.296, and for the coefficient of specialization, 0.301: the two values are almost identical. Higher moments –measuring skewness and kurtosis – confirm that the two frequency distributions are virtually identical in form and also reveal that 'level of diversification' is a normally distributed variable for these cities.

The results of this analysis do not suggest any systematic relationship between level of diversity and the factor of location. Cities of every level of diversity, from highly diversified to highly specialized, are found in every region. There is no general tendency for cities to be geographically segregated on the basis of their indexes of specialization. Accordingly, we conclude that location does not significantly influence the level of diversification attained by any particular city.[1]

The analysis does suggest, however, that there may be a significant relationship between level of diversification and city size. For example, of the topmost 20 per cent of all cities according to population size, a majority (specifically, thirty-four out of fifty-four cities) also belong to the 'most diversified' 20 per cent. This suggests a positive correlation between level of diversification and city size: the larger the city, the more diversified its profile of employment.

There are good theoretical grounds for expecting such a relationship. As a city expands in population, it becomes an increasingly large market

1 This conclusion is supported by a variety of statistical tests. However, because the results are uniformly negative, details of these tests are omitted.

and may therefore produce a growing number of different goods and services formerly obtained from elsewhere (or not consumed at all). As this process of substituting locally produced goods and services for imports continues, the growing city comes more and more to be a microcosm of the national urban economy as a whole. Thus 'increased city size brings greater industrial diversification' (Thompson 1965, 147). If this theory of import substitution is valid, and if we assume that all cities have similar histories of diversification, then a cross-sectional analysis at any time should show larger cities to be more diversified than smaller ones.

At a rather general level of data aggregation, the theory of import substitution can be tested by calculating the Spearman coefficient of rank correlation between level of diversification and city size. For this calculation, the cities are ranked according to their Gini indexes in ascending order –that is, number one is the most diversified city. With respect to their sizes, however, the cities are ranked in descending order, with number one the largest center. Given these orderings, a positive correlation is expected if the theory of import substitution is valid.

Two technical comments must be made. First, the variable 'city size' is measured not by population, but rather by the size of the labor force, because military personnel are excluded from employment data, but not from counts of total population. Thus total population is tallied on a different basis than the labor force statistics on which the Gini measures are based. Even though only a few cities have a large proportion of the total population directly attributable to the military presence, it was thought best to define 'city size' in a manner consistent with the data used to measure industrial diversification. (There is a very strong correlation between total nonmilitary labor force and total city population. The labor force, as here defined, almost always lies between 35 and 40 per cent of the total population. Prominent military centers such as San Antonio and Norfolk-Portsmouth, however, have distinctly low ratios.)

Second, the Spearman rank correlation coefficient is used in preference to the Pearson product-moment correlation coefficient because the latter should be used only when both variables are normally distributed. In the present case the Gini index values are normally distributed but the city sizes emphatically are not. In general, the variable of city size is perhaps best described as forming a truncated zeta distribution (see chapter 11). Under these conditions, the nonparametric Spearman coefficient is the appropriate one to use.

The rank correlation, R, between the Gini indexes and city size is 0.59.

TABLE 4.3
Cross-classification of 268 cities by size and level of industrial diversification

		Specialized				Diversified
		LEVEL OF DIVERSIFICATION				
		V	W	X	Y	Z
Large	A	1	4	4	11	34
	B	6	8	8	18	13
CITY SIZE	C	8	14	15	14	3
	D	16	14	13	6	4
Small	E	23	13	14	4	0

If diversification is measured by means of the coefficient of specialization instead of the Gini index, the marginally higher value of 0.61 is obtained. Although these results are statistically significant when evaluated by means of the standard *t*-test, the associated level of explanation is disappointingly low. The coefficient of determination, R^2, is at best 0.37: only 37 per cent of the variation in level of diversification can be accounted for by variation in city size. City size is not the only factor influencing the level of industrial diversification.

Before considering a second factor, however, we can cast additional light on the relationship between diversification and city size by examining our data in a more disaggregated form. Table 4.3 presents a quintile-by-quintile cross-classification of the 268 cities in the form of a five-by-five contingency table. Each row in this table contain one-fifth of the total number of cities, as does each column. For ease of reference the rows and columns are lettered from A to E and from V to Z, respectively. A chi-squared test applied to the entire table yields a statistically significant result, but visual inspection of the individual cell values in the table reveals that the association between economic diversity and city size is not equally strong on all levels or 'strata' of either the city size variable or the diversification variable. The variation from stratum to stratum in the strength of the relationship can be evaluated by applying the Kolmogorov-Smirnov test separately to each row and column in turn.[2]

2 The Kolmogorov-Smirnov test compares the shape of the distribution of an ordered series of frequencies (here, cell counts in a particular row or column) to the shape of the corresponding 'ideal' distribution derived from an appropriate null hypothesis. In the present case the ideal or expected frequencies are assumed to be those employed in the standard chi-squared test of the same data. For details of the Kolmogorov-Smirnov test see Siegel (1956, 47–52) and Norcliffe (1982, 102–10).

From the ten Kolmogorov-Smirnov tests thus performed, seven significant results are obtained. Specifically, rows A, B, D, and E, and also columns V, Y, and Z, exhibit sequences of cell frequencies that differ significantly (at the 95 per cent confidence level) from the sequences that would occur if city size and industrial diversification were completely unrelated. Moreover, the levels of statistical significance attained in these tests are highest for rows A and E and for columns V and Z – that is, the outermost rows and columns in the contingency table. This analysis demonstrates that the overall positive association between city size and industrial diversification is accounted for largely by the characteristics of the smallest cities (approximately 50,000–125,000 people) and the largest cities (750,000 and more), with cities of intermediate size conforming only weakly. In short, small cities show a marked tendency to be highly specialized and large cities are almost always highly diversified, but the level of diversification among cities of intermediate size is much more variable and, accordingly, unpredictable.

One possible explanation of this pattern is that the rate at which a growing city becomes more diversified is not constant. Instead, there may be a long period of equilibrium at a low level of diversification, a relatively short period of rapid change, and finally a long period of equilibrium at a high level of diversification. The short period of rapid change may fall anywhere within the long time interval between the 'milestones' of (roughly) 100,000 and 1 million. Cities in this size-range will therefore vary widely in levels of diversification, with some diversifying soon after passing the 100,000 mark and other 'late bloomers' remaining specialized until they are on the threshold of a million inhabitants. Regrettably, it is impossible to test this hypothesis of varying rates of change with data currently available.

THE ROLE OF FUNCTIONAL DIFFERENTIATION

A city's level of industrial diversification appears to be influenced by the functional character of its employment profile as well as by its overall size. The initial evidence for this claim comes from a consideration of the functional character of the cities occupying the extremes of the continuum of diversification. Table 4.4 shows the functional classes – as established in chapter 3 – to which the fifteen most diversified and the fifteen most specialized cities belong. The fifteen most diversified cities are either mixed or nodal. In contrast, the fifteen most specialized centers form a heterogeneous group in which mixed and modal cities are poorly represented. Their places are taken by manufacturing cities and by

TABLE 4.4
The most diversified and the most specialized North American cities according to the Gini index

Rank	City	Gini index	Functional type
Most diversified			
1	Philadelphia, Pennsylvania	0.089	Mixed
2	St Louis, Missouri	0.091	Mixed
3	Columbus, Ohio	0.093	Mixed
4	Los Angeles, California	0.100	Mixed
5	Baltimore, Maryland	0.107	Mixed
6	Kansas City, Missouri	0.110	Mixed
7	Portland, Oregon	0.112	Mixed
8	Denver, Colorado	0.119	Nodal
9	Toronto, Ontario	0.122	Mixed
10	New Haven, Connecticut	0.126	Mixed
11	Boston, Massachusetts	0.127	Mixed
12	Cincinnati, Ohio	0.137	Mixed
13	Syracuse, New York	0.147	Mixed
14	San Francisco, California	0.148	Nodal
15	Tacoma, Washington	0.148	Nodal
Most specialized			
1	Sudbury, Ontario	0.525	Mining
2	Steubenville, Ohio	0.472	Manufacturing
3	Las Vegas, Nevada	0.468	Resort
4	Flint, Michigan	0.462	Manufacturing
5	Midland, Texas	0.451	Mining
6	Columbia, Missouri	0.449	University
7	Sault Sainte Marie, Ontario	0.446	Manufacturing
8	Bryan–College Station, Texas	0.440	University
9	Rochester, Minnesota	0.432	Nodal
10	Anderson, Indiana	0.426	Manufacturing
11	Tallahassee, Florida	0.422	University/government
12	Odessa, Texas	0.410	Mining
13	Sydney, Nova Scotia	0.409	Mining
14	Pueblo, Colorado	0.406	Mixed
15	Petersburg, Virginia	0.404	Mixed

several types of 'exceptional' centers, including one resort, two university towns, the unique university-cum-government city of Tallahassee, and all four mining towns.[3]

3 The fifteen most diversified cities are generally large; the fifteen most specialized are relatively small. If the 268 cities are ranked in descending order of size, the average of ranks for the former group is 27.9; for the latter, 207.5.

In order to examine systematically the role of functional differentiation, rank sums and average ranks according to the Gini index values were calculated for each functional class of cities. The results are shown in Table 4.5. Two distinct hypotheses appear worth testing: that the 41 exceptional cities, taken as a single group, are significantly more specialized than the 227 normal centers; and, at a finer level of disaggregation, that significant differences in levels of diversification exist among the seven functional classes of cities when these classes are compared with one another simultaneously.

The first hypothesis was tested by means of the Mann-Whitney sum-of-ranks test. This test asks, in effect, whether the exceptional cities are clustered toward the specialized end of the scale of diversification more than might be expected if cities were allocated randomly to the rank positions along this scale. In the present case, a highly significant result is obtained: with random allocation of ranks, the observed degree of bunching would occur, on average, only once in every fifty million trials. We therefore reject the hypothesis of randomness and conclude that the exceptional cities, taken as a group, are indeed significantly more specialized than the normal centers.[4]

The second hypothesis was examined by means of the Kruskal-Wallis technique for analysis of variance. This is a nonparametric test that may be described as an extension of the Mann-Whitney test to situations involving more than two classes. The Kruskal-Wallis test yields a statistic which, under the hypothesis of randomness in the assignment of cities to classes, has the same frequency distribution as chi-squared with $k - 1$ degrees of freedom, where k is the number of classes considered in the test (Siegel 1956, 184–93). For the data of Table 4.5, the value of the Kruskal-Wallis test statistic is 78.18. With six degrees of freedom, the critical value of chi-squared at the 99.9 per cent confidence level is only 22.46, so we again have a highly significant result. The seven functional classes of cities differ significantly in the degree of diversification of their employment profiles.

Given significant differences among the functional classes, the data on average ranks (Table 4.5) may be used to place the various classes in an ordered sequence along the continuum of diversification. Mixed cities are the most diversified type, followed closely by nodal cities. After a noticeable gap, government centers, manufacturing cities, and resorts are

4 The Mann-Whitney test, also known as the Wilcoxon test, is the nonparametric counterpart of the parametric *t*-test for comparison of two means. Here the numbers of normal and exceptional cities are sufficiently large that the method of normal approximation must be used. For details see Alder and Roessler (1964, 141–50).

TABLE 4.5
Sum of Gini ranks and average Gini rank for each functional class of cities

Functional class	Number of cities	Sum of Gini ranks	Average Gini rank[a]
Normal cities			
Manufacturing cities	69	11,756	170.4
Mixed cities	80	7,773	97.2
Nodal cities	78	8,490	108.8
All normal cities	227	28,019	123.4
Exceptional cities			
Government centers[b]	20	3,219	161.0
University towns	12	2,858	238.2
Resort centers	5	905	181.0
Mining towns	4	1,045	261.3
All exceptional cities	41	8,027	195.8

a Rank number 1 is assigned to the most diversified city in the urban system. Hence, the lower the average rank value, the more diversified the cities of a given class.
b For certain tests the functional classes of cities must be mutually exclusive. In these cases Tallahassee is counted only as a university town and not also as a government center. No other city is a member of more than one class.

clustered along the midsection of the continuum. Another gap follows, and finally the university and mining towns are found near the specialized end of the scale. These statements refer to each class as a whole, and there is variation in diversification within each class. Nevertheless, the progression from mixed cities to mining towns is a valid generalization that provides useful insights into the relationship between industrial diversification and functional type.

Recalling the discussion of functional classification in the preceding chapter, it is satisfying to discover that mixed cities are the most diversified of the seven functional classes. First, given that complete diversification is defined by the average employment profile of the urban system as a whole, we would expect the most diversified class to be found among normal rather than exceptional cities, since normal cities comprise 85 per cent of all cities. Second, mixed cities occupy a unique position among normal centers, with significant amounts of employment in both manufacturing and trade. They are, as we have mentioned earlier, 'the most normal of the normal centers.' It is therefore entirely appropriate that they should emerge from the present analysis as the most diversified functional class.

Nodal cities, as a class, are distinctly more diversified than manufacturing cities. While many manufacturing centers produce a wide variety of manufactured goods, a considerable number of them concentrate on one particular kind of product, thus drawing the manufacturing class as a whole toward the specialized end of the scale. This unbalanced type of structure is much less likely to occur among nonmanufacturing activities, which produce 'goods' (health care, education, retail facilities, and so forth) that cannot be physically transported over large distances and therefore must be made available everywhere. Within a nodal city, the manufacturing sector itself might be highly specialized but, by definition, this sector forms only a minor part of the city's total employment structure. Nodal cities, in short, are dominated by employment in precisely the kinds of activities that tend to be spatially ubiquitous. It is not surprising, therefore, that nodal centers turn out to be much more diversified than manufacturing cities.

Occasionally, of course, a nonmanufacturing activity will exhibit an exceptionally high percentage in a particular city's labor force. If this occurs in mineral extraction, education, personal services, or public administration, the result is an exceptional city – recall the criteria established in chapter 3. However, if it occurs in some other nonmanufacturing activity, the result is a nodal city with unusually high specialization. This is rare. The only noteworthy example, in fact, is Rochester, Minnesota, home of the world-famous Mayo Clinic. Employment in health services comprises 26.0 per cent of Rochester's labor force, more than three times the system-wide average of 7.8 per cent and almost twice the value for the city ranking second in this activity (Kingston, Ontario, at 14.2 per cent). Rochester's Gini index is an impressive 0.432 (Table 4.4). Only 8 of our 268 cities are more specialized than this, and Rochester is far and away the most specialized center in the nodal class.

We have now seen that the level of industrial diversification in North American cities is associated with two distinct factors: namely, city size and functional type. In a mechanistic sense these factors may be regarded as 'causes' of the observed levels of diversification, though causality is a metaphysical concept not open to empirical demonstration. Some writers prefer to give causality an evolutionary or 'process' interpretation, treating the three elements of employment structure, size, and industrial diversification as interdependent manifestations of a single general process of urban system development. In this view, these variables do not act causally on one another but rather hold equal status as the joint effects of causes buried in the details of each city's unique history. However, supporters of the historical, or 'vertical' conception also recognize mechanistic, or 'horizontal' interconnections among various manifestations of the general evolutionary process. Thus, regardless of one's position concerning causality, it is useful to have discovered that the variables of city size, functional type, and industrial diversification are significantly interrelated.

We are left, however, with one conspicuous loose end. Diversification is related both to city size and to functional type, but are the latter two variables directly related to each other? This question may be addressed by using the method employed in testing the relationship between diversification and functional type. First, using the Mann-Whitney test, it is found that exceptional cities, taken as a group, are not significantly different in size than normal cities. Second, however, a Kruskal-Wallis test of size differences among all seven functional classes yields a result that is significant at the 99 per cent confidence level. Detailed examination of the data shows that the Kruskal-Wallis test is successful because the mining and university towns are notably smaller than cities of the other types. But the remaining exceptional centers – resorts and government cities – are not especially small, and hence exceptional cities as a group are not significantly smaller than normal towns. The relationship between city size and functional type, while not insignificant, is much weaker than the relationships between each of these two factors and the level of industrial diversification.[5]

DIVERSIFICATION WITHIN MAJOR EMPLOYMENT SECTORS

In the preceding sections of this chapter we have examined the level of diversification of the entire labor force of each city. But the general concept of diversification can also be applied to *particular sectors* of the labor force. Probably the most fundamental distinction is that between manufacturing and nonmanufacturing industries. Indeed, early studies of diversification were concerned exclusively with manufacturing, apparently on the grounds that the nonmanufacturing sector was unimportant to a city's economic viability or potential for growth (Tress 1938; Rodgers 1957). Avoiding the 'industrialist bias' implicit in this one-sided approach,

5 In a similar study of seventy-nine Canadian cities for 1951, 1961, and 1971, no significant relationship between city size and functional type was found, although both variables were associated with differences in diversification (Marshall 1981). This illustrates the point that two variables that display significant correlations with a third variable need not be significantly correlated with each other. For Canadian cities, the effects of city size and functional type on diversification are, statistically speaking, mutually independent.

we now briefly discuss the results obtained when levels of diversification for manufacturing and nonmanufacturing activities in our 268 cities are measured separately.

Measurement of industrial diversification within a given sector involves redefining the 'total labor force' to be the sum of all employment within that sector only. Workers in other activities are simply irrelevant to the calculations. Thus, in measuring diversification within the manufacturing sector, the 'total' labor force of each city is defined as the sum of all employment in manufacturing, and the system-wide definition of 'complete diversification' is the average percentage distribution of employment within that sector. In short, the manufacturing sector – which, on average, makes up only 25.4 per cent of the total labor force in all activities (see Table 3.2 above) – is treated as 100 per cent in this portion of the analysis. In similar fashion, the relevant standard of reference for the analysis of the nonmanufacturing sector is the sum of all employment in the nonmanufacturing activities.

Apart from these basic changes, the approach is the same as that described at the beginning of this chapter. Either the Gini index or the coefficient of specialization may be used; the two sets of results thus obtained, for reasons already given, will be almost perfectly correlated with each other. In the present case the two sets of index values have a rank correlation coefficient of 0.98 in both the manufacturing and the nonmanufacturing sectors. By coincidence, this value is the same as the correlation obtained when the basis of measurement for each index is the entire labor force.

Since manufacturing activities tend to be spatially sporadic and nonmanufacturing activities (except for mineral extraction) ubiquitous, we should expect the general level of specialization to be high in manufacturing and low in nonmanufacturing. As Table 4.6 shows, this expectation is borne out by the facts. No matter which technique of measurement is used, the average index value for the 268 cities is approximately three times as large in manufacturing as in nonmanufacturing. Moreover, the average index value for all industries taken together lies between the values obtained for the manufacturing and nonmanufacturing sectors taken separately. Thus, if an individual city's index for its entire labor force has an above-average value – signifying an above-average level of overall specialization – the structure of the city's manufacturing sector is probably responsible. Among cities having a Gini index greater than 0.267 for the entire labor force, more belong to the manufacturing group than to any other functional class.

In the case of data pertaining to the manufacturing sector, the

TABLE 4.6
Measures of diversification within major employment sectors

	Gini index		Coefficient of specialization	
	Average	Standard deviation	Average	Standard deviation
All industries	0.267	0.079	18.3	5.5
Manufacturing only	0.410	0.138	31.7	11.9
Non-manufacturing only	0.145	0.062	10.7	4.7

coefficient of rank correlation between the level of diversification (using the Gini index) and city size (represented by the total labor force in all activities) is 0.53. The corresponding figure for nonmanufacturing activities is 0.50. These coefficients are statistically significant at the 99 per cent confidence level and support the view that import substitution operates with about equal strength in both sectors. In both sectors, moreover, these significant correlations between diversification and size are largely the result of the characteristics of the very largest and the very smallest cities in the study. The level of diversification is almost always high among large cities and low among small cities but varies widely among cities of intermediate size (roughly speaking, those of 125,000–750,000 population). In this respect the structure of employment in the manufacturing and nonmanufacturing sectors closely resembles the structure in the labor force as a whole (see Table 4.3).

In the manufacturing sector there is no significant overall relationship between level of diversification and the seven categories of the functional classification (Kruskal-Wallis test). Thus the diversity of manufacturing does not differ appreciably between cities classed as manufacturing centers and other types of cities. In some manufacturing centers the manufacturing sector itself is highly specialized (e.g. Flint, Michigan; Steubenville, Ohio; and Sault Ste Marie, Ontario), while in others it is markedly diversified (e.g. Toledo, Ohio; Lowell, Massachusetts; and Kitchener-Waterloo, Ontario). But equally, some cities not classed as manufacturing centers have a highly specialized manufacturing sector (e.g. Baton Rouge, Louisiana; Pueblo, Colorado; and Chicoutimi, Quebec), while in others the manufacturing sector is diverse (e.g. Baltimore, Kansas City, and Toronto). In short, the 'level of commitment' to manufacturing, as measured by the percentage of a city's total labor force devoted to manufacturing, is unrelated to the level of diversification attained by this sector.

For the nonmanufacturing activities, in contrast, the overall relationship between level of diversification and functional classification is highly significant – the Kruskal-Wallis test yields a result significant beyond the 99.9 per cent confidence level. Of course, the four classes of exceptional cities possess anomalous profiles in their nonmanufacturing sectors by definition. A city is simply not classified as being 'exceptional' unless it deviates sharply from the system-wide percentage in the relevant nonmanufacturing activity. Indeed, a separate test shows that the levels of diversification within the nonmanufacturing sector do not vary significantly among the three classes of normal cities. The variation becomes significant only when normal and exceptional cities are combined in a single, all-inclusive test. Thus the relationship between the functional typology and the levels of diversification within the nonmanufacturing sector is essentially an expression of the fundamental distinction between normal and exceptional types of towns.

Diversification and time

A natural extension of the above static analyses involves the study of changes in levels of diversification through time. Where long periods of time are concerned, such work is severely hampered by definitional changes and other limitations inherent in the census data on employment, but for short periods it is possible to assemble a body of data that uses identical categories of industry in all years. Given year-to-year consistency in the data, valid statements can be made about the behavior of indexes of diversification through time.

A start has been made in this direction in a recent study of seventy-nine Canadian cities for the years 1951, 1961, and 1971 (Marshall 1981). The study was motivated by the fact that import substitution is fundamentally a theory about process, and therefore requires a longitudinal analysis for proper evaluation. The fact that city size and level of diversification are significantly correlated in a cross-sectional study, though highly suggestive, does not logically entail that individual cities become increasingly diversified as they grow larger. It is quite apparent – though often overlooked – that some historical depth is required if questions about processes of change are to be adequately addressed.

In the Canadian study the labor force of each city was broken down into twenty-eight categories of activity (as compared with twenty in the earlier sections of this chapter) and Gini indexes were calculated for every city for each of the census years 1951, 1961, and 1971. The system-wide average profile changes slightly from year to year. For example, average employment in textile manufacturing fell from 1.9 per cent of the system-wide profile in 1951 to 1.2 per cent in 1961 and finally to 0.9 per cent in 1971. Conversely, education rose dramatically, from 2.7 per cent in 1951 to 4.2 per cent ten years later and to 7.5 per cent in 1971. In other words, the employment structure that defines complete diversification is a 'moving target' (Marshall 1981, 324). Thus, in the unlikely event that a city should experience absolutely no alterations in its employment profile over a ten-year period, its Gini index would still change. By the same token, constancy in level of diversification can be achieved only by appropriate small changes in the city's industrial structure.

For the decade 1951–61, the mean rate of change in the size of the labor force in the seventy-nine Canadian cities was 35.7 per cent and the mean rate of change in the Gini index value was −12.3 per cent. The corresponding figures for 1961–71 were 42.1 per cent and −8.9 per cent, respectively. In both decades, therefore, the average city grew larger (positive rate of change in size of labor force) and became more diversified (negative rate of change in Gini index). These results are consistent with the theory of import substitution.

The cities were subdivided into functional classes by means of a method similar to the procedure described in chapter 3. Because the total number of cities in the study was relatively small, the exceptional centers were grouped together under 'special activities.' The complete classification thus consisted of four categories: manufacturing, mixed, nodal, and special activities. The ensuing analysis revealed that the average Gini index value declined in both decades in all four functional classes, as shown in Figure 4.2. Therefore import substitution appears to be a general process, not restricted to particular functional classes of towns.

Looking at individual cities, I found that seventy-three of the seventy-nine cities grew larger during the 1951–61 decade, and sixty-six (or 90 per cent) of these became more diversified. In the following decade seventy-four cities increased in size, and again sixty-six of these became more diversified. Thus the vast majority of the cities grew both larger and more diversified, thereby behaving in accordance with our theoretical expectations.

In the few cities that increased in size but became *less* diversified, the increase in specialization can generally be attributed to pronounced employment growth in a single activity. For example, the military town of Trenton, Ontario (an 'exceptional' center), saw increased employment in

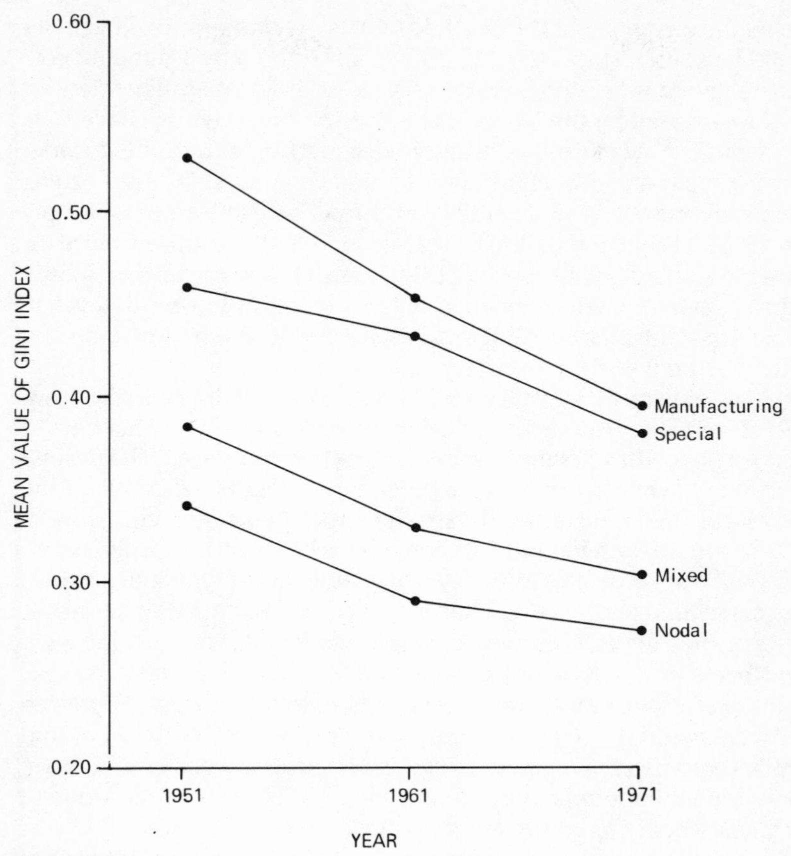

Figure 4.2
Changes in the average level of industrial diversification for four types of cities in Canada, 1951–71. The falling values of the Gini index indicate increasing diversification over time.

'federal administration' – actually, the air force – during the 1950s, from 22.0 to 29.7 per cent of its total labor force; its Gini index rose from 0.435 to 0.495. In St Thomas, Ontario, a major investment by Ford Motor Co. raised employment in the manufacturing of transportation equipment from 4.2 per cent of the city's labor force in 1961 to 12.5 per cent in 1971. St Thomas's Gini index accordingly rose from 0.337 to 0.356.

Just the opposite mechanism appears to be involved in the very few cases where a city declined in size but nevertheless became more diversified. The decline in size is accounted for by a specific activity that represents the backbone of the city's economy. As this specialty becomes relatively less important during its decline, the city's employment structure converges on the system-wide average profile. The clearest example is provided by the Nova Scotia coal-mining town of Glace Bay, the only center in which the labor force declined in both of the decades examined. Employment in mineral extraction comprised 50.7 per cent of this town's labor force in 1951, falling to 38.7 per cent in 1961 and 24.1 per cent in 1971. The latter figure is still large when compared with the system-wide average of 1.0 per cent, but clearly the progressive loss of coal-mining employment implies a decrease in the overall level of specialization. Glace Bay's Gini indexes for the years in question were 0.685, 0.597, and 0.464, respectively.

There was only one case of a city declining in size and becoming more specialized: namely, Moose Jaw, Saskatchewan, 1961–71. (This type of change does not contradict the theory of import substitution.) The decline in size resulted primarily from the loss of approximately 250 jobs in the railroad yards. The increase in specialization was associated with absolute growth in both federal and nonfederal civil service employment.

Additional tests indicated that rates of change in the Gini index values were not correlated with rates of change in city size, or with city size at the beginning of each decade. However, some significant relationships were observed between the rates of change in the Gini values and the four functional classes of cities. Some reversals of relative status between decades were noted. For example, during the 1950s, both nodal cities and mixed cities were becoming increasingly diversified at significantly faster rates than the exceptional cities; but during the 1960s, the latter group of centers was diversifying at the greater rate.

In addition to providing support for the theory of import substitution, this longitudinal study makes it clear that the employment profiles of Canadian cities have tended to converge toward the system-wide average profile over time. If this convergence is true for North American cities in general, we may infer a gradual weakening of the structural distinctions that separate the various functional classes of cities. However, this may not be equally true for all classes of cities. The distinctions that identify exceptional types of cities – mining towns, government centers, resorts, and university towns – will probably tend to remain relatively sharp. In normal centers, however, similarities among cities may eventually become more significant than differences. If present trends continue, functional type will become a less and less important determinant of level of diversification, whereas city size will come to play an increasingly significant explanatory role.

Structural convergence undoubtedly reflects the ubiquity of the most rapidly growing activities of recent years – health care, education, office employment, and various other tertiary activities. If types of work that are found everywhere grow rapidly, while other, geographically sporadic types grow slowly or not at all, employment profiles must gradually become more and more alike. It will be interesting to see how far this process of convergence advances.

Industrial diversity and economic health

We conclude this chapter with a brief discussion of the general relationship between industrial diversification and a city's overall economic health. The basic principle is that the likelihood of a city's falling on hard times is inversely proportional to the diversification of its economy: the more diversified the city, the more easily it should be able to survive regional or national recession or depression. Conversely, the more specialized the city, the more vulnerable its residents will be to economic hardship in times of stress. For ease of reference we will call this relationship the 'principle of vulnerability.'

The principle of vulnerability first attracted widespread interest in the United Kingdom during the 1940s (Smailes 1943). At that time, many British politicians, planners, and academics were concerned about the possible deleterious effects of an expected postwar economic slump on the country's urban system, especially on the major centers of Victorian industrialization in the Midlands, the northern counties, and South Wales. Although diversification in these old industrial cities had apparently never been measured, it was widely accepted that in many cases their economic profiles were, to use Smailes's term, 'ill-balanced,' with high levels of specialization, particularly in manufacturing.

It became a virtual axiom of postwar British planning than an unduly specialized urban economy could only lead to unemployment, poverty, and perhaps even social unrest, and the perceived remedy was to promote diversification by every reasonable means. This belief can be discerned in various attempts to discourage excessive concentration of new employment growth in the Greater London area. Decentralizing new employment away from London could, in addition to relieving the southeastern region of some of the pressures of rapid growth, also reduce specialization elsewhere. Thus decentralization and diversification might be made to go hand in hand. Similar sentiments have been expressed in other European countries – notably the Netherlands – and occasionally in the United States and Canada.

The theoretical basis of the principle of vulnerability is easy to understand. If a city is heavily dependent on one industry, and if the market for that industry's products fails, it stands to reason that the city will soon be experiencing economic difficulties. Workers in the major industry will be laid off; and, even though they may receive some income from unemployment insurance, their spending power will be sharply reduced. The resulting decline in the amount of money circulating within the local economy will seriously affect the retail and service sectors. Stores that were barely viable before the disaster struck may have to close down; others will make ends meet by reducing the number of sales clerks and other personnel. These repercussions will cause unemployment to rise still higher. Some families will leave town in search of greener pastures. Others will go heavily into debt. Only the recovery of the market – possibly under the stimulus of government intervention – or the introduction of new sources of employment can prevent the downward spiral of decay from ultimately destroying the town.

In a city with a diversified economic structure, however, the situation is distinctly less gloomy. First, unless there is a massive economic slump on the scale of the Depression of the 1930s, it is unlikely that all industries will experience unfavorable market conditions simultaneously. Hence, workers laid off by one firm may be able to find work with a firm of a different kind. Second, the impact of even the complete closing of a single factory in a diversified city is likely to be much smaller, in relative terms, than the effect of a cutback in the major enterprise within a highly specialized center. Third, a diversified city's labor force usually encompasses a wider variety of skills than that of a specialized city, and therefore a diversified city has a better chance of attracting replacements for any firms that go into decline. In short, there is much to be said for 'not putting all your eggs in one basket.'

Given our earlier finding that levels of industrial diversification are positively correlated with city size, an important implication of the principle of vulnerability is that, generally speaking, large cities should be less susceptible to economic reversals than small cities. Large cities, of course, possess several defenses against economic decline, including greater political clout than small centers, a mass market for goods and services of all kinds, and, almost certainly, a steadier supply of entrepreneurial talent (Thompson 1965, 21–4). A diversified economy should be added to this list as one of the several factors that make large centers less vulnerable than small centers to the vicissitudes of economic life. But no city, however large or industrially diversified, is ever immune from absolute decline. For example, Cleveland, St Louis, Pittsburgh, and even New York each experienced an absolute loss of population during the 1970s – and we are speaking here of the metropolitan area as a whole, not merely of the central city. And let us not forget Rome, which fell from a population of 750,000 at the time of Christ to a comparatively meager 40,000 in the Middle Ages. Who, in the age of Augustus, would have thought such a fate possible? Clearly it is folly to believe, as some have done, that there is a critical threshold of size above which a city's future growth is forever assured.

To end this chapter on a forward-looking note, the principle of vulnerability offers opportunities for the testing of specific hypotheses that relate industrial diversification to various indicators of economic health. One might, for example, measure the economic health of each city over a selected period of time by means of the average rate of unemployment, the per capita volume of unemployment insurance payments, the number of plant closings, or other similar variables. Relationships between these sorts of data and measures of diversification could then be sought. To the best of my knowledge, work of this kind has not yet been undertaken. The hypotheses in question, however, seem eminently worthy of consideration.

Summary

This chapter has been concerned with the study of industrial diversification. A city is diversified if its employment profile resembles that of the urban system as a whole. The more a city's employment structure deviates from the system-wide average profile, the more specialized the city is said to be.

Two suitable indexes for measuring diversification were described in detail: namely, the coefficient of specialization and the Gini index of concentration. If these two indexes are applied to the same urban system, the resulting sets of index values will be almost perfectly correlated with each other.

For North American cities it was shown that levels of diversification are significantly related to city size and to the functional classification described in chapter 3. In general, the larger the city, the more diversified is its economic structure. This finding is consistent with the theory of import substitution, but it was noted that the observed general relationship tends to break down among cities of intermediate size. With regard to functional type it was found that the seven functional classes could be arrayed along the scale of increasing specialization in a definite order: mixed cities, nodal cities, government centers, manufacturing centers, resorts, university towns, and mining towns.

Levels of diversification were also measured separately for the manufacturing sector and the nonmanufacturing sector. On average, the level of specialization in the manufacturing sector is three times as high as in nonmanufacturing. In both sectors, the levels of diversification are positively correlated with city size. It was also found that the level of diversification achieved by the manufacturing sector is unrelated to whether or not a city is classified as a manufacturing center.

A longitudinal analysis of Canadian cities showed that almost all centers, taken individually, do become increasingly diversified as they grow larger in size. This historical analysis lends strong support to the theory of import substitution.

The chapter concluded by drawing attention to the hypothesis that specialized cities are more vulnerable than diversified cities to economic hardship in times of recession. It was noted that formal testing of this hypothesis awaits future research.

5

Central place theory: the spatial structure of local trade

The basic question to which central place theory is addressed may be stated as follows. What would be the spatial structure of an urban system if towns and cities functioned solely as providers of consumer goods and services to surrounding tributary areas?

In order to appreciate the importance of this question, let us recall certain general points made in earlier chapters. First, the development of urban systems has been dominated historically by the influence of two major categories of economic activity: trade and manufacturing. Second, the category of trade may be divided into two distinct components: local and long-distance. Combining these two statements, we may say that local trade, long-distance trade, and manufacturing are the three essential pillars of the urban economy – the three sources of support without which urban systems, as we know them, would not exist.

In terms of these three major sources of support, central place theory must be seen as a theory of local trade. It is not concerned with long-distance trade, and it is not concerned with manufacturing. Nor is it concerned with what might be termed the 'lesser urban functions' – that is, special activities such as mineral extraction, tourism, and military training. Indeed, the single most important introductory point about central place theory is that it is not, and was never intended to be, a general theory of the spatial structure of urban systems. It is a *partial* theory, concerned only with the role of towns as centers of local trade.

Given that central place theory is only a partial theory, why then is it usually regarded as one of the most important topics in urban systems analysis? The answer is that local trade is a more ubiquitous source of urban support, on a worldwide basis, than any other activity that towns perform. As indicated in chapter 1, the role of a town as a provider of goods and services to consumers in its surrounding region comes closest to expressing the fundamental essence of urbanism as a way of life. Thus, while central place theory is not concerned with all aspects of urban support, it does deal with the particular kind of activity that has had the greatest formative influence on the spatial structure of the typical urban system.

Writing at the beginning of the present century, the German economic historian Carl Bücher described the spacing of towns:

> If we take a map of the old German Empire and mark upon it the places that, up to the close of the Middle Ages, had received grants of municipal rights – there were probably some three thousand of them – we see the country dotted with towns at an average distance of four to five hours' journey in the south and west, and in the north and east of seven to eight. All were not of equal importance; but the majority of them in their time were, or at least endeavoured to be, the economic centres for their territory, leading just as independent an existence as the manor before them. In order to form a conception of the size of these districts, let us imagine the whole country evenly divided among the existing municipalities. In this way each town in southwestern Germany has on the average forty to somewhat over fifty square miles, in the central and northeastern parts between sixty and eighty-five, and in the eastern from somewhat over one hundred to one hundred and seventy. Let us imagine the town as always situated in the centre of such a section of country, and it becomes plain that in almost every part of Germany the peasant from the most distant rural settlement was able to reach the town market in one day, and be home again by nightfall. (Bücher 1901, 119–20)

Bücher's comments endorse the view that a town's most basic task is to serve as a market center for its surrounding district. They also identify a mechanism capable of explaining the spacing of local market towns: namely, the requirement that every peasant could reach one within a day's round trip, on foot or at best on horseback. Market towns had thus to be fairly closely spaced. The single-day market journey is generally assumed to have been the norm throughout Europe – and in eastern North America before the introduction of the internal combustion engine. Thus the basic spatial pattern of the original network of small market towns on both sides of the North Atlantic can be related to pre-automobile transportation technology.

Bücher did not make any further contribution to the study of the spacing of towns; the geographer Walter Christaller (1893–1969), writing during the 1930s, presented the theory of central place patterns in a more fully articulated form (Christaller 1933). As Christaller remarked many years later, his theory, at least initially, was not well received (Christaller 1968, 99). His fellow geographers tended to regard it as economics, while economists dismissed it as geography! That the theory addresses a geographical problem is beyond doubt; yet, even today, not all geographers find Christaller's work entirely to their liking. Some find central place theory cold and unappealing because of its highly abstract character. Others, in contrast, regard it as one of the brightest jewels in geography's crown. To a large extent, such differences of opinion are based on varying philosophical views of the relationship between the process of theory-building and the conduct of empirical work. If one accepts that a theory is intended not to *reproduce* reality (in the sense of creating an exact replica) but rather to provide a conceptual framework that renders the world of our experience more intelligible, central place theory looks like a gift from the gods.

Christallerian central place theory

THE ASSUMPTIONS

Christaller's theory is built on a set of assumptions that describe the character of an imaginary world in which theory and 'reality' are one and the same. As with all good theories, the assumptions are essentially simplifications of selected aspects of everyday experience. They are designed to provide an uncluttered arena for the working out of the specific problem in hand. The making of assumptions is by no means a reprehensible activity; indeed, it is an essential, if sometimes unrecognized, component of all theorizing. Care should be taken, however, to preserve as much plausibility as possible. One is reminded of the story of the shipwrecked economist who approached the problem of opening his sole can of beans by declaring: 'First, I assume a can-opener'! The assumptions in central place theory are not that far-fetched, though there is probably no part of the real world in which they are fulfilled to the last detail.

The first assumption is that the urban system takes shape on a hypothetical type of landscape known as an isotropic plain. This is a level, featureless, agricultural plain on which a rural population is distributed at a constant density. Formally, the term *constant density* signifies that a planar cell of fixed dimensions that contains x inhabitants when centered on a randomly chosen point in the plain will contain $x \pm e$ inhabitants when centered on any other point, where e is vanishingly small compared to the magnitude of x. In addition, the isotropic plain is everywhere equally fertile, the farmers are all equally competent, they possess identical tastes and preferences for consumer goods and services, and disposable income per person is constant. Finally, movement across the plain is equally easy in all directions from all points, with the cost of transportation being directly proportional to the distance traveled. Because of this homogeneity with respect to movement, an isotropic plain is sometimes referred to as a 'transport surface.'

In some branches of location theory, an isotropic plain may initially be defined as being unpopulated. In central place theory, however, the presence of a market for consumer goods and services is essential, and a veneer of rural population is therefore regarded, in effect, as part of the definition of the isotropic plain. If we like, we can first assume an empty plain and then argue that farmers will distribute themselves evenly in response to the plain's physical uniformity. This produces the desired result by a slightly different route. Further, some writers consider the isotropic plain to extend outward to infinity in all directions; others assume merely that it is extremely large. Infinite extent does not seem necessary to Christaller's argument. But the plain should be extensive, and the urban system should develop somewhere in its interior, so that boundary effects can safely be ignored.

The second assumption is that all urbanizing forces except local trade are absent. This 'insulating' assumption eliminates the effects of alternative causes of town growth. The major causes thus neutralized are long-distance trade and modern large-scale manufacturing, but less common activities such as mining, tourism, and military training are also removed. This assumption is less unrealistic than it may appear. Although real cities, especially large ones, tend to be multifunctional, the great majority of urban centers in many parts of the world are dependent far more on local trade activity than on any other function. This is the normal state of affairs, indeed, in regions dominated by agriculture. Such regions, moreover, are often plains, or at least areas of low relief – for example, the Paris Basin, the Ganges Valley, and the American Midwest. In short, many areas that come tolerably close to meeting the assumption that towns are supported solely by local trade also tend to satisfy the assumption of an isotropic plain.

The third assumption is that the prevailing economic system is one of pure competition. Individual transactions are too small in value to affect the market price of any commodity; there is no governmental control of prices or of the quantities of products made available; entrepreneurs are free to enter and leave the market at will; and there is no price-fixing or any other form of collusion either among sellers or among buyers

(Leftwich 1960, 23–7). However, in contrast to the nonspatial classical theory of pure competition in economics, Christaller's argument recognizes that the competitive position of each supplier is affected by location. Since every possible site for a central place is unique, each village or town enjoys a monopoly in its own vicinity, and the suppliers within the town may be regarded as spatial monopolists rather than pure competitors. Christaller's theory thus modifies the assumption of pure competition by explicit consideration of location.

The fourth assumption states that, as the cost of acquiring a good or service increases, the quantity demanded by consumers declines. Although this assumption is often taken for granted, cost and quantity demanded could be related in some other way, and hence it is advisable to make the inverse relationship explicit. In reality this assumption is true for virtually all goods; goods of this type are termed 'normal goods' in economic analysis. The fourth assumption simply means that our hypothetical economy contains only normal goods.

The fifth assumption is that the amount of revenue accruing to each central place per unit of time is directly proportional to the size of its tributary area. This assumption does not entail, as some have supposed, that all consumers invariably patronize the nearest central place at which their current shopping needs can be fulfilled. The idea that central place theory requires a strict 'nearest center' assumption is a myth. So far as the theoretical argument is concerned, consumers are free to visit towns more distant than their nearest center provided that such 'longer than necessary' shopping trips cancel each other out. For example, a family residing five miles from town *A* and ten miles from town *B* may shop regularly in town *B* so long as it is assumed that a second family, one living closer to *B* than to *A*, travels regularly to shop in *A*. Sets of self-neutralizing exchanges involving more than two towns are also possible. Moreover, shopping trips of this sort are not in some sense 'irrational.' They are merely motivated by concerns other than a desire to minimize the distances covered. In summary, the revenue received by each central place is the amount that would be generated if all consumers minimized their travel distances, but minimization of distance is not necessary for the generation of that particular amount.[1]

1 Studies completed in Iowa have shown that many rural consumers do not always shop at the closest available outlet (Golledge, Rushton, and Clark 1966; Rushton, Golledge, and Clark 1967). However, these studies did not consider the effects of such behavior on the revenues of central places, and therefore the extent of self-cancellation among longer-than-necessary trips remains unknown. See also Webber (1971, 21–3) for additional commentary on the non-essential character of the nearest-center assumption.

The sixth and final assumption requires a preliminary discussion of the concept of threshold. As used in central place research, this term denotes either the minimum number of customers or the minimum dollar volume of revenue required to keep a particular type of retail store or service establishment in business. If the threshold is not equaled or surpassed by the available number of consumers, the store's costs will exceed its revenues and the enterprise will fail. The 'costs' in this context are defined as including a reasonable level of profit, termed 'normal profits.' Now different establishments of the same kind in the real world will probably not have identical thresholds, but in theory a single threshold value is assumed for each business type. Accordingly, we may conceive of all the different types of stores and services as being ranked in descending order on the basis of their threshold requirements. At the top of the list are found 'high-order' types, such as exclusive fashion boutiques and the most specialized kinds of medical services. At the bottom are found 'low-order' types, such as gas stations, taverns, and convenience stores.

The sixth assumption can now be stated as follows. The various suppliers of goods and services on the isotropic plain will agglomerate in space in such a way that, if a central place contains a particular type of business with threshold *t*, it will also contain all other types having thresholds equal to or smaller than *t*. Despite its apparent stringency, this assumption is by no means unrealistic. It means, in essence, that the types of stores and services present in a particular town must not be absent from any town of larger size. On the whole, the real world obediently follows this rule, though exceptions are known to occur. The sixth assumption merely prohibits all exceptions within the theoretical models.

For ease of reference, the six Christallerian assumptions are here summarized in a numbered list.

Assumption 1. The urban pattern takes shape on an isotropic plain.

Assumption 2. All urbanizing forces other than central place activity (i.e. the provision of consumer goods and services to the population surrounding each urban center) are absent.

Assumption 3. The prevailing economic system is one of pure competition, but the influence of location is also taken into account.

Assumption 4. The economy contains only normal goods: that is, as the cost of acquisition increases, the quantity demanded declines.

Assumption 5. Each central place's revenue is proportional to the size of its market area.

Assumption 6. If a central place contains a type of store or service with threshold *t*, it also contains all other types having thresholds equal to or smaller than *t*.

THE ARRANGEMENT OF POINTS OF SUPPLY

Suppose that a single retail store is opened at some arbitrary point on the isotropic plain. For the moment it does not matter what kind of store or service outlet this happens to be. Whenever a farmer comes to shop in this store, he or she pays not only the in-store purchase price of the goods selected but also the cost of traveling from the farm to the store and back. The in-store purchase price is a fixed cost that is the same for all consumers, but the cost of transportation is variable and depends on the distance between the farm and the store (assumption 1). The total cost incurred by the consumer is the sum of the fixed and variable costs and must rise as the distance between farm and store increases. Accordingly, the quantity of goods demanded declines with increasing distance from the store (assumption 4). At a certain distance a point is reached at which the demand is zero, since farmers at this point are not willing to bear the transportation cost required to visit the store. Beyond this distance, too, the level of demand is zero. Moreover, the point of zero demand will be equally distant from the store in every direction (assumption 1). Therefore the market area of our isolated store will be perfectly circular, with the store situated at the center of the circle (Figure 5.1, panel A).

The radius of this market area is termed the *range* of the store. For any given level of rural population density, the smallest possible value of the range is determined by the size of the store's threshold. If the range is not great enough to capture at least the threshold number of consumers, the store will not be economically viable and will presumably go out of business.

It is useful to make a formal distinction between the two possible situations in which the store will remain viable. First, the market area may be exactly large enough to satisfy the store's threshold, and no larger. In this case the store earns only normal profits and the market area is termed the threshold market area. Second, the market area may be larger than is required in order to keep the store in business. In this case the market area itself is given no special name, but the store is said to earn excess profits. By definition, excess profits are simply profits over and above the level of the normal profits provided by the threshold market area. The distinction between normal and excess profits will prove important as our discussion proceeds.

Now suppose that additional stores of the same type as the first store are established at various points on the plain. So long as these stores stay well clear of one another, their market areas will remain circular (Figure 5.1, panel B). In addition, the size of each market area will be limited only

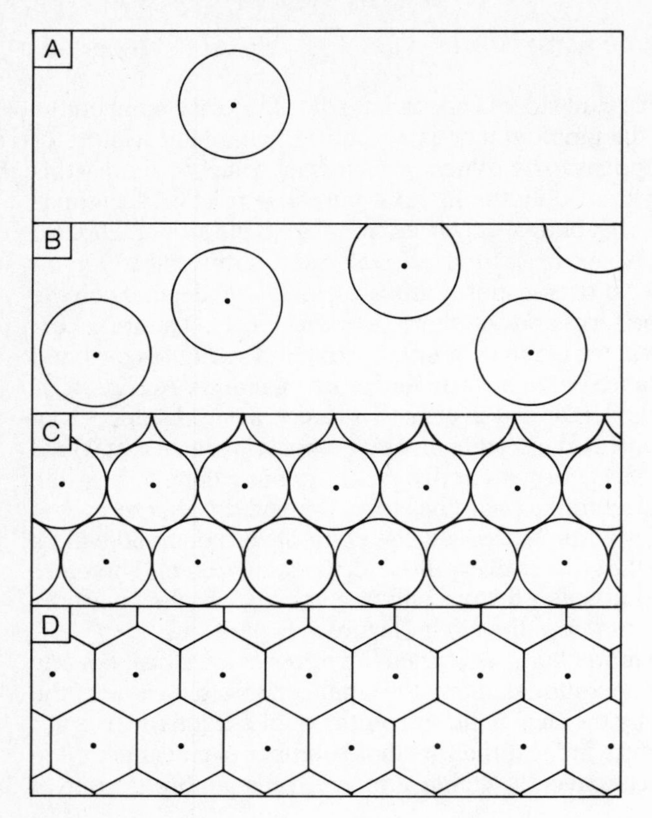

Figure 5.1
The development of a triangular lattice of supply points on an isotropic plain: (A) an isolated supplier; (B) several suppliers; (C) many suppliers; (D) spatial equilibrium

by the maximum distance that consumers are willing to travel. The range of each store will be at its upper limit, and excess profits will be earned. However, since the economic system permits free entry and exit of firms (assumption 3), and since the unserved areas lying beyond the territories claimed by established stores offer attractive opportunities to entrepreneurs, more and more stores will appear on the plain. Eventually the stores will begin to compete directly with one another for territory, and this spatial competition will reduce excess profits to the vanishing point. As equilibrium is approached, the plain becomes filled with market areas of exactly threshold size. These market areas are packed together as closely as possible, which means that each of them is symmetrically surrounded by six others (Figure 5.1, panel C).

At this point, a vitally important feature of the final equilibrium solution has already been attained – the locations occupied by the central places (i.e. the sites of the stores) form a perfect triangular lattice. Each store is equidistant from six others, and any three adjacent stores form the vertexes of an equilateral triangle. This regular arrangement of the points of supply is a fundamental property of all central place models developed on Christallerian principles. The triangular lattice is caused, of course, by the packing of the maximum possible number of suppliers into the plain, each supplier being surrounded by a circular market area of threshold size.

It is customary to make one further adjustment to the pattern thus far developed. Circular market areas leave small, quasi-triangular interstitial territories that do not fall within the market area of any specific store, and it has become traditional to eliminate these interstitial territories by replacing the circles with regular hexagons (Figure 5.1, panels C and D). Two technical points must be noted. First, the interstitial zones eliminated by the introduction of hexagons are commonly described as 'unserved' areas, but this label is highly misleading. Since each store's potential maximum range is likely to be somewhat larger than the radius of its threshold market area, the supposedly 'unserved' farmers are not 'out of range' at all, and their patronage is properly regarded as representing excess profits for the three nearest stores. Second, if hexagons are introduced without any change being made in the spacing of the stores, the resulting hexagonal market areas will be larger than threshold size, because a hexagon is larger than the circle inscribed within it. Therefore, if we wish to restrict each store to normal profits, the stores must be moved slightly closer together in order to reduce the hexagons to threshold size. By general agreement this final step produces the equilibrium spatial solution (Figure 5.1, panel D). However, Christaller himself did not insist that excess profits be eliminated.

The use of hexagons may be justified on esthetic grounds by postulating that the market areas should form a space-filling tessellation. Among all regular polygons only hexagons, squares, and equilateral triangles can fulfill this requirement; and among these three permissible shapes the hexagon departs least from the ideal circular form. However, from the point of view of understanding the spatial structure of Christaller's models, the hexagonal shape of the market areas is only a secondary feature – almost, in fact, a distraction. As noted above, the most important characteristic of the theoretical pattern is the triangular lattice formed by the central places themselves. It really does not matter whether one uses hexagons or circles for the market areas, so long as the triangularity of the network of supply centers is preserved. In short, it is the geometry of the *points*, not of their associated areas, that is the key to the remainder of the argument.

THE $K = 3$ MODEL

We now address the task of creating a system of central places that contains all the different types of retail stores and consumer services present in the economy. Each distinct type of store or service – for example, hardware store, bank, beauty parlor – is known as a central function, often abbreviated to the single word 'function' where the context ensures that the meaning is clear. It is convenient to regard all the different central functions as being ranked according to the magnitude of their threshold requirements. Let there be n different functions in our system, and let them be ranked in ascending order so that rank number 1 is assigned to the function with the smallest threshold and rank number n is assigned to the function with the largest threshold.

In order to derive a rational solution that gives all consumers access to all types of central functions, we introduce the tactical assumption that the first function to appear on the plain is function n, the type having the *highest* threshold. (This is, of course, a distortion of history. Nevertheless, in real life the earliest urban settlements founded in each region generally do become established as the points of supply for the highest-threshold functions offered at any given time.) Suppliers of function n will form a triangular lattice on the plain in accordance with the argument of the preceding section. Let the points comprising this initial lattice be known as A centers. Since the economy permits unrestricted entry of firms (assumption 3), spatial equilibrium is attained when the A centers are situated as close to one another as possible. The size of each A center's market area is determined by the threshold of function n. The suppliers of this function earn only normal profits.

Now consider the function having the second largest threshold requirement. Following the lead of the suppliers of function n, suppliers of this second function will locate themselves in the A centers (assumption 6), thereby maximizing their accessibility to consumers. By definition, the members of this second set of suppliers have a smaller threshold than the suppliers of function n, yet they serve the same market area; that is, their

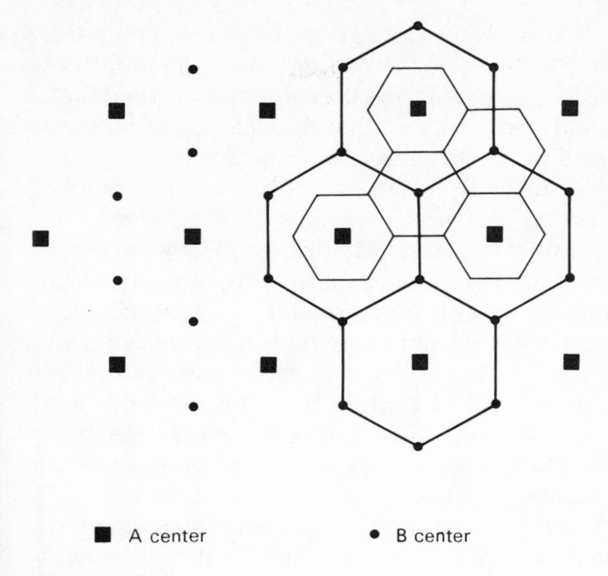

Figure 5.2
The emergence of low-order centers (B centers) within the interstices of the triangular pattern formed by the high-order centers (A centers)

market area is defined not by their own threshold but by the larger threshold of function *n*. Suppliers of the second function are therefore in a position to earn excess profits.

The same reasoning applies to succeeding functions with progressively smaller thresholds. Given assumption 6, suppliers of all functions will continue to locate in the A centers. In due course, therefore, each A center will provide every central function available on the plain. In addition, as successive functions are added, the volume of excess profits earned will increase as threshold requirements decline.

The increase in excess profits, however, does not continue indefinitely. As functions with progressively lower thresholds are added to the system, a function is bound to be reached for which the interstitial excess profits are themselves just large enough to support additional suppliers of this same function – that is, suppliers other than those located in the A centers. Let this critical function be function *p*. Its distinguishing feature is a threshold market area just small enough that it can be placed on the plain not only around the A centers but also *between* the A centers in the manner displayed in Figure 5.2. Function *p* is therefore supplied not only from the A centers but also from an entirely new set of interstitial locations. These new central places are termed B centers. Each B center is situated precisely at the midpoint of a triangle formed by three adjacent A centers (Figure 5.2). The market areas for function *p*, around both the B centers and the A centers, are only of threshold size. It follows that the suppliers of function *p*, like the suppliers of function *n*, earn only normal profits.

Function *p*, which marks the entry of the set of B centers, is termed a hierarchical marginal function. (In some versions of the theory, the argument is developed in terms of hypothetical single-good firms, in which case the term *hierarchical marginal good* is used.) The A centers can provide every function that the economy has to offer, but the B centers, because of their limited market areas, cannot provide any function having a greater threshold than that of function *p*. Accordingly, there is a sharp distinction between A centers and B centers in terms of functional complexity. These two kinds of centers, in fact, represent two orders in a *hierarchy* of central places. This distinctive hierarchical structuring within the urban system is a fundamental feature of Christallerian theory.[2]

The argument used above to account for the appearance of the B centers is also used to justify the subsequent appearance of lower orders of C centers, D centers, and so forth. Continuing down the list of functions below function *p*, the A and B centers are jointly adopted as supply points until another function is reached – let this be function *r* – for which the excess profits are again just large enough to support additional suppliers besides those taking up positions in the pre-existing centers. Function *r* is the hierarchical marginal function for the set of C centers, each of which provides function *r* and all functions having still smaller thresholds. Farther down the list, another hierarchical marginal function is encountered when the geometry permits the emergence of D centers; still another marks the appearance of E centers, and so forth. The total number of hierarchical orders that will emerge is indeterminate, but the construction of the system obviously stops when no more functions remain to be added to the plain.

Since the market areas for each hierarchical marginal function are only of threshold size, suppliers of these particular functions earn only normal

2 Christaller's original discussion of the theoretical rationale for distinct hierarchical orders was incomplete and vague. Early treatments in English made some headway (Ullman 1941; Dickinson 1947, 53–62), but the first comprehensive explanation was provided by Berry and Garrison (1958a). Christaller's ideas required pruning and clarification before a coherent theory emerged – much more so than is generally recognized. In this respect, Berry and Garrison stand to Christaller very much as Playfair stood to Hutton.

profits. The suppliers of function *n*, at the top of the list, are also in this position. In effect, function *n* is the marginal function for the highest order in the hierarchy. As we have seen, all functions except the hierarchical marginal functions are theoretically able to earn excess profits in varying degree. This does not necessarily mean, however, that some types of stores and services in the real world are intrinsically more profitable than others. In the first place, the hypothetical excess profits available to a particular function might be absorbed in the real world by an expansion in the scale of the individual establishments.[3] In the second place, although very little is known about the actual profit levels of individual stores, it is clear that profitability is determined by several factors in addition to the spatial structure of the urban system. Quality of service, effectiveness of advertising, and variations in consumer tastes and preferences come immediately to mind. On the whole, the theoretical distinction between normal and excess profits probably has little meaning for real-world entrepreneurs. However, there is a very real constraint embodied in the simple fact that small market areas cannot support high-threshold functions. Regardless of profit levels, this basic spatial constraint places an upper limit on the variety of goods and services that can be offered at each level of the hierarchy.

Figure 5.3 shows the theoretical pattern of central places and market areas for an urban system consisting of four hierarchical orders. Christaller called this pattern the Versorgungsprinzip, or 'marketing principle' model, but in English it is generally known as the $K = 3$ model. The symbol K denotes the ratio between the sizes of the market areas on any two adjacent levels in the system. In the model under consideration, this ratio is exactly three to one for every pair of adjacent levels, a fact that may readily be verified by careful study of Figure 5.3. Accordingly, the market area of an A center is three times as large as that of a B center, nine times as large as that of a C center, twenty-seven times as large as that of a D center, and so forth. As we will see in the next section, other values of K are possible besides 3. It is also possible to construct models in which the value of K is not a constant. Geometrically speaking, the $K = 3$ pattern

3 Such an expansion would represent an increased threshold for the central function, because the expanded establishments could not be sustained by the pre-expansion threshold. If excess profits are indeed absorbed in this way, the thresholds of all functions in the system will tend to converge on the thresholds of the hierarchical marginal functions. In effect, the sizes of establishments will become neatly adapted – in an almost Darwinian sense – to the sizes of market areas provided by the geometry of the system.

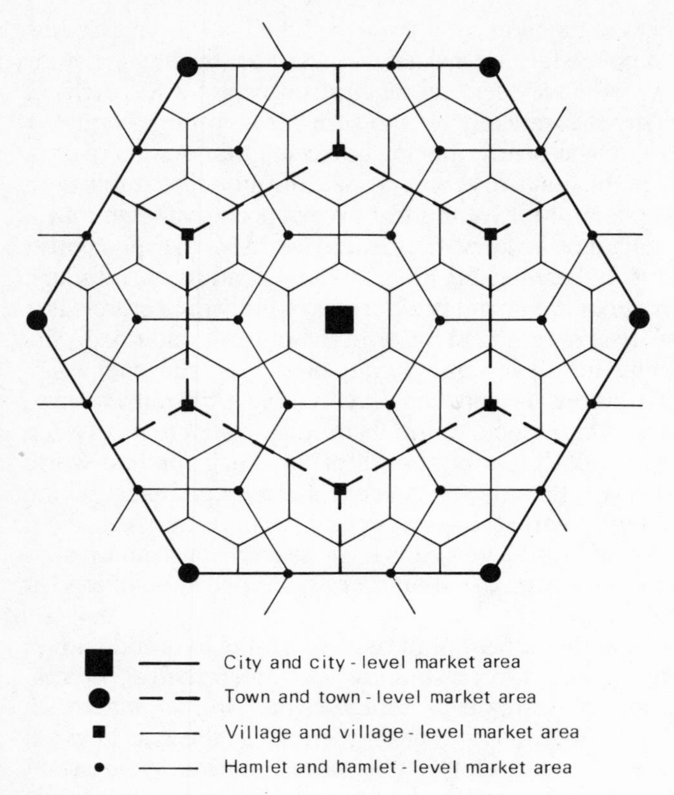

Figure 5.3
Christaller's $K = 3$, or 'marketing principle' model. Four hierarchical orders of central places and market areas are shown.

may be viewed as the simplest member of a whole family of closely related models.

One of the most striking characteristics of the $K = 3$ model is the fact that the lattice of central place locations is always perfectly triangular, regardless of how many orders of centers are added to the system. This regularity is obviously true for the A centers taken alone, but it is not immediately obvious that the triangular structure of the lattice will be preserved when the lower orders of B centers, C centers, and so forth, are placed successively on the plain. In formal geometrical terms we may state the matter as follows. If, in a perfect triangular lattice, a new lattice point is added at the midpoint of every triangle formed by three adjacent pre-existing points, the resulting lattice of old and new points taken together is also perfectly triangular. Thus, as each successively lower order of central places is added to the plain, the *arrangement* of the centers (i.e. their positions in relation to one another) remains unchanged. It is only the *density* of the pattern that is altered. An important consequence of this regularity is that, for any central place except those belonging to the highest order, the three nearest neighbors of higher rank are equidistant from the center in question and form the vertexes of an equilateral triangle (Figure 5.3). In the final analysis, this arrangement of 'one lower surrounded by three higher' is the fundamental element in the spatial structure of the $K = 3$ pattern.

THE $K = 4$ AND $K = 7$ MODELS

We now consider additional representatives of the family of models in which the $K = 3$ pattern is the simplest member. In particular, we examine models in which K takes the values 4 and 7, respectively.

The explanation of the $K = 4$ model is identical to that of the $K = 3$ model up to the point at which the A centers are distributed uniformly across the plain. Assumption 1 – that the plain is isotropic – is then modified by the introduction of an auxiliary assumption, that the appearance of each order of central places is accompanied by the construction of a network of major transportation facilities connecting all the centers. These transport lines might be either railroads or highways or a combination of both. The important point is that they represent routes of especially easy movement, thus distorting the perfectly isotropic character of the plain.

It is then proposed that the most rational pattern of locations for the second set of centers – the B centers – is not at the midpoints of the triangles formed by the A centers, as in the $K = 3$ model, but at the midpoints of the routes that now connect the A centers to one another. Accordingly, each B center in the $K = 4$ model lies exactly halfway between two A centers, as illustrated in Figure 5.4.

With the B centers thus positioned, a further set of transport links is constructed to connect them with one another, so that once again every pair of adjacent central places on the plain is joined by a line of especially easy movement. Next, a set of C centers is added at the midpoints of the routes linking every pair of centers of higher rank. Each C center thus lies either halfway between two B centers or halfway between a B center and an A center. New transportation links and new orders of central places are

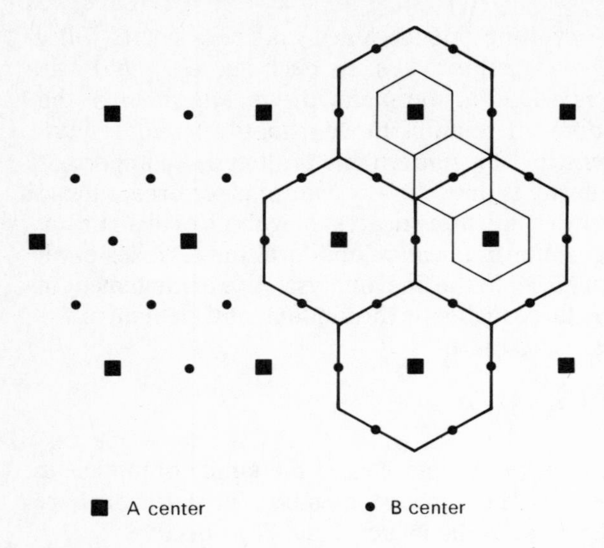

Figure 5.4
In the $K = 4$ model, each low-order center (B center) is situated exactly halfway between two neighboring high-order centers (A centers). Compare with Figure 5.2.

added according to the same pattern until every function in the economy is present on the plain. The resulting spatial arrangement of centers and market areas is shown for four orders in Figure 5.5.

This second model was termed the Verkehrsprinzip, or 'traffic principle' model by Christaller. It is also known as the $K = 4$ model because, as may be calculated from Figure 5.5, the market areas on any given level of the hierarchy are exactly four times as large as those on the level immediately below. The ratio of market area sizes is therefore four to one instead of three to one – this is the essential difference between the $K = 3$ and $K = 4$ patterns. In other respects the two models are quite similar. In particular, in both models the lattice of supply points remains perfectly triangular as successively lower orders of central places are added to the system. All market areas on every level of the hierarchy in both models are therefore hexagonal.

In addition to the marketing and traffic principles of organization, Christaller also proposed a third principle, the Absonderungsprinzip, or 'separation principle.' In both of the models already described, many central places lie exactly on the boundaries of the market areas of centers

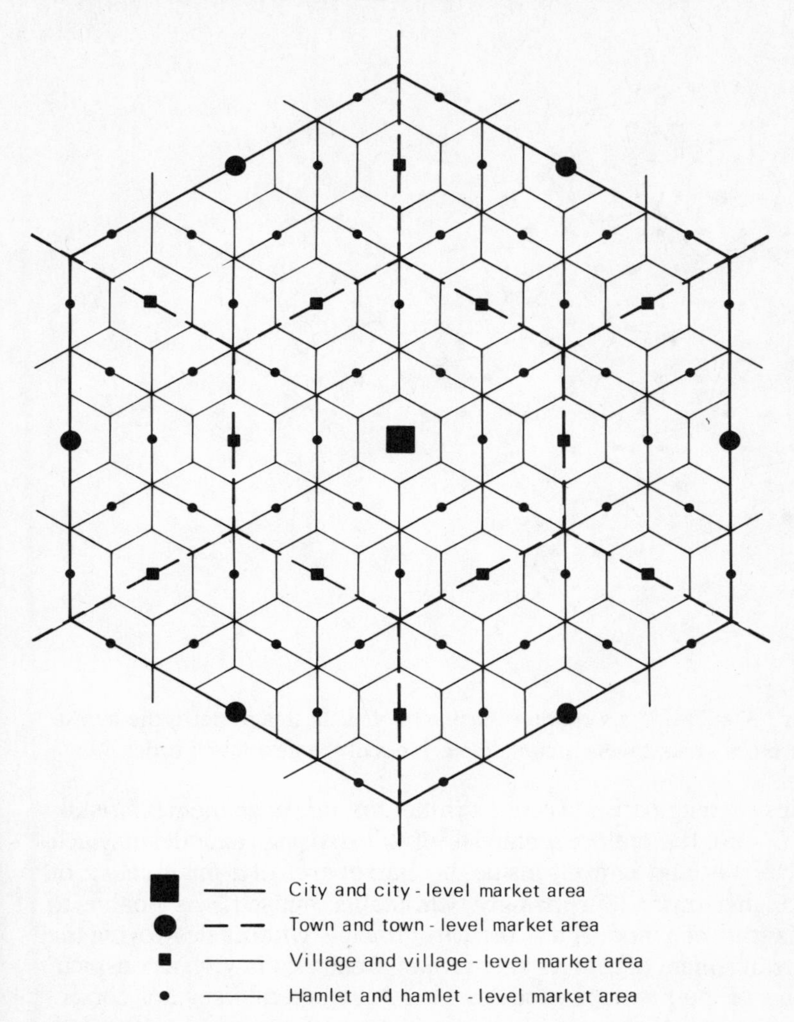

Figure 5.5
Christaller's $K = 4$, or 'traffic principle' model. Compare with Figure 5.3.

of higher rank (Figures 5.3 and 5.5). Christaller suggested that the divided economic allegiance of towns situated in these boundary locations would not accord well with the goal of effective political administration, and he pointed out that regional and provincial seats of government usually exercise their jurisdiction over whole towns of lower rank, with political

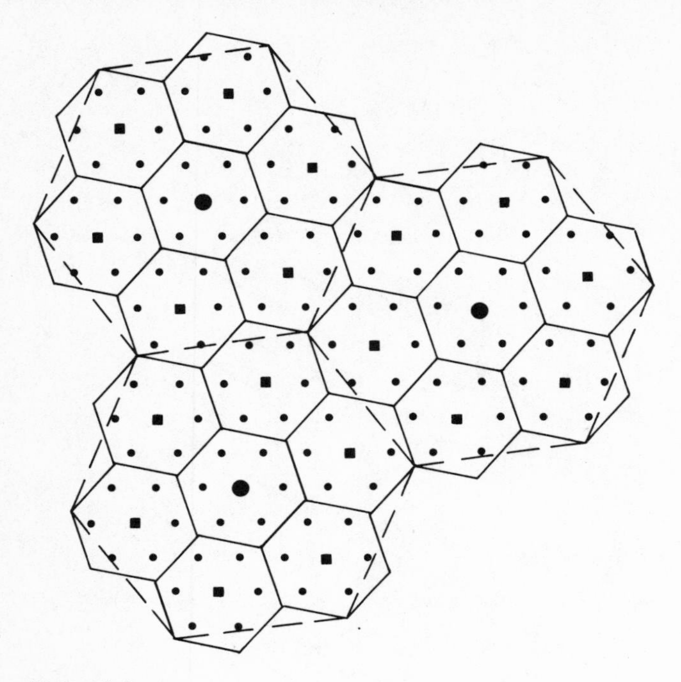

Figure 5.6
Christaller's $K = 7$, or 'separation principle' model. In this pattern, the market area of each center contains six complete centers of the next lower order.

boundaries running *between* centers rather than *through* them (Christaller 1966, 77–80). It therefore seemed desirable to design a model in which each central place lay entirely inside the market area of a single center of the next higher order. Surprisingly, Christaller himself was unable to draw a diagram of a model based on this principle without destroying the regular arrangement of centers and market areas that is such a conspicuous feature of the previous models. There are actually several correct solutions. The simplest, one version of which was first provided by Lösch (1954, 132), is shown in Figure 5.6.

The ratio of market area sizes between adjacent hierarchical levels in this third model turns out to be seven to one. Hence this pattern is known as the $K = 7$ model. It resembles the previous models in its use of hexagonal market areas and in the fact that the lattice of supply points remains triangular as each new order of central places is added to the plain.

Can additional hierarchical models be devised that incorporate values of K other than 3, 4, and 7? A definite answer can be given, but we will not concern ourselves with the mathematical explanation of it. We assume, following the pattern set by the three models already presented, that the central places are invariably required to form an unbroken triangular lattice of points. Even under this strict constraint, the number of theoretically possible values of K is infinite! The first ten values in this infinite series are as follows: $K = 3, 4, 7, 9, 12, 13, 16, 19, 21$, and 25. The entire series can be generated by the following formula:

$$K = x^2 + xy + y^2, \tag{5.1}$$

where $x \leqslant y$ and both x and y are non-negative integers.[4]

There is no evidence at present that values of K greater than 7 are of any importance for empirical work. Large values of K are unlikely to occur under real-world conditions. In the first place, given free entry of firms (assumption 3), real-world conditions will tend to favor maximization of the number of different sizes of market areas carved out by the different kinds of entrepreneurs, and it is clear that the number of different sizes of market areas within the urban system declines significantly as the value of K increases. In the second place, entrepreneurs are not very likely to choose locations in a manner that would lead to a pattern having a high value of K. The arguments that justify the $K = 3$ and $K = 4$ arrangements are not difficult to understand and might conceivably form the basis of conscious locational strategies, but the same cannot be said for other models. As we have noted, Christaller himself was unable to draw even the $K = 7$ pattern in its correct geometrical form.

It is noteworthy that different values of K can occur simultaneously within a single hierarchical model (Marshall 1977a; Parr 1978). Figure 5.7 shows the simplest possible case, with the two smallest values of K combined in a single urban system. This pattern was generated by allowing the traffic ($K = 4$) and marketing ($K = 3$) principles to alternate as successive orders of central places were added to the system. Evidence presented in chapter 7 suggests that mixed structures of this kind are more common in the real world than patterns in which the value of K remains constant.

4 This formula was derived by Dacey (1964, 1965) in connection with a problem related to the structure of the Löschian landscape (see chapter 8). The same equation defines the theoretically possible values of Christaller's K. The value $K = 1$ (corresponding to $x = 0$, $y = 1$) is ignored.

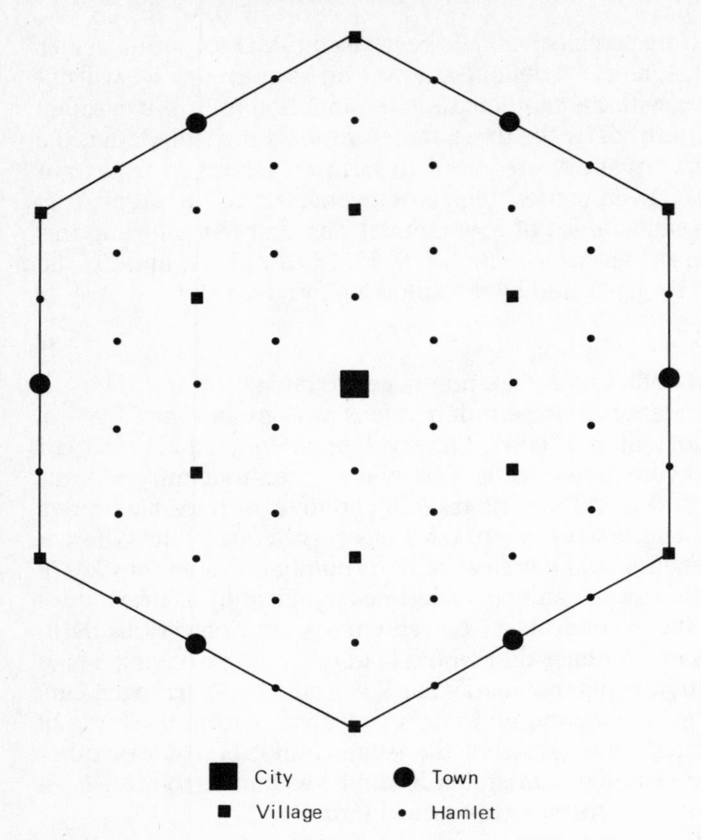

Figure 5.7
A variable-K central place model that incorporates elements of both the $K = 3$ and $K = 4$ patterns. In the interest of clarity, market areas of towns, villages, and hamlets are omitted.

GENERAL CHARACTERISTICS OF HIERARCHICAL MODELS

Although the various central place models described up to this point differ in certain particulars, they also have several important characteristics in common, which may be regarded as the definitive properties of the general concept of a hierarchy of settlements. Four specific diagnostic features of hierarchical structuring may be identified: (1) discrete stratification of centrality; (2) incremental baskets of goods; (3) a numerical pyramid in the frequency of centers in successive orders; and (4) a spatial pattern involving interstitial placement of successive orders. In this section we will discuss each of these four necessary conditions in turn.[5]

First, the term *centrality* denotes the degree to which an urban center functions as a central place. Population size can be used as an approximate indicator, but centrality is usually measured by counting the numbers of retail and service establishments of different kinds present in each center. (Technical aspects of the measurement of centrality are discussed in detail in chapter 6). The phrase 'discrete stratification' means that differences in centrality *within* each hierarchical order are small whereas differences *between* orders are clearly marked. In the theoretical models, differences within each order are zero, since all centers of a given order provide identical arrays of goods and services. Differences between orders are significant as a result of clear distinctions in the relative sizes of market areas.

It is a matter of common observation that, as the size of towns increases, so also does their complexity as retail and service centers. However, it is not immediately apparent whether the increases in size and functional complexity within the urban system are smooth and gradual, or whether, as the concept of a hierarchy implies, there are distinct 'plateaus' on which towns tend to be concentrated. In the past this critical question has often been ignored, and the word 'hierarchy' has been used to signify nothing more than the fact that large centers are less numerous than small ones (e.g. Florence 1955, 115–18; O.D. Duncan et al 1960, 56–69). It needs to be emphasized that the notion of discrete levels is an integral part of all Christallerian models. As Raymond Murphy has remarked, 'The idea of distinct classes with real breaks between them is rooted in Christaller's work and is in fact inherent in his theory' (Murphy 1966, 96).

Second, the presence of incremental baskets of goods means that each order of central places provides all the functions available at lower hierarchical levels plus a set of additional functions that lower levels cannot provide because of the limited size of their market areas. In effect, a center of high order is surrounded not by a single market area but by a nested series of market areas, one for each of the hierarchical levels on which it operates. A distinct set of functions or 'basket of goods' is

5 Some years ago I described seven diagnostic features, three of which I now believe to be either unnecessary or else implicitly contained within the four here discussed (Marshall 1969, 23–8).

associated with each market area in this concentric series (see Figures 5.3, 5.5, and 5.6).

Third, a hierarchy is a 'pyramid' in that the number of central places in any given order is smaller than the number in the order immediately below. If we arbitrarily set the number of centers in the highest order at 1, the relative frequencies, F, of centers in successively lower orders in the three basic Christaller models run as follows: in the $K = 3$ model: $F = 1, 2, 6, 18, 54, 162, \ldots$; in the $K = 4$ model: $F = 1, 3, 12, 48, 192, 768, \ldots$; and in the $K = 7$ model: $F = 1, 6, 42, 294, 2058, \ldots$ These sequences of numbers are dependent on the specific geometrical procedures used in the construction of the individual models. The general rule is simply that the number of centers on each level increases as we descend from the top to the bottom of the hierarchy.

The final feature – interstitial placement of orders – does not easily lend itself to precise definition. Low-order centers occupy the interstitial spaces between centers of higher rank, but this statement is too general to be of much use in empirical work. Most researchers seem to agree that a real-world central place system should not be regarded as a Christallerian hierarchy unless (among other things) it 'looks like' one or another of the models described above. However, it remains unclear how the degree of resemblance between reality and the model might best be evaluated.

Suppose that we are studying two separate urban systems, X and Y, which occupy areas of similar size and shape, and suppose that both of these study areas come acceptably close to meeting the six basic Christallerian assumptions described earlier. Moreover, suppose that both systems satisfy the conditions of discrete stratification of centrality, incremental baskets of goods, and a pyramid of town frequencies. The two systems, in other words, appear to be similarly structured. But when we make maps of the two areas, a startling difference is revealed. System X (Figure 5.8, panel A) looks agreeably like the $K = 3$ model and may therefore be regarded as satisfactorily explained by the Christallerian theory. But system Y (Figure 5.8, panel B) looks nothing like any of our models! Of course, system Y in this example has been deliberately designed as an extreme case. If we actually encountered such a system we should probably respond by describing it as hierarchical in structure but non-Christallerian in geographical expression. Its pattern on the ground certainly could not be explained by invoking Christallerian principles. Difficulties of this sort, while not widespread, do pose a problem in case studies of actual urban systems, and at present there is no general agreement as to how this type of problem might be solved.

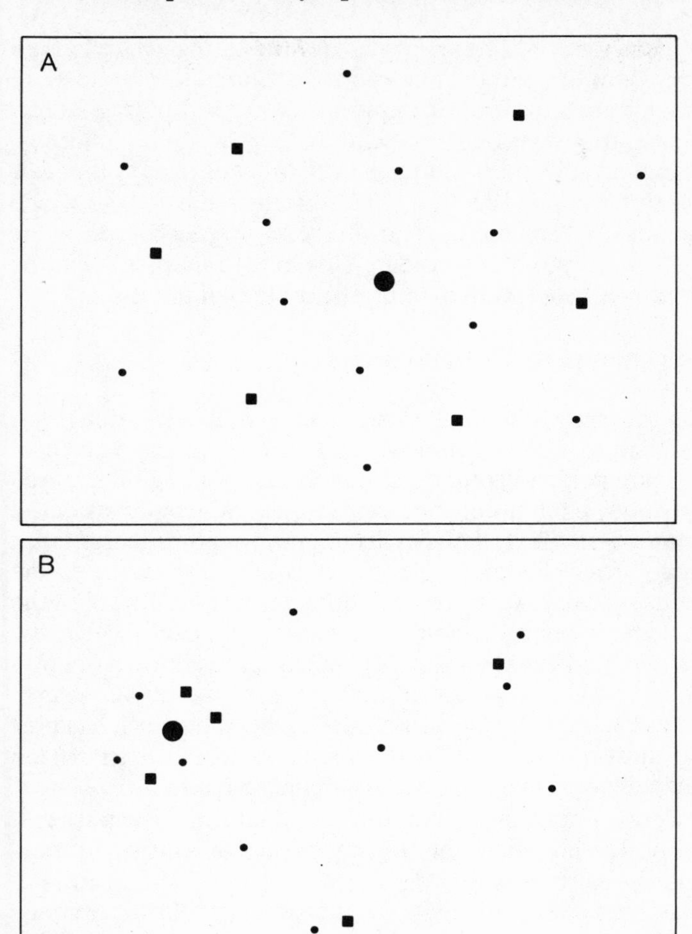

Figure 5.8
Two different hypothetical spatial arrangements of nineteen urban centers belonging to three distinct hierarchical orders: (A) a pattern that resembles the $K = 3$ model; (B) a pattern unlike any Christallerian model

The Christallerian concept of hierarchical structuring embodies an explicit and inescapable spatial dimension. In Christaller's theory of settlement, the relative functional complexity of central places simply cannot be divorced from their relative locations in geographic space.

The above diagnostic features stand as a checklist of criteria that can be applied to real urban systems in order to assess the extent to which they are hierarchically structured. The utility of these criteria in the context of a specific case study is considered in detail in chapter 7. For the moment, however, we must continue our exploration of theory.

Squares instead of hexagons: the Gourlay grid

We noted earlier that only hexagons, squares, and equilateral triangles – the only space-filling regular polygons – can be market areas in central place models having perfect geometrical uniformity. Among these three space-filling polygons, equilateral triangles are generally not considered to be serious contenders. First, they depart too far from the ideal circular shape. Measured from the midpoint of an equilateral triangle, the distance to the most remote point on the perimeter is twice as far as the distance to the closest point. Second, central places positioned at the midpoints of a space-filling tessellation of triangles do not form a uniform lattice of points. To achieve complete regularity, an additional center must be placed at each point where six triangles meet, but such a center would have no market area of its own. For these reasons, triangular market areas have never been regarded as a plausible alternative.

Squares, however, are a different matter. First, intuition suggests that squares should be readily adaptable to the rectilinear systems of land division and road construction that dominated the historical process of rural settlement over most of central and western North America (Pattison 1957; H.B. Johnson 1976). Second, central places positioned at the midpoints of squares form a complete square lattice of points with no unsightly gaps. For both historical and esthetic reasons, therefore, square market areas are worthy of consideration.

Arguing by analogy with Christallerian theory, we adopt the convention that, as successively lower orders of central places are added to the plain, the pattern formed by all orders considered simultaneously must always be a perfect square lattice. This ensures that the network of market areas for each incremental basket of goods is a regular space-filling tessellation. A central place belonging to any given order in a hierarchy also operates on all lower levels and possesses a separate market area for each level on which it operates.

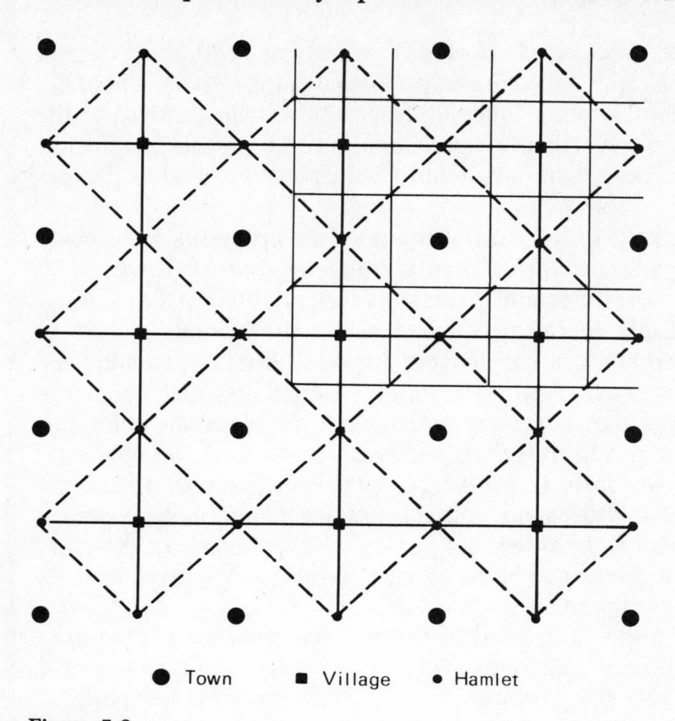

Figure 5.9
The simplest possible hierarchical structure based on square market areas. Each lower-order center is situated at the midpoint of the square formed by the four closest higher-order centers. The resulting value of K is 2. For clarity, market areas of hamlets are shown only at top right.

The simplest hierarchy based on square lattices is shown in Figure 5.9. It is easy to calculate that the market areas on each level in this system are exactly twice as large as those on the level immediately below. Hence the result is a fixed-K hierarchy in which the value of K is 2. Inevitably we are led to ask what values of K are possible with square market areas. It can be shown that the possible values are generated by the following simple formula:

$$K = x^2 + y^2, \qquad (5.2)$$

where, as before, $x \leq y$ and both x and y are non-negative integers. The ten smallest values yielded by this formula are $K = 2, 4, 5, 8, 9, 10, 13, 16, 17$, and 18. (As before, the value $K = 1$ is ignored.)

Remarkable as it may seem, a model belonging to this family was proposed by Robert Gourlay (1778–1863) as early as 1822, more than a century before Christaller's theory was published. Gourlay was a political reformer who sought, among other things, to introduce changes in the administration of land grants in Upper Canada (now southern Ontario). He viewed his model as a rational scheme for the planting of urban centers in virgin territory.[6]

It seems very likely that Gourlay intended to apply his theoretical scheme of town location to the standard nineteenth-century township of thirty-six square miles; the model is certainly well adapted to that setting. The standard township was a square block of land measuring six miles on each side, with the boundaries, at least in the ideal situation, running due north–south and due east–west. Two series of roads, also oriented to the cardinal points of the compass, were laid out at intervals of one mile, thus dividing the township into thirty-six 'sections' of 640 acres. Each settler was granted a square farm comprising a quarter-section, or 160 acres. Families were large in those days, and it is not unreasonable to assume a population of seven or eight persons per farm. The density of the rural population would therefore be about thirty persons per square mile once the land was fully occupied.

If the Gourlay model is applied to this setting, each township has a market town situated at its geometrical center, as shown in Figure 5.10. The boundary of this town's market area coincides with the political boundary of the township, giving each market town a trade area of thirty-six square miles. (Compare this figure with those given for Germany in the quotation from Bücher near the beginning of this chapter.) In terms of population, the market town serves about 1,000 rural inhabitants. Encircling the town is an orderly series of eight villages, each serving a market area of four square miles and a rural population of about 120. The central town itself, of course, also functions on the village level. Accordingly, the ratio of market area sizes in this system is nine to one, giving K the value of 9.

In Christallerian terms, the pattern illustrated in Figure 5.10 follows the Absonderungsprinzip, or 'separation principle.' Every village –

6 The model was incorporated in a folded map that served as the frontispiece to Gourlay's *Statistical Account of Upper Canada*, published in England in 1822. However, no verbal description of the model was included in either of the two volumes then published. Gourlay stated: 'My own method of laying out the waste lands of the crown ... [will be fully explained] ... in my third volume' (Gourlay 1822, i–ii). The promised third volume, however, never appeared. For discussions of Gourlay's career in general, see Craig (1963, 93–100) and Milani (1971).

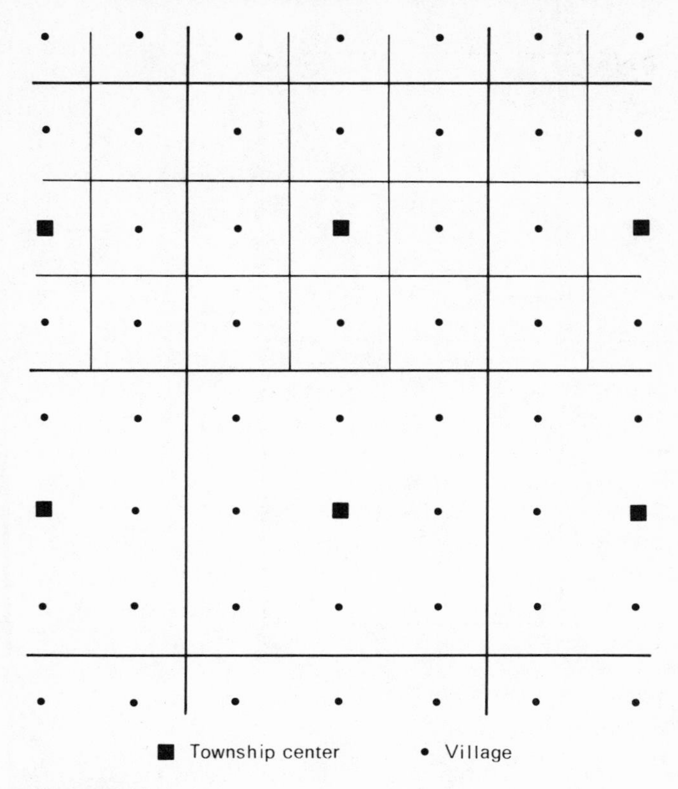

Figure 5.10
The pattern of central places proposed by Robert Gourlay in 1822. Each township contains one higher-order center and eight lower-order centers. The value of *K* is 9. Market area boundaries for the lower-order centers are shown only in the upper half of the diagram. SOURCE: adapted from Gourlay (1822, frontispiece)

indeed, every village's complete market area – lies wholly inside the market area of a single town. Each set of nine central places therefore forms a self-contained unit, creating an arrangement that accords well with the township's status as a distinct political entity at the lower end of a hierarchy of administrative areas. If the Gourlay model is extended to higher orders, it is easy to visualize the ideal 'county' as a unit consisting of nine townships arranged in three rows of three, with the market town of the central township promoted to the rank of 'county seat.'

The Gourlay scheme, for reasons unknown, was not adopted by the

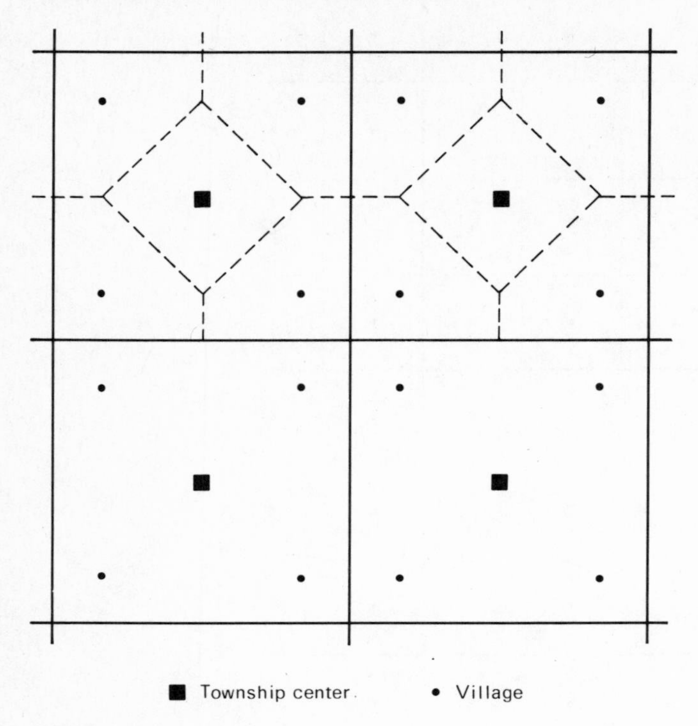

Figure 5.11
The arrangement of central places proposed by Thomas Adams in 1917 and attributed by him to Sir William Van Horne. The lattice of supply points is not completely uniform, and hence the lower-level market areas (shown in the upper half of the diagram) are of two different shapes. The village-level market area is slightly larger for each town than for each village (in the ratio 8:7). SOURCE: Adams (1917, Figure 9), with market area boundaries added

authorities of the day. In fact, it seems to have been quickly forgotten. However, a similar model appeared ninety-five years later in a treatise written by the prominent British planner Thomas Adams (1917, 54). Adams made no mention of Gourlay's work and attributed the pattern in question to the well-known 'railroad general,' William Cornelius Van Horne. As Figure 5.11 indicates, this Van Horne model is actually the Gourlay model with exactly half of the small villages removed. As a result, the network of supply points on the village level is not a uniform square lattice, thus violating our convention that all lattices should be complete.

It might be argued, however, that a total of five central places per township – including the central market town – is quite enough, especially for an era in which disposable income per rural family could not be expected to be high. We do not know if Van Horne would have defended the model in these terms. We also do not know if he was aware of Gourlay's contribution when he designed his own ideal scheme.

Under certain conditions it is unreasonable to insist, as we have done up to now, that the lattice formed by all central places considered simultaneously should be geometrically complete. In particular, departures from perfection should be expected in areas where central places are strongly attracted to a set of major transportation routes that link only the higher-order centers and not, as in Christaller's $K = 4$ model, the lower-order centers as well. Throughout the western interior of North America, for example, the spatial structure of the settlement pattern has been significantly influenced by the alignment of the principal railroad lines. Research carried out by Berry in Iowa provides a representative illustration of the resulting effects (Berry 1967, 38–40). If we call the four lowest orders of central places, in descending order, cities, towns, villages, and hamlets, and if we disregard the hamlets (which are randomly scattered and generally in decline), we find that the sequence of centers along the main east–west rail lines in Iowa tends to run as follows: city – village – town – village – city. The same sequence is repeated between the city-level centers along the roads and railroads running north and south, as shown in Figure 5.12. In consequence, large quadrilaterals are formed – ideally, squares or rhombuses – with city-level centers at their corners and with regular sequences of towns and villages along their sides. But within each large quadrilateral there is normally no more than a single center of village rank, whereas nine centers would be required in order to produce an uninterrupted lattice of supply points. A rational explanation of the observed pattern is found in the fact that the plain is not isotropic. The population has become concentrated close to the main transportation lines that form the edges of each quadrilateral, leaving the interior zones with a low density of demand that is capable of supporting only a single center of village rank.

Unlike Christaller's models, which were intended to aid our understanding of a pattern that had existed in Bavaria for centuries, the designs proposed by Gourlay and Van Horne were planning models, intended to serve as blueprints for the initial settlement of virgin lands. As it happens, these models seem never to have received any official backing in either Canada or the United States. More recently, however, the general

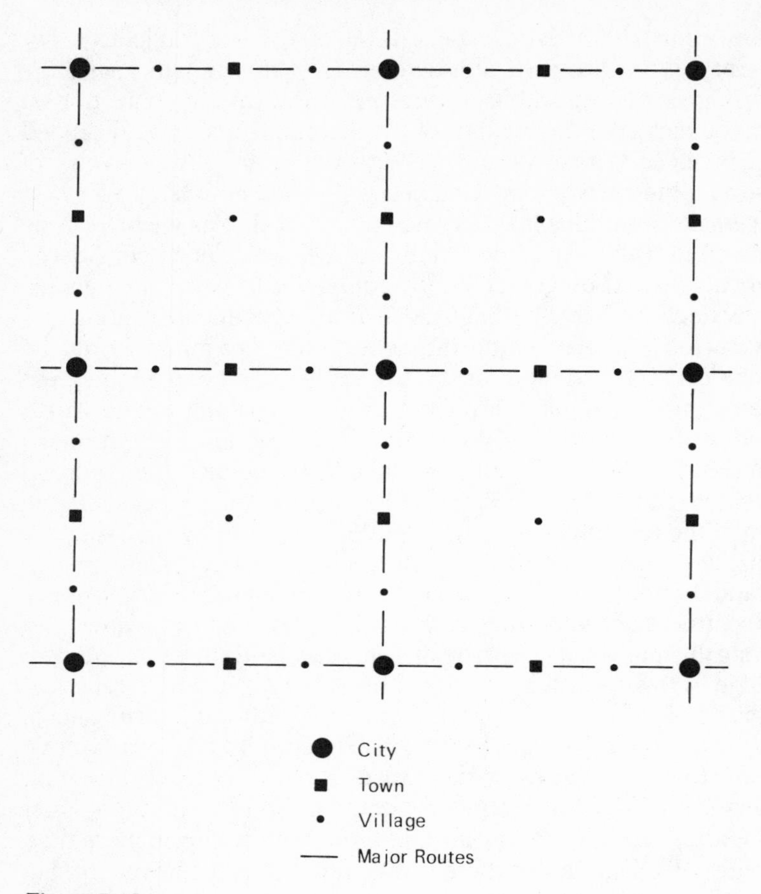

Figure 5.12
Schematic representation of the arrangement of central places in relation to major transportation routes in the American Midwest. After Berry (1967, 39)

concept of hierarchical structuring has been used in the planning of new systems of central places in other areas, including the Guadalquivir Delta in Spain (Enggass 1968), the Lakhish Plain in Israel (Berry 1967, 131), and the Ijsselmeer polders in the Netherlands (Takes 1960; Constandse 1963; Van Hulten 1969). Central place theory has also been taken into account in the controversial 'key settlement' policy favored by some rural planners in England and Wales (Cloke 1979). Finally, there is the particularly intriguing case of the plans drawn up in Nazi Germany during the Third

Reich for 'regularization' of the spatial pattern of market towns in Poland (Lösch 1954, 133; Kaniowna 1963; Golachowski 1964). According to Carol (1970, 68), Christaller himself took an active part in making these plans. However, the scheme was abandoned following the defeat of Germany by the Allies at the end of the Second World War.[7]

There is no doubt that the principle of hierarchical structuring can serve as a valuable tool in the field of regional planning. It is only because the hospitable areas of the earth's surface are now more or less fully settled that opportunities for the practical application of central place theory on a large scale do not often arise. But maybe some day, when we journey to the stars ...

Hierarchical structure in algebraic form

Within the total population (urban and rural combined) inhabiting a central place system, a certain number of individuals comprise what may be termed 'the suppliers.' The suppliers, as this term is used in this section, include not only persons actually employed in providing retail goods and services to the population at large but also dependents of these retail and service workers (i.e. spouses, children, and perhaps other relatives). Thus suppliers comprise that portion of the total population directly supported by the performing of central place activities. The market served by these suppliers, of course, is the total population within the system, including the suppliers themselves. The relationship between the suppliers and their market is therefore one that exists not between two mutually exclusive groups of individuals but between part and whole. By analyzing the manner in which part and whole are related, we can construct a model that expresses the sizes of central places as an algebraic function of the external market area populations that they serve (Dacey 1966; Beckmann and McPherson 1970; Parr 1970; Marshall 1975a).

We begin with a simple situation in which the landscape contains no

7 Hudson has examined the early development of service centers on the Great Plains (Hudson 1979, 1985). Conscious planning of settlement was involved, but not co-ordinated planning by a public authority. Instead, it was a case of corporate planning carried out by rival railroad companies under conditions of fierce competition. In general, the resulting network of central places was overbuilt. The subsequent sorting process, in which some towns thrived while others stagnated, may have reflected the natural operation of central place principles within the urban network after the original planners and promoters had left the scene.

central places above the level of hamlets. (Places of higher rank will be added shortly.) Each hamlet contains 60 inhabitants and serves an umland containing a rural population of 240 persons.[8] We assume also that we are dealing with a 'pure' central place system – i.e. all non-central urban activities such as manufacturing and mining are completely absent. Our immediate task is to derive a mathematical expression that relates the suppliers in this simple system to the population they serve.

The relationship between the suppliers and their market involves an interesting infinite regression. To begin with, a certain number of suppliers is required in order to cater to the needs of the 240 inhabitants of the umland. Let this initial number of suppliers be denoted by h. But now an additional number of suppliers is needed in order to meet the demands created by the initial group of h suppliers; and then more suppliers are required in order to cater to the demands of these additional inhabitants; and so forth, ad infinitum. The sum of this infinite series is the total population of a hamlet, or sixty persons.

The situation may be expressed symbolically as follows. Let p denote the total population of a hamlet; let r denote the population of a hamlet-level umland; let h denote the number of suppliers (including dependents) required in order to satisfy the needs of r consumers; and let k (not the K of Christallerian theory) denote the ratio between a given number of suppliers and the number of consumers they are capable of serving (the ratio k is always less than 1). Then, by definition,

$$h = kr \tag{5.3}$$

and

$$p = h + kh + k^2h + k^3h + \ldots \tag{5.4}$$

By substitution we obtain

$$p = kr + k^2r + k^3r + k^4r + \ldots \tag{5.5}$$

We can also write

$$kp = k^2r + k^3r + k^4r + k^5r + \ldots \tag{5.6}$$

If we now subtract equation 5.6 from equation 5.5, all terms except one on the right-hand side of the resulting equation vanish, and we are immediately left with

$$p - kp = kr. \tag{5.7}$$

8 The German word *umland* means 'surrounding land.' The word 'umland' is used freely in English and is synonymous with 'market area.'

Finally, by rearrangement of terms:

$$k = p/(p + r). \tag{5.8}$$

Using our assumed values of 60 and 240 for p and r, respectively, we see that the ratio k in this example has the value 0.200: the system is in equilibrium when there is one supplier (including dependents) for every five people. Alternatively, one-fifth of the total population is supported by providing central place services for the benefit of all.

The logical next step is to extend the above analysis to central places of greater than hamlet rank. At this step, however, two features make our task somewhat more complex. First, every central place above the hamlet level possesses not one umland but a nested series of umlands, one for each hierarchical level on which the center operates. To each umland there corresponds a distinct bundle of goods and services consisting of those central functions that are incremental on the hierarchical level in question. It will be convenient to introduce a new symbol, n, to denote both the various levels of the hierarchy and the corresponding bundles of goods. The levels are conveniently numbered in ascending order: thus, for the hamlets, $n = 1$; for the villages, $n = 2$; and so forth. A central place belonging to the nth hierarchical order supplies n distinct bundles of goods to n distinct umlands. The nth bundle of goods is supplied to the nth-level umland but not to any larger umland. This fundamental concept of nested umlands must be taken into account as we extend the mathematical model to centers above the hamlet level.

Second, as soon as the urban system is considered to contain two or more hierarchical orders, the ratio k may be a variable rather than a constant. Specifically, a distinct (although not necessarily unique) value of k may be associated with each different bundle of goods. We should expect, indeed, that the value of k will usually vary from one bundle of goods to another. For example, the incremental bundle of goods on the village level typically contains about ten times as many central functions as the hamlet-level bundle, and it would be absurd to assume that k is identical for two bundles of goods differing so greatly in size. Moreover, even if two bundles of goods contain approximately equal numbers of central functions, their different functions justify the expectation that k will vary. In what follows, therefore, we assume that each bundle of goods gives rise to a distinct value of k.

We now modify our notation in order to accommodate the presence of different hierarchical levels and different bundles of goods. Let p_n denote the total population of a central place belonging to the nth level of the hierarchy; for example, the population of a hamlet is denoted by p_1, that of a village by p_2, and so forth. Let r_n denote, in similar fashion, the population of an nth-level umland. Let k_n denote the ratio between a given number of suppliers of nth-level goods and the number of consumers they are capable of serving; the value of k_n is necessarily less than 1 for all values of n.

For the hamlets, by analogy with equation 5.5, we obtain

$$p_1 = k_1r_1 + k_1{}^2r_1 + k_1{}^3r_1 + k_1{}^4r_1 + \ldots \tag{5.9}$$

The first term on the right-hand side of equation 5.9 represents the 'export-oriented' component of each hamlet's population – the component supported by supplying goods and services to customers who live outside the hamlet. The remaining terms, as in equation 5.5, cover the infinite series of ever-diminishing secondary effects. Now let x_n denote the 'export-oriented' component of a central place of greater than hamlet rank. As each center above the hamlet level supplies n different bundles of goods to n distinct umlands, we obtain

$$x_n = k_1r_1 + k_2r_2 + k_3r_3 + \ldots + k_nr_n. \tag{5.10}$$

Or, using standard summation notation:

$$x_n = \sum_{i=1}^{n} k_ir_i. \tag{5.11}$$

The total population of this center, given that it also supplies its own inhabitants with n bundles of goods, is then given by

$$p_n = x_n + x_n \sum_{i=1}^{n} k_i + x_n \left(\sum_{i=1}^{n} k_i\right)^2 + \ldots \tag{5.12}$$

And this, by analogy with equations 5.5, 5.6, and 5.7, reduces to

$$p_n - p_n \sum_{i=1}^{n} k_i = x_n. \tag{5.13}$$

By substitution from equation 5.11, we then obtain

$$p_n - p_n \sum_{i=1}^{n} k_i = \sum_{i=1}^{n} k_ir_i. \tag{5.14}$$

And finally, by rearrangement of terms,

$$p_n = \sum_{i=1}^{n} k_ir_i \Big/ \left(1 - \sum_{i=1}^{n} k_i\right). \tag{5.15}$$

TABLE 5.1
Some population relationships for central places in part of southern Ontario

Type of place	Number of places	Total population (1971)	Average size of center	Average external population served
Rural[a]	–	45,853	–	–
Hamlets[b]	83	8,013	97	441
Villages	15	16,200	1,080	2,565
Towns	4	12,800	3,200	11,678
Cities	2	28,011	14,006	41,433
Total	104	110,877	–	–

SOURCE: Results reported here are drawn from the case study described in detail in chapter 7.
a Includes not only dispersed population but also inhabitants of nucleated centers that perform no central functions.
b For 36 of the 83 hamlets, no published population figures are available. Since the minimum size for inclusion in the census is 50, these 36 hamlets are assumed to contain an average of 25 persons each.

Given a set of population totals for central places and their associated umlands, equation 5.15 can be solved to yield a set of values of k_n for the urban system in question. However, the values of k_n must be obtained sequentially, beginning with k_1 and proceeding in ascending order. Owing to the algebraic structure of the model, k_{n-1} must be known before k_n can be calculated.

Under real-world conditions, both the central places and the umlands on any particular hierarchical level may be expected to vary in population size. The model, however, assumes uniformity on each level. Unfortunately, therefore, it is necessary to make use of average values in order to suppress any variation in the empirical data. As an illustration, we now apply equation 5.15 to the average population values provided in Table 5.1. The values of k_n thereby obtained for the four bundles of goods in this example are: hamlet-level goods, $k_1 = 0.180$; village-level goods, $k_2 = 0.221$; town-level goods, $k_3 = 0.085$; and city-level goods, $k_4 = 0.100$.

In interpreting these results we must take into account the fact that the various bundles of goods are unequal in size. For example, the hamlet-level bundle of goods in the study area described by Table 5.1 contains only three incremental functions, whereas the village-level bundle contains thirty-two. This being the case, it would be inappropriate to compare the calculated k-values with each other directly. Consider a fixed total market consisting of 1,000 consumers. According to the calculated values of k_1 and k_2, this market is provided with hamlet-level goods by 180 suppliers (including the suppliers' dependents) and with village-level goods by 221 suppliers. But the 221 village-level suppliers provide thirty-two different central functions; the 180 hamlet-level suppliers, three.

The different bundles of goods may be compared more equitably if we calculate, for each bundle, the number of central functions that could theoretically be provided for 1,000 consumers by an arbitrary supply force of 100 persons (including dependents). On the village level, for example, 1,000 consumers can be provided with thirty-two functions by 221 suppliers; therefore 100 suppliers could provide, on a simple pro rata basis, 14.5 functions to the same market. The results of these pro rata calculations are shown for all four bundles of goods in Table 5.2. In spite of the rough method, the results strongly suggest that the average productivity of suppliers tends to increase as we ascend the levels of the hierarchy (though perhaps not without limit). Two possible explanations have been suggested (Parr 1970, 235–6). First, suppliers of high-order goods may be able to take advantage of economies of scale that are simply not available to suppliers of low-order goods. Second, the volume of consumer expenditure per unit of time may decline as one moves from low-order goods to high-order goods. Both of these mechanisms imply that a supply force of any given size would be able to provide more high-order functions than low-order functions.[9]

Although equation 5.15 effectively captures the concept of a nested series of umlands, each associated with a different bundle of goods, the model is limited by the assumption of a 'pure' central place system in which all urban centers are supported solely by performing central place activities. In reality, of course, all centers except the smallest are likely to derive some portion of their economic support from noncentral activities such as manufacturing and wholesaling. As a general rule, the relative importance of noncentral activities within the urban economy increases as we ascend the levels of the hierarchy. Even the smallest centers, moreover, are not necessarily 'pure,' since hamlets and villages often contain significant numbers of retired persons – in particular, retired

9 Parr's discussion was based on the behavior of the *k*-values themselves; he did not consider the role of variation in the sizes of the bundles of goods (Parr 1970, 235–6). As Table 5.2 indicates, the raw *k*-values do tend to decline as we ascend the hierarchy, though somewhat erratically. The trend of increasing productivity is more convincingly revealed when 'size of bundle' is explicitly taken into account.

TABLE 5.2
Characteristics of the hierarchical bundles of goods in the southern Ontario study area

Bundle of goods	Number of incremental functions	Value of k	No. of functions that 100 persons could provide for 1,000 consumers
Hamlet	3	0.180	1.7
Village	32	0.221	14.5
Town	22	0.085	25.9
City	23	0.100	23.0

SOURCE: Data in second column taken from case study in chapter 7.

farmers – whose presence inflates the urban population but does not rest on the performance of central place activities. On every hierarchical level, therefore, central place activity may support something less than the total population of each urban center.

On the theoretical level this problem can be resolved as follows (Marshall 1975a, 640–1). Let d_n denote, for centers of the nth hierarchical order, the average number of persons (including dependents) who are not supported by performing central place activities. Persons included in d_n live within an urban center of the nth order, not in that center's umland. In other words, d_n forms part of p_n, the latter being the total population of an nth-order center.

Equation 5.11 now becomes

$$x_n = \sum_{i=1}^{n} k_i r_i + \sum_{i=1}^{n} k_i d_n. \tag{5.16}$$

Notes that x_n no longer identifies the 'export-oriented' component of the central place's population – the component that serves only external consumers. It now identifies the component supported by supplying central goods and services not only to the external portion of the market (i.e. the umland population) but also to d_n, the internal population that is not supported by performing central place activities. We can also write, by analogy with equation 5.12,

$$p_n - d_n = x_n + x_n \sum_{i=1}^{n} k_i + x_n \left(\sum_{i=1}^{n} k_i \right)^2 + \ldots \tag{5.17}$$

The left-hand side of equation 5.17 represents the portion of the center's total population that *is* supported by performing central place activities. Equation 5.17 reduces to

$$p_n - d_n = x_n \Big/ \left(1 - \sum_{i=1}^{n} k_i\right). \tag{5.18}$$

And finally, by substitution from equation 5.16 and by rearrangement of terms, we obtain

$$p_n = \left(\sum_{i=1}^{n} k_i r_i + d_n\right) \Big/ \left(1 - \sum_{i=1}^{n} k_i\right). \tag{5.19}$$

Equation 5.19 is identical to equation 5.15, except for the addition of d_n to the numerator on the right-hand side. This small difference, however, acknowledges that central place activities do not necessarily support the entire urban population on each hierarchical level.

The application of equation 5.19 to real urban systems is unfortunately hampered by the difficulty of measuring d_n, particularly for lower hierarchical orders, since the appropriate kinds of data – such as statistics on types of employment and on age structure – are not often readily available for small centers. Some progress can be made, however, through the use of educated guesses based on field experience and on whatever fragmentary evidence is available. Consider, for example, the hamlet-level centers in the system summarized in Table 5.1. The typical hamlet contains no economic activity other than a handful of central place functions, but field experience indicates that a substantial proportion – probably more than 50 per cent – of the total hamlet population is composed of retired people and of households whose working members commute to jobs in larger centers. If we assume, for the sake of illustration, that 60 per cent of the average hamlet's population should be allocated to the term d_n, equation 5.19 gives k_1 a value of 0.072, which is much lower than the value obtained from equation 5.15 (Table 5.2). It may be argued, however, that the smaller value lies closer to the truth, since not all residents of the average hamlet are supported by employment in the local retail and service outlets.

As research proceeds, it is to be hoped that reliable estimates of the values of d_n for the various levels of the hierarchy in different regions will be obtained. Possibly the most effective method for constructing these estimates will be to carry out detailed case studies of selected centers on each hierarchical level, the results then being averaged to produce a representative value. However, the average value of d_n on any given level may not be constant from region to region. On the contrary, the extent to which both d_n and k_n vary across space should be a major target of future research. Meanwhile, the model represented by equation 5.19 stands as a convenient algebraic statement of the functional relationships that link suppliers of central place goods and services to internal and external markets within a hierarchically structured urban system.

Summary

Central place theory is concerned with the provision of goods and services at the retail level to the final consumer. Under ideal physical and economic conditions – described by the six assumptions on which central place theory is based – the most rational spatial pattern is a *hierarchy* of urban centers. Hierarchical structuring is characterized by four diagnostic properties: (1) discrete stratification of centrality; (2) incremental baskets of goods; (3) a numerical pyramid in the frequency of centers in successive orders; and (4) interstitial placement of the orders. The number of geometrically distinct arrangements of centers that satisfy these four criteria is, in theory, infinite. Among this wealth of possibilities, the three classic Christallerian patterns known as the $K = 3$, $K = 4$, and $K = 7$ models are the simplest and also the most likely to occur under real world conditions.

It is possible to design a family of hierarchical models on the basis of market areas that are square instead of hexagonal. Gourlay's contribution in this connection was published more than a century before Christaller's work appeared.

The final section described an algebraic model that expresses the population sizes of central places as a function of the umland populations served by the various levels of centers. This model is constructed so as to take account automatically of the fact that every central place above the hamlet level possesses not one umland but a nested series of umlands. In short, the model expresses the relationship between central place populations and market area populations in a manner that faithfully captures the essence of the concept of hierarchical structuring.

When applied to empirical data, the algebraic model produces a set of numerical estimates of the ratio that relates the number of suppliers (including dependents) on each hierarchical level to the number of persons supplied. The available evidence indicates that the value of this ratio declines with increasing hierarchical rank, suggesting a tendency for per capita productivity of suppliers to increase as we ascend the levels of the hierarchy.

6

Bridging the gap: techniques linking theory with observations

The purpose of this chapter is to describe some important operational procedures that permit application of central place theory in the study of real urban systems. Three topics are addressed: first, measurement of urban centrality; second, delimitation of umlands; and third, use of data on intercity flows to investigate relationships of dominance and subordination among cities. The chapter as a whole bridges the gap between the theoretical ideas discussed in chapter 5 and the case study presented in chapter 7.

Measuring urban centrality

THE GENERAL CHARACTER OF CENTRALITY

The simplest approach to the measurement of centrality is to assume that a town's importance as a central place is adequately represented by the size of its population. This assumption is reasonable, since both the volume and the variety of goods and services offered to consumers are, in general, strongly correlated with city size. In the case of large cities – as a rule of thumb, those above 50,000 inhabitants – total population probably measures centrality as reliably as any other variable.

Among smaller centers, however, population size can give an exaggerated impression of a town's importance as a central place. For example, the addition of a manufacturing plant may inflate a town's resident population without stimulating a commensurate increase in its ability to attract shoppers from outside. Or a small center, located within commuting distance of a large town, may take on the role of dormitory: the additional population of commuters and their dependents tends to increase, at least initially, more rapidly than retail and service facilities. In these kinds of situations, population size may not provide a trustworthy measure of centrality. Accordingly, measurement of the centrality of a small center is based usually on the number and variety of retail stores and consumer services. Data on population are not ignored, but take second place.

The concept of centrality refers to an urban center's role in serving the needs of *external* retail and service customers – customers who do not reside within the center itself (Christaller 1966, 17–18; Preston 1971b, 1975). The total demand for central goods and services that is satisfied by a given town falls into two parts: internal and external. The internal component represents goods and services purchased by the town's own residents. The external component is the measure of the town's centrality. By definition, the internal and external components of total demand are mutually exclusive:

$$D = C + L, \tag{6.1}$$

where D is the total demand satisfied by the town, C (centrality) is the external demand, and L (local demand) is the volume of demand attributable to the residents of the town itself.

It follows than an ideal measure of urban centrality should incorporate an acceptable procedure for separating the effects of external and internal demands. If reliance is placed exclusively on population statistics, no logical basis for achieving such a separation can be devised. As we shall see, however, use of data on retail and service outlets does permit a plausible separation.

It is instructive to consider the method originally used by Christaller to measure centrality in Bavaria during the 1930s. The key to his approach lies in the rarity at that time in Bavarian private households of the telephone, which was almost exclusively a tool of commerce. Although firms with central place functions were perhaps not the only types of businesses possessing telephones, Christaller contended that an acceptable measure of a town's influence over its surrounding region could be obtained by finding the difference between the number of telephones in the town and the number that could be expected on the basis of the 'telephone density' of the region as a whole:

$$Z_z = T_z - E_z(T_g/E_g), \tag{6.2}$$

where Z_z is the centrality of a town, T_z is the number of telephones present in the town, E_z is the population of the town, T_g is the number of telephones in the region as a whole (the county or administrative district in which the town was situated), and E_g is the region's (county's) total population. The expression $E_z(T_g/E_g)$ provides a pro rata value for the expected number of telephones in the town. This value is then subtracted from the observed number, T_z, in order to yield the desired measure of centrality (Christaller 1966, 147).

Christaller's procedure makes it abundantly clear that centrality is not to be equated with the total or absolute importance of a town. Centrality measures a town's importance relative to its surrounding region. In effect, the concept is one of 'excess capacity.' A town contains a certain number of telephones: some fulfill internal needs, and only those in excess, if any, cater to the needs of the surrounding area. Such data would nowadays be inappropriate; but the concept of centrality as a measure of external importance remains unchanged.

CENTRAL FUNCTIONS AND FUNCTIONAL UNITS

In the analysis of data that describe the retail stores and services in central places, a distinction is made between central functions and functional units. A central function (normally abbreviated to 'function') is a recognizably distinct type of retail or service business, such as 'hardware store,' 'bank,' or 'beauty parlor.' A functional unit, in contrast, is an individual occurrence of any function. To take a simple example, consider the case of a hypothetical hamlet with two gas stations and a general store. This hamlet possesses two functions ('gas station' and 'general store') and three functional units.

A functional unit is not identical to an establishment, the latter being defined as a physically distinct place of business. Most establishments perform only a single function and therefore consist of a single functional unit. However, some establishments perform more than one function, and these special cases can provide valuable information. For example, in many central places of hamlet size, a gas station and a general store are housed together on the same premises under a single owner. Because 'gas station' and 'general store' are distinct types of business, often occurring independently, it is appropriate to regard the combination as consisting of two separate functional units, even though housed under one roof. Another example is the combination of a furniture store and a funeral home. This seemingly incongruous pair is now relatively rare, but it can still be found in small central places that are remote from metropolitan influences. The combination originated at a time when furniture dealers, who once manufactured much of what they sold and who worked mainly in wood, branched out into the sideline of manufacturing caskets, and thence into the complete business of undertaking. Here, too, a single establishment may reasonably be treated as two functional units.

When using the concepts of 'function' and 'functional unit' to prepare an inventory of central place activities, one pays no attention to the ownership of the stores and services in question. For instance, suppose that a successful local businessman in a particular town owns a lumberyard, a furniture store, and two separate restaurants. Suppose, moreover, that this entrepreneur operates all these outlets under a single corporate identity: 'J. Holzhacker and Sons, Incorporated – Lumber, Home Furnishings, and Good Eats.' Only a single firm is involved, but we regard each outlet as a separate entity. Thus Mr Holzhacker's activities are recorded as four distinct functional units representing three different central functions. Our basic unit of analysis, in short, is the functional unit rather than the firm. We are concerned with the visibly distinct components of the commercial landscape, not with underlying legalities.

In drawing up a list of activities that qualify as central functions in any particular area, one should treat an activity as a central function if it exists primarily to satisfy the demands of the final consumer. Central functions fall into two broad classes: provision of tangible goods and provision of services. The first class consists of what we ordinarily understand by the term 'retail stores': grocery stores, hardware stores, clothing stores, and so forth. The second class contains a wide variety of outlets specializing in consumer services, including barber shops, beauty parlors, banks, restaurants, dentists' offices, and public libraries. There is no universal catalog of central functions that can be used as a general checklist in all parts of the world. The specific set of functions found in any particular area is obviously a product of the manner in which the retail and service sector has evolved, and a list that is appropriate for Iowa or Saskatchewan may prove quite unworkable in Belize, in Bengal, or even in Bavaria. In every case, however, all activities that qualify for inclusion as central place functions have a recognizable identity and an orientation toward the needs of the final consumer.

Paradoxically, two common elements of the urban landscape possess both of the above features but are often excluded from consideration: namely, places of worship and schools. Their exclusion seems to be based partly on their 'institutional' rather than 'commercial' character and partly on the fact that the geographical patterns of occurrence do not always appear closely related to city size. Commercial activities have always been the primary focus of concern in central place studies; but some institutional functions, such as hospitals and public libraries, are normally included. Thus the 'institutional' argument for exclusion does not seem to be decisive. However, the spatial distribution of places of worship and schools appears different in kind from that of other consumer-oriented activities. For example, a church may occur in isolation in the open countryside where no nucleated settlement exists (though one may have existed there in the past), and a district high school serving a scattered rural population may be located in a small village rather than a large town. The factors influencing the locations of places of worship and schools tend to be social and political in character rather than economic, perhaps placing these functions outside the normal scope of central place research. But there are good arguments on both sides, and the final decision appears to be essentially a matter of the investigator's personal preference.

The primary source of data for a central place study is fieldwork. However, the cost of carrying out an exhaustive census of retail and service functions at first hand is often prohibitively high, and considerable reliance must accordingly be placed on secondary sources. The most useful published materials are the following:

1. *Local business directories*. These are published for specific towns by the local Chamber of Commerce, Planning Board, or Retail Businessmen's Association. They are sometimes combined with promotional literature designed to attract new investment (especially in manufacturing).

2. *General commercial directories*. These are compiled by the internationally known market research firm of Dun and Bradstreet Inc. They include manufacturers and wholesalers as well as retailers, but the entries are classified by type, and it is not difficult to separate the retail outlets from the other categories. Coverage of retailers is comprehensive, but coverage of service facilities is fragmentary. The Dun and Bradstreet directories contain confidential financial information and are therefore not freely available to the general public. However, they are usually accessible to qualified researchers on presentation of appropriate credentials.

3. *Specialized directories*. These provide information concerning a single function, usually at the state or national level. An example is the *Canadian Medical Directory*, which includes a list giving the location of every non-military physician and surgeon currently practicing anywhere

in Canada. Similar directories are available in Canada for hospitals, banks, newspapers, public libraries, and beer and liquor stores.

4. *Telephone directories*. Not every retail and service outlet can be identified unambiguously in telephone directories, but these volumes are particularly useful for services that tend to be incompletely covered in other sources – lawyers, dentists, and funeral homes are examples. Both the white and the yellow pages should be used.

Experience indicates that the above types of sources, used in combination, can provide the central place researcher with an almost exhaustive body of functional data. Nevertheless, there are several good reasons why fieldwork should also be undertaken.

First, fieldwork provides an opportunity for spot-checking apparent anomalies in the published data. For example, the directories may indicate that a certain fairly large town lacks a hardware store, even though hardware stores are present in many other smaller settlements. Does this town really have no hardware store, or are the directories incorrect? Fieldwork will supply the answer.

Second, a full understanding of an area's central place network requires not only an inventory of functional data but also, as will be explained in chapter 7, a knowledge of patterns of consumer travel behavior. In brief, one needs to know the extent of each town's tributary area, and direct field investigation is the most reliable approach to this problem.

Third, fieldwork is the best way to become familiar with special features of the study area that may affect the structure and functioning of the central place system. Under this heading come such things as the influence of physiography, the details of the area's road network, and the role of noncentral activities such as manufacturing and tourism.

Once the data on functions and functional units have been collected, it is convenient to display them in an orderly fashion by preparing a table in which each row represents a different central place and each column represents a different central function. The value entered in each cell of this table indicates the number of units of a particular function located at a specific central place. A table of this kind for a small hypothetical study area is shown as Table 6.1. The area represented by this table contains ten central places (labelled A through J), eight central functions (P through W), and a total of sixty-six functional units. For ease of reference, places are ranked by population and functions by numbers of central places in which they occur. Summing horizontally along any individual row yields

TABLE 6.1
Occurrence of functional units in a hypothetical set of central places

Central places	Population[a]	Functions P	Q	R	S	T	U	V	W	Total
A	2,600	5	4	2	3	2	1	2	1	20
B	800	3	2	2	1	0	1	0	0	9
C	600	2	3	1	2	1	0	0	0	9
D	500	2	1	2	1	1	0	0	0	7
E	360	1	2	1	0	1	0	0	0	5
F	280	2	1	0	1	0	0	0	0	4
G	240	1	2	1	0	0	0	0	0	4
H	200	1	1	1	0	0	0	0	0	3
I	190	2	1	0	0	0	0	0	0	3
J	180	1	1	0	0	0	0	0	0	2
Total	5,950	20	18	10	8	5	2	2	1	66

a For central places; the study area is assumed to contain 12,000 inhabitants, including rural population.

the total number of functional units of all types present in a particular central place. Summing vertically down a column gives the number of units of a particular function occurring in the entire region. The table as a whole represents the raw material for the calculation of each place's centrality.

THE DAVIES FUNCTIONAL INDEX

The standard approach to the measurement of centrality in recent years has been based on an index devised by W.K.D. Davies in his studies of the urban centers of South Wales (Davies 1967, 1970). The Davies index in its original form is a measure not of centrality (i.e. external importance) but of the total functional importance of a central place. Following Lukermann (1966) and Preston (1971b, 1975), we will use the term *nodality* to denote the total functional importance of a place. In this section we will first consider the Davies index in its original form and then introduce a modification that transforms this index from a measure of nodality into a measure of centrality.

The Davies functional index rests on three axioms. First, every functional unit present in a central place contributes toward that place's nodality. Second, all functional units representing the same central function are equal in weight. And third, units that represent commonly occurring functions (for example, gas stations and grocery stores) must be given less weight than units that represent rarely occurring functions (for example, bookstores and piano tuners). More specifically, the weight given to any functional unit is assumed to be inversely proportional to the frequency with which units of the function in question occur within the study area as a whole.

The first axiom simply affirms that the data assembled should be, as far as resources permit, complete. A number of early central place studies deliberately avoided the task of preparing a complete inventory and placed their faith in what may be termed the 'indicator functions' approach. A few functions were selected arbitrarily to serve as taxonomic criteria, and the urban centers were then assigned to hierarchical orders on the basis of the numbers of these diagnostic functions that they possessed (Smailes 1944; Duncan 1955). This approach is too presumptive. Without analyzing the patterns of occurrence of *all* functions, we cannot decide which activities are reliable indicators for each hierarchical level. Moreover, a single set of indicators will probably not prove uniformly reliable across a large area containing regional variation in population density. For these reasons it is now generally accepted that the approach using indicator functions is of very limited value.

The implications of the second axiom are somewhat more subtle. On the one hand, we know that the assumption of equal weight is empirically false, since we can observe differences in quality among stores or service outlets of any given type (Potter 1980). On the other hand, it is difficult to formulate satisfactory guidelines that take differences in store quality into account. In particular, the physical condition of an establishment is not a trustworthy indicator of quality of service. The inhabitants of the rural regions where most central place research is carried out are generally more concerned with substance than with style, and many small-town stores that may appear second-rate to the metropolitan visitor offer good-quality merchandise, fair prices, a high level of personal attention, and even a cheerful proprietor. To award demerit points for unimpressive physical appearance is to go skating on thin ice. All things considered, it seems reasonable to adopt the twofold auxiliary assumption that outlets of genuinely inferior quality are relatively few in number and distributed randomly both among central functions and among central places. To the extent that these conditions are fulfilled, the axiom that all functional units of the same type are equal in weight can be accepted without fear of introducing any systematic bias.

The third and final axiom is concerned with the fact that central

TABLE 6.2
Nodality scores calculated from data in Table 6.1

Central places	Functions								Total nodality
	P	Q	R	S	T	U	V	W	
A	25.0	22.2	20.0	37.5	40.0	50.0	100.0	100.0	394.7
B	15.0	11.1	20.0	12.5	–	50.0	–	–	108.6
C	10.0	16.7	10.0	25.0	20.0	–	–	–	81.7
D	10.0	5.6	20.0	12.5	20.0	–	–	–	68.1
E	5.0	11.1	10.0	–	20.0	–	–	–	46.1
F	10.0	5.6	–	12.5	–	–	–	–	28.1
G	5.0	11.1	10.0	–	–	–	–	–	26.1
H	5.0	5.6	10.0	–	–	–	–	–	20.6
I	10.0	5.6	–	–	–	–	–	–	15.6
J	5.0	5.6	–	–	–	–	–	–	10.6
Total	100.0	100.2	100.0	100.0	100.0	100.0	100.0	100.0	800.2

functions differ widely in their frequency of occurrence. For example, the two-county area that we examine in detail in chapter 7 contains 215 gas stations (the most frequently occurring function) but only forty-two drugstores and only a single music store. Clearly it would be misleading to give each gas station the same numerical weight as the solitary music store. It can be argued, in fact, that the drawing power of the music store is 215 times as great as the drawing power of a single gas station. Accordingly, the weights used in the calculation of nodality scores should be in the ratio 215 to 1. The weights applied to all other functions should reflect their relative frequencies in a similar manner.

The third axiom implies, in effect, that the numerical weights should be defined as follows:

$$W_i = 100/T_i, \qquad (6.3)$$

where W_i is the weight assigned to each functional unit of function i, and T_i is the total number of units of function i within the study area as a whole. For example, if there are forty-two drugstores in the entire study area, the weight of a single drugstore is 2.38 nodality units (i.e. 100 divided by 42). The weight of a single functional unit is termed the *location coefficient* of the function in question (Davies 1967). The value of 100 is arbitrary, ensuring simply that the location coefficient of each function is inversely proportional to the frequency of occurrence of the corresponding functional units.

Once the location coefficient of every different function has been calculated in this way, the numbers of functional units of each type in each central place are multiplied by the appropriate location coefficients in order to obtain the amounts of nodality conferred on each place by each function. These amounts are then added together to produce an overall nodality score for each center, termed by Davies the center's functional index. Thus the formula for a central place's functional index may be expressed as

$$F_j = \sum_{i=1}^{N} E_{ij}W_i, \qquad (6.4)$$

where F_j is the functional index of central place j, N is the number of different functions in the study area, E_{ij} is the number of functional units of function i in central place j, and W_i is the location coefficient of function i.

It is not necessary to calculate the individual location coefficients, W_i. By combining the above two formulae we can write

$$F_j = 100 \sum_{i=1}^{N} E_{ij}/T_i, \qquad (6.5)$$

where T_i, as before, is the total number of units of function i within the study area as a whole. This is the most convenient formula to use in practical work.

Table 6.2 illustrates the calculation of functional indexes for the set of hypothetical central places shown in Table 6.1. The values in the body of Table 6.2 are the amounts of nodality contributed by each function to each central place. For example, the system as a whole contains twenty units of function P (Table 6.1), and hence the location coefficient of function P is 5.0 (100 divided by 20). Central place A possesses five units of this function and therefore receives 25.0 points toward its overall functional index (Table 6.2). When all the nodality points are added together, central place A attains a final nodality score of 394.7 points.

If the results are accurate, each column in Table 6.2 should sum to 100, this being the total weight allotted, by definition, to each central function. (Minor discrepancies due to rounding can be ignored.) Similarly, the grand sum of the overall functional index values should be, in this case, 800. More generally, this sum can be obtained by multiplying N, the number of different functions included in the analysis, by 100.

Functional indexes derived in this manner take account of variation among central places in terms of numbers of both functions and functional units. Because the weights assigned to the various functions are inversely proportional to their frequencies of occurrence, a single unit of a rare function can add more to a center's nodality than several units of a ubiquitous function, which accords with our intuitive expectations. The Davies technique measures the nodality of central places more realistically than counts of either functions or functional units alone. For example, if we compare central places A and E in Tables 6.1 and 6.2, we see that center A has twice as many functions and four times as many functional units as center E. Its functional index, however, is more than eight times as large, because it possesses a number of functional units representing relatively rare functions.

Davies considered modifying his approach by using data on the numbers of persons employed by each central function instead of data on the numbers of functional units. With this change, the individual worker takes the place of the individual functional unit as the basic element in the analysis. The calculations thus take into account the possibility that functional units may vary in size (as measured by employment). However, published data on employment at the level of functional units are not readily available. Davies addressed this problem by collecting employment data in the field. Using these data, he compared functional indexes based on employment with those based on numbers of functional units. The nodality scores of the largest central places tended to be relatively higher when employment data were used, and this was taken to reflect the fact that, for several functions, the average number of workers per functional unit tended to increase in the larger centers. However, when the central places were ranked separately according to the two sets of functional indexes, rank differences were very minor. Davies concluded that the differences between the two approaches were too small to justify the greatly increased effort required to obtain the necessary employment data for large areas (Davies 1967).

The Davies functional index is a measure of nodality rather than of centrality – it measures the total functional importance of a center rather than the 'excess capacity' available to serve customers living outside the center. Here we can be guided by the logic of Christaller's original telephone index. What we require is a method for estimating self-oriented nodality – the amount of nodality that a central place would possess if it served only itself. The difference between this amount and the observed total nodality will provide an estimate of the place's centrality, or its importance to external customers.

Estimates of self-oriented nodality can be obtained on a pro rata basis as follows. We begin by relating the grand sum of the functional indexes of all central places to the total population of the study area. Let us assume, for example, that the study area represented in Tables 6.1 and 6.2 contains 12,000 inhabitants – including, of course, rural population as well as the 5,950 persons who live in the area's ten central places (Table 6.1). As we have seen, the grand sum of the functional indexes in this study area is 800 nodality units (Table 6.2). Accordingly, there is an overall ratio of 800 nodality units to 12,000 inhabitants, or one nodality unit for every 15 inhabitants. We apply this ratio to the population of a central place in order to obtain an estimate of that place's self-oriented nodality. Finally, we obtain the required estimate of centrality by simple subtraction. Consider, for example, central place A in Table 6.2:

Population of central place A = 2,600.

Ratio of nodality points to total population within the study area as a whole = 800/12,000 = 1/15.

Therefore, self-oriented nodality of center A = 2,600(1/15) = 173.3.

Observed nodality of A (Table 6.2) = 394.7.

Therefore, centrality of A = 394.7 − 173.3 = 221.4.

Since we are dealing with relatively large numbers of consumers, the amount of 'pressure' exerted by each consumer (whether rural or urban) on the stores and services located in the central places can be treated as a constant factor. In effect, we assume that consumers exhibit a homogeneous pattern of demands. Accordingly, simple proportionality can be used to estimate the volume of demand exerted by any given number of consumers. If a central place possesses a higher nodality score than it requires in order to satisfy the demands of its own population, the excess is assumed to reflect demands exerted by external customers. The entire approach, in fact, is completely analogous to the technique employed in Bavaria by Christaller.

It might be thought that the type of adjustment described above should be carried out independently for each function, by dealing with the cells in Table 6.2 one at a time instead of making a single adjustment to the nodality score of each central place as a whole. Actually these two possible approaches produce identical results. Since every function accounts for the same total amount of nodality in the study area as a whole (100 points), it makes no difference whether we carry out a series of separate adjustments, which must then be summed, or a single overall adjustment after the total nodality of each center has been obtained. From

TABLE 6.3
Nodality and centrality scores for central places of Table 6.2

Place	Population	Nodality	Internal demand	Net centrality
A	2,600	394.7	173.3	221.4
B	800	108.6	53.3	55.3
C	600	81.7	40.0	41.7
D	500	68.1	33.3	34.8
E	360	46.1	24.0	22.1
F	280	28.1	18.7	9.4
G	240	26.1	16.0	10.1
H	200	20.6	13.3	7.3
I	190	15.6	12.7	2.9
J	180	10.6	12.0	−1.4

SOURCE: Calculated from Tables 6.1 and 6.2 by method described in text.

a computational standpoint it is much easier to make a single adjustment to the total nodality score than to adjust the scores for each function individually.

In formal terms our index of centrality may be written as follows:

$$C_j = F_j - 100N(P_j/P_t), \tag{6.6}$$

with C_j the centrality index of central place j, F_j the nodality score (Davies functional index) of central place j, N the number of different functions in the study area, P_j the population of central place j, and P_t the total population, rural and urban combined, in the study area as a whole.

The results of applying this modification to all the hypothetical central places in Table 6.2 are shown in Table 6.3. Two interesting features, both of which also occur under real-world conditions, may be noted. First, the ranking of places according to their centrality scores is not necessarily identical to the ranking of the same places according to their total nodality. For example, center F (Table 6.3) has a higher nodality score than center G, but center G has the greater centrality. Second, it is possible for a central place's centrality score to be a negative value, as in the case of center J. This means simply that the place in question is not theoretically capable of meeting even its own needs, let alone those of external consumers. In fact, negative centrality is common in the real world among central places containing fewer than one or two hundred inhabitants.[1]

The scores obtained above are in no sense absolute values. They are entirely dependent on the size and composition of the study area. If we make our study area larger we will add more central places, more functional units, and perhaps even more central functions to our body of data, and these additions are bound to affect the values of the nodality and centrality indexes. Measures of the type here described are meaningful only within the limits of whatever study area we have defined. Direct comparison with values obtained in other areas, even if identical methods have been used, is unwarranted.

The delimitation of umlands

GENERAL CONSIDERATIONS

Regardless of the technique employed, the delimitation of an urban center's market area always involves the idea of relative dominance. In essence, the umland of a given town is not the entire area within which the town's influence as a central place can be detected. It is the more limited region within which the influence of the town is stronger than that of any other town of comparable size and functional complexity. In the typical case, the influence of a town as a retail center declines in strength as we move outward from the town into the surrounding area. Sooner or later a point is reached where this declining influence is equaled by the attractiveness of a competing center. At this point a consumer is pulled equally strongly – or equally feebly – toward each of the competing towns, and the location is therefore known as a point of indifference.

By definition, a point of indifference belongs simultaneously to the umland boundaries of two or more competing centers. It does not follow, however, that the ability of each town to attract customers stops abruptly at a point of indifference. It is normal, in fact, for a town to draw shoppers from beyond the encircling points of indifference, though perhaps not in great numbers. Each point of indifference marks a transition from one zone of relative dominance to another. It is these zones of relative

1 Negative centrality, however, does not imply that a central place attracts absolutely no external consumers. A hamlet containing a single general store may have negative centrality, yet the store may have regular customers among nearby farmers. Our estimates refer to net centrality, not to the actual numbers of external patrons.

dominance that define the network of market areas of the towns in question (Bracey 1956; Whitelaw 1962; Carter and Davies 1963).

It follows that the delimitation of an umland requires the identification of points of indifference along routes radiating in all directions from the town under investigation. A line connecting these points of indifference represents the desired umland boundary. The problem of defining an umland boundary may thus be viewed as a problem of determining the locations of points of indifference.

We defined an umland, above, as the area within which the influence of a particular central place is stronger than that of any other center of comparable size and functional complexity. The phrase 'of comparable size and functional complexity' requires explanation. Recalling our discussion of central place theory in chapter 5, we should anticipate that all urban centers except the very smallest will be found to have more than one umland. Specifically, each center will have a separate umland for each level of the urban hierarchy on which it operates. The phrase 'of comparable size and functional complexity' therefore means 'functioning on the same hierarchical level.' For example, if we are interested in town-level umlands, the study must include all places of town rank or higher. Moreover, for centers of higher rank than towns, only the town-level umlands will be of concern, not the umlands served by these places at higher hierarchical levels. To rephrase our definition, a central place's umland on a particular hierarchical level is the area within which the influence of that center is stronger than the influence of any other central place that operates on the same hierarchical level.

A meaningful delimitation of umlands therefore requires that the central places under consideration first be classified into hierarchical orders. It requires that, at the very least, centers be separated into two groups: namely, places on or above a certain hierarchical level, and places below this level. This relationship between umland delimitation and hierarchical classification is not always fully appreciated, possibly because many studies of umlands focus on only one or two cities and therefore do not appear to be concerned with a *system* of towns. Even in a study of a single city, however, some form of hierarchical classification is implicit: the city's market area can be defined only by considering the influence of nearby competing centers (Dickinson 1930; Jones 1950; Macaulay 1954; O'Farrell 1965). In short, a central place's umland cannot be delimited unless one first identifies a set of nearby competitors of equivalent hierarchical rank.

In the remainder of this chapter we will take it for granted that, for any particular town, competitors of equivalent hierarchical rank can be identified without difficulty. The task of carrying out a complete hierarchical classification of an area's central places is considered in detail in chapter 7.

Three approaches to the delimitation of umlands are currently in use: namely, the field survey method, the analysis of newspaper circulation, and a method based on the Reilly gravity model. We now discuss each of these three approaches in turn.[2]

FIELD SURVEYS

In the field survey approach, information concerning habitual patterns of consumer movement is collected by means of direct inquiries in the field. This information is then used to delimit umlands by application of the following definition. Suppose that *X* and *Y* are two urban centers, *Y* being the larger of the two. Further, let *G* be a good or service that is available in *Y* but not in *X*. Center *X* lies within the umland of center *Y* with respect to good *G* if the residents of *X* obtain this good more frequently from *Y* than from any other center.

An obvious implication of this definition is that a small urban center may lie simultaneously within the umlands of several larger places, the relationships being based on different goods and services in each case. Such a set of relationships, of course, is entirely consistent with the concept of an urban hierarchy. Indeed, quite apart from the connection between consumer movement and the delimitation of market areas, field investigation of shopping behavior is a necessary step in the classification of central places into hierarchical orders (see chapter 7).

If the study area is relatively small, the researcher may be able to carry out a field survey that incorporates a precisely defined sampling framework (e.g. Murdie 1965; Tarrant 1967). With a large area, however, a formal survey is usually impracticable, and an informal approach must be used. In fact, for the purposes of umland delimitation and hierarchical classification, an informal survey is generally quite satisfactory, since the investigator does not seek to make any statistical inferences about the characteristics of consumers. The objective of the study is simply to identify the 'orientation' of each nucleated settlement with respect to the

2 A fourth technique, not discussed here, is based on the analysis of rural bus routes around market towns (Green 1950, 1953, 1958, 1966). This approach has merit in many parts of Europe, especially where automobile ownership remains limited, but it is not applicable in North America.

shopping opportunities offered by larger centers. Experience indicates that these orientations can almost always be determined unambiguously through interviews with persons who, by virtue of their positions within the community, possess a good knowledge of local shopping habits. These persons include bank managers, town clerks, editors and publishers of newspapers, and presidents of chambers of commerce. In the smallest centers, where these kinds of officials are not found, the most helpful informant usually turns out to be the owner of the local general store. As the work proceeds, the information gathered in each center can be checked against the responses received elsewhere, and in this way a comprehensive picture of the normal pattern of consumer movement in the area can be assembled.

It has been found useful to obtain advance publicity for field activities by arranging for a suitable announcement to appear in the local newspapers, including weekly papers as well as dailies. If possible, this announcement should take the form of a short article that identifies the investigator, describes the aims of the study, and outlines the possible value of the results to local merchants (see the section below entitled 'The Uses of Umlands'). Presentation of a copy of such an article can be an effective method of 'breaking the ice' in cases where a potential respondent's initial reaction to the request for an interview is one of hesitation. The article can also serve as a supplementary means of identification in the event that the investigator's activities are questioned by local authorities.[3]

Sometimes a small settlement, located approximately halfway between two competing centers of higher order, shows 'divided loyalty,' with roughly equal amounts of patronage going to each of the higher-order places. Although it might be possible to prove, in such cases, that one of the two higher-order centers exerts a slightly stronger force of attraction than the other, it would be somewhat misleading to allocate the low-order settlement wholly to the stronger center's sphere of influence. A more realistic solution is to regard the low-order center as lying precisely on the point of indifference between the high-order places. Thus, the low-order center lies on the boundary that separates the two higher-order umlands. This is by no means an 'untidy' solution. On the contrary, it is exactly this

3 On one occasion I was leading a field party of undergraduates in Renfrew County, Ontario, when we were challenged by the police at gunpoint. We later learned that two armed robberies had been committed that same afternoon in nearby villages by a gang whose getaway car matched the description of our field vehicle. I am sure that some of the students on that field trip promptly abandoned all thoughts of pursuing careers in geography!

state of affairs that is predicted by the $K = 4$ model. Indeed, in both the $K = 3$ and $K = 4$ models (though not in the $K = 7$ pattern), the umland boundaries on any given hierarchical level pass through the centers that belong to the level immediately below. Accordingly, occurrences of this sort in the real world should be cause for satisfaction, not dismay.

NEWSPAPER CIRCULATION

If the towns under investigation are sufficiently large, their umlands may be defined by analysis of newspaper circulation records. As one researcher has noted: 'Newspaper circulation is closely related to cultural and commercial regions of urban dominance. For trading and recreational activities, a person tends to travel to the city from which his newspaper comes, and conversely he prefers a paper which carries advertisements and news from the city which he visits most frequently' (Menefee 1936, 66).

In other words, regardless of which is the cause and which is the effect, the geographical pattern of newspaper circulation and the extent of a city's sphere of influence as a local trade center are closely correlated. In fact, the 'newspaper approach' has been used for umland delimitation in both North America and Europe (particularly in France) for more than fifty years (Park 1929; Park and Newcomb 1933; Haughton 1950; Gauchy 1955; Chatelain 1957; Preston 1979).

Newspapers fall into two distinct classes on the basis of their frequency of publication. If publication takes place on at least four days of each week, the label 'daily' is applied. Almost all other newspapers are published only twice a week or less; they are called 'weeklies.' It is quite common for a newspaper to commence publication as a weekly and become a daily when the town grows large enough to justify the change.

If we seek to delimit the umland of a small town that publishes a weekly paper but not a daily, and if – as is often the case – the areal extent of this small town's umland is governed partly by competition from a large center that publishes a daily paper but not a weekly, we run into difficulties because the circulation patterns of daily and weekly newspapers are not directly comparable. For this reason, the newspaper method is normally restricted to towns that publish dailies.

It does not follow, of course, that all towns that publish a daily paper belong to a single hierarchical order. Therefore, if we seek to delimit the umland of a particular daily newspaper town, we should not assume that all nearby towns that also publish dailies are necessarily equivalent in hierarchical rank: some may be of lower rank. We are once again brought face to face with the concept of 'comparable size and functional complexity': hierarchical classification involves more than the presence or absence of daily newspapers.

Let us assume, however, that we have already identified all central places that operate on (or above) a specified hierarchical level; let us assume also that all these places are daily newspaper towns. This set of centers may be termed the 'target set.' Official records of the circulation totals of daily newspapers are maintained in such a way that the umland of each city can be defined as a region of relative newspaper dominance: the umland of any city is the area receiving more newspapers from that city than from any other city in the set. The delimitation is accomplished by assigning each small settlement listed in the circulation records to the city that supplies it with the greatest number of newspapers.[4]

Umland boundaries delimited on the basis of daily newspaper circulation are shown for a portion of southern Ontario in Figure 6.1. Although the five cities of Windsor, London, Kitchener-Waterloo, Hamilton, and St Catharines all belong to the same hierarchical order, their umlands vary widely in size. London's tributary area is exceptionally large, whereas that of St Catharines – constrained by the lakes, the international border, and the influence of Hamilton – is very small. The umlands of London and Kitchener-Waterloo are elongated toward the north, reflecting the absence of significant competition in that direction. The umland of Kitchener-Waterloo is 'pinched out' in the north before it reaches Lake Huron. This brings London into direct competition with metropolitan Toronto (a center of higher order) in the area south of the Bruce Peninsula (Figure 6.1).

THE REILLY GRAVITY MODEL

Suppose that there are two towns, X and Y, separated by a distance of D_{xy} miles. It is assumed that X and Y belong to different hierarchical orders and that the residents of X, the lower-order center, shop regularly in Y. Given these initial conditions, the Reilly gravity model is based on the idea that the flow of shoppers linking X and Y may be treated as being analogous to the force of gravitational attraction that exists between physical objects, as described by Newtonian mechanics. Specifically, the flow of trade from X to Y may be expressed by the following formula:

4 Although circulation records can be obtained from individual publishers, a unified source of data is maintained for North American newspapers by the Audit Bureau of Circulations, or ABC. As with the Dun and Bradstreet business directories, the ABC newspaper data are usually accessible to qualified scholars for research purposes.

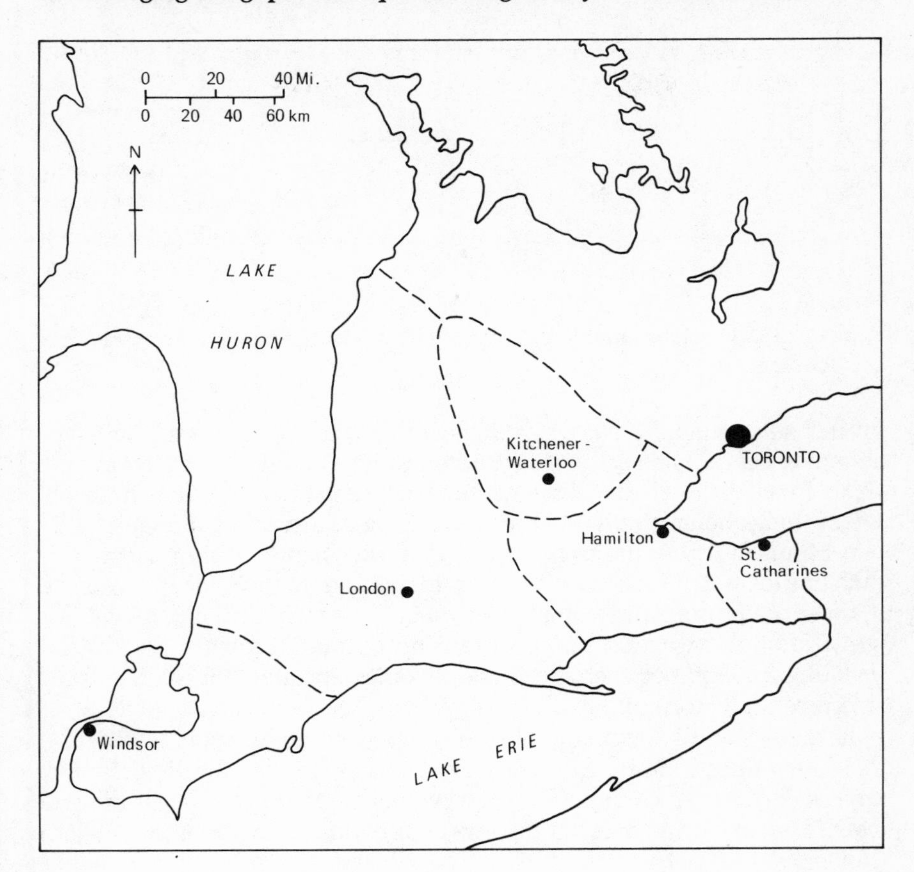

Figure 6.1
Umlands of major cities in southwestern Ontario, based on daily newspaper circulation patterns. Note wide variation in sizes and shapes.

$$F_{xy} = G(P_x)^m(P_y)^m(D_{xy})^{-n}, \tag{6.7}$$

where F_{xy} is the flow of trade from X to Y, measured, for example, by the dollar volume of sales or by the number of shoppers per unit of time; P_x is the population, or 'mass,' of town X; P_y is the population of town Y; D_{xy} is the distance from X to Y; and G, m, and n are constants: G is a proportionality factor (the analog of the universal gravitational constant in Newtonian physics), and m and n are exponents attached to mass (i.e. population size) and to distance, respectively.

Equation 6.7 is, in effect, a generalized form of Isaac Newton's famous

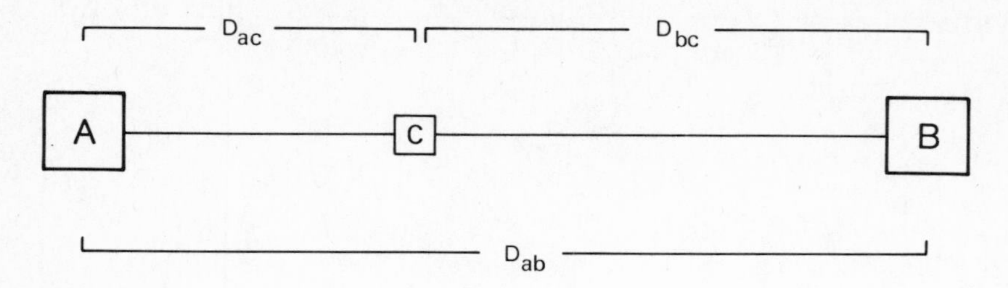

Figure 6.2
A trio of urban centers suitable for use in calibrating the Reilly gravity model. See text for explanation.

inverse square law. It expresses the fact that a town's attractiveness as a shopping center is directly proportional to its size and inversely proportional to the distance that shoppers must travel in order to reach the town. Other things being equal, a larger town is more attractive than a smaller one. Similarly, other things being equal, a town nearby is more attractive than one located farther away. Accordingly, a small town located close to a prospective consumer may exert just as much drawing power or 'gravitational attraction' as a large center situated farther afield. The resulting tradeoff between mass and distance accounts for the algebraic structure of the formula.

In the original Newtonian inverse square law the exponents m and n took the values 1 and 2, respectively. However, following Reilly's own approach, we will not impose a priori values on these exponents but will instead regard m and n as parameters to be estimated from empirical data (Reilly 1929, 48–50).

Let us now broaden the context and consider a case involving three towns rather than two; their relative positions are indicated in Figure 6.2. Two of the towns, denoted by A and B, are assumed to be high-order centers, equivalent in hierarchical rank, but not necessarily equal in size. The third town, C, is a low-order center located at an arbitrary point along the direct route connecting A and B. It is assumed that residents of C patronize the stores and service outlets in both A and B. Using the gravity formula given above, but with the subscripts modified to suit the new context, we obtain the following two equations:

$$F_{ca} = G(P_c)^m(P_a)^m(D_{ac})^{-n} \tag{6.8}$$

and

$$F_{cb} = G(P_c)^m(P_b)^m(D_{bc})^{-n}. \tag{6.9}$$

These two equations can be combined into a single expression by the simple procedure of dividing F_{ca} by F_{cb}:

$$\frac{F_{ca}}{F_{cb}} = \frac{G(P_c)^m(P_a)^m(D_{ac})^{-n}}{G(P_c)^m(P_b)^m(D_{bc})^{-n}}. \tag{6.10}$$

G, the proportionality constant, and P_c, the 'mass' of the low-order center, now vanish as a result of cancellation, leaving a relatively simple formula to work with:

$$\frac{F_{ca}}{F_{cb}} = \frac{(P_a)^m(D_{bc})^n}{(P_b)^m(D_{ac})^n}. \tag{6.11}$$

The left-hand side of equation 6.11 is simply a ratio (that is, a dimensionless quantity) that describes the relative 'popularity' of towns A and B from the standpoint of town C. This ratio can be given a numerical value by collecting appropriate information on the amounts of trade flowing from town C to towns A and B, respectively. Using Texas as their study area, Reilly and his associates evaluated this ratio through the use of data pertaining to the numbers of persons holding charge account privileges in major stores in the larger cities (Reilly 1929). In this way the numerical value of the ratio was calculated for 255 separate three-center cases of the type show schematically in Figure 6.2.

The basic data required for the right-hand side of equation 6.11 can be obtained relatively easily. Reliable population figures are available in the reports of the national census, and the distances separating the various towns can be measured on large-scale maps. Reilly used actual over-the-road mileages in his calculations, but later work has suggested that, in most regions, the correlation between over-the-road mileages and straight-line distances is so high that the latter can be used without significantly affecting the value of the subsequent results (Nordbeck 1964; Olsson 1965, 57–8).

The remaining elements in equation 6.11 – namely, the exponents m and n – presented Reilly with something of a problem. He was evidently unaware that the best-fitting values of these exponents could be derived from his empirical data simultaneously by means of multiple regression analysis. He recognized, however, that if the value of one exponent were known, the value of the other exponent could easily be found. He therefore assumed that the value of m, the exponent on the populations, was fixed at 1. (He attempted to justify this assumption by showing that total external trade is directly proportional to city size, but this argument is not convincing.) Given the value of 1 for m, equation 6.11 can be

TABLE 6.4
Frequency distribution of exponent of distance in Reilly gravity model as applied to cities in Texas

Value of exponent	Number of cases
0.00–1.50	45
1.51–2.50	87
2.51–3.50	35
3.51–4.50	24
4.51–5.50	15
5.51–6.50	14
6.51–7.50	6
7.51–8.50	5
8.51–9.50	12
9.51–10.50	5
10.51–11.50	3
11.51–12.50	4
Total	255

SOURCE: Reilly 1929, 49–50.

solved in order to produce a value of n for each group of three towns. Reilly's data for the cities of Texas permitted the solution of 255 separate equations. The resulting frequency distribution of the 255 values of n is shown in Table 6.4.

The next step is to select a representative value of n to serve as the exponent on distance in a general statement of the gravity equation. If the 255 values of n in Table 6.4 are averaged, the result is 3.5. However, Reilly evidently felt – justifiably so – that simple averaging was inappropriate in view of the marked positive skewness of the frequency distribution. He therefore opted for the modal value of n (i.e. the most frequent case), which he took to be exactly 2. By a remarkable coincidence, this is the same value as the exponent on distance in the original Newtonian inverse square law.

Setting the values of m and n at 1 and 2, respectively, Reilly's gravity equation may now be written in what has come to be regarded as its standard form:

$$\frac{F_{ca}}{F_{cb}} = \frac{P_a}{P_b}\left(\frac{D_{bc}}{D_{ac}}\right)^2. \qquad (6.12)$$

This standard form expresses the quantitative relationships believed to exist among trade flows, populations, and distances in an *ideal* situation – where the relative magnitudes of the flows of trade are influenced only by distance and city size. In the real world, of course, other factors may come into play – most significantly, the alignment of political boundaries (Mathieson 1957, 1958), but also variations from city to city in quality of service, effectiveness of advertising, and the general ambiance of the major shopping districts. Presumably the influence of these additional factors accounts for the observed variation in real-world values of n, as exemplified by Table 6.4. Like Christaller's models of the arrangement of central places, Reilly's gravity model is intended not to be a replica of reality but to provide a framework of ideas within which to pursue an understanding of real situations. Reilly's model has sometimes been referred to as 'the *law* of retail gravitation'; however, although it summarizes an important general tendency, it does not describe an immutable relationship to which all cities inevitably conform.

Up to this point I have not sought to relate Reilly's model to the general problem of delimiting umlands. The relationship is really quite simple. Suppose we are faced with a situation similar to the one shown schematically in Figure 6.2. We do not know, initially at least, where the point of indifference between A and B is situated. We do know, however, that customers living precisely at this point are attracted with equal force to both A and B. In other words, the point of indifference – Reilly's 'breaking-point' – lies where F_{ca} is exactly equal to F_{cb}. Accordingly, the gravity equation for the point of indifference may be written as follows:

$$\frac{P_a}{P_b}\left(\frac{D_{bc}}{D_{ac}}\right)^2 = 1. \tag{6.13}$$

There need not be a settlement of any kind located precisely at the point of indifference. However, if such a settlement did exist, its inhabitants could be expected to behave in accordance with equation 6.13.

We also know, of course, that the sum of D_{bc} and D_{ac} is equal to D_{ab}, the total distance between A and B (Figure 6.2). Equation 6.13 thus becomes:

$$\frac{P_a}{P_b}\left(\frac{D_{bc}}{D_{ab} - D_{bc}}\right)^2 = 1. \tag{6.14}$$

And finally, by rearrangement of terms, we obtain

$$D_{bc} = \frac{D_{ab}}{1 + \sqrt{P_a/P_b}}. \qquad (6.15)$$

Because the populations of A and B and the distance from A to B are known (or can easily be determined), equation 6.15 can be used to find the distance from town B to the breaking-point. By repeated application of the formula, the locations of a series of breaking-points around any specified city can be established. These points can then be connected in order to provide a closed umland boundary.[5]

In any particular application of equation 6.15 it is convenient to designate the larger city as A and the smaller city as B. The equation then gives the distance from the smaller city to the point of indifference. For example, suppose that cities A and B are ninety miles apart and contain populations of 250,000 and 160,000 inhabitants, respectively. By inserting these values into the gravity formula we obtain:

$$D_{bc} = \frac{90}{1 + \sqrt{\dfrac{250{,}000}{160{,}000}}},$$

from which it is easily calculated that the breaking-point lies forty miles from city B and fifty miles from city A. Note that the larger city, A, has the greater 'reach.' This accords well with our intuitive expectations and with common experience.

The Reilly model has been tested at various times against results obtained by the use of field surveys (Converse 1949, 1953; R.B. Reynolds 1953; Mathieson 1957, 1958; Jung 1959; Schwartz 1962; Wagner 1974). In general, these tests have shown that equation 6.15 yields reliable results, but in some cases the field evidence indicates that the correct value of n for a particular pair of competing cities is not equal to 2. However, such evidence does not necessarily invalidate Reilly's claim that the modal value of n is 2. After all, as Table 6.4 indicates, Reilly's own results

5 The connecting of the breaking-points must be done with caution. If the gravity model is used to assign umlands to several cities simultaneously, the use of straight lines to connect adjacent breaking-points will leave some territory unallocated. Ingenious geometrical rules can handle this problem, but the impression of accuracy that they create is inconsistent with the model's general character as a rough-and-ready working tool. It seems more appropriate to allocate the left-over territory subjectively, linking each pair of adjacent breaking-points with a line that includes an 'elbow' at some point along its course.

showed that the value of n exhibits considerable variation. Accordingly, the discovery of a particular case in which the value of n is not equal to 2 does not imply that Reilly's model is generally false. It implies only that the cities in question represent a nonmodal case: most probably, a case in which the flows of trade are significantly influenced by factors other than distance and city size.

For any given pair of cities, the position of the breaking-point as determined by Reilly's formula is fairly insensitive to variation in the value of n. Consider, for example, the hypothetical case described above in which, given 2 as the value of n, the breaking-point was found to lie 40 miles from city B. If the value of n is reduced to 1, the breaking-point lies 35 miles from B (and 55 miles from A). If the value of n is increased to 6, the breaking-point lies 43 miles from B (and 47 miles from A). In this particular example, therefore, any value of n between 1 and 6 places the breaking-point within a short, 8-mile stretch of the 90-mile route connecting the two cities. A similar degree of stability can easily be demonstrated for other examples. All things considered, it seems reasonable to accept Reilly's evidence concerning the shape of the frequency distribution of n and, correspondingly, to accept the apparent modal value, $n = 2$, as the appropriate exponent to use in practical applications of the model.

In the three methods of umland delimitation described above, overall accuracy is inversely proportional to the effort involved in their use. Field surveys give results that are entirely accurate by definition (i.e. this is the method by which 'truth' is defined), but the necessary fieldwork is both expensive and time-consuming. The Reilly method is the least accurate but the easiest to apply. The newspaper circulation method is intermediate on both counts. All three techniques, however, can be relied on to produce satisfactory results.

THE USES OF UMLANDS

Umlands have significant value in both theoretical and practical contexts. Their theoretical importance is related both to central place theory and to the long-standing distinction in geography between uniform and nodal regions. Their practical value is related to certain aspects of market research and also to the broader field of regional planning.

With respect to central place theory, the delimitation of umlands casts light on the manner in which a real-world pattern of central places and market areas departs from the regular geometry postulated in the theory.

Meaningful delimitation of a system of umlands requires that the central places in question be first classified into hierarchical orders. Let us assume that all places that operate on or above a particular hierarchical level in a specific region have been identified and that a network of umlands has been delimited for this set of places. As in the theoretical central place models, these umlands will be nonoverlapping and will exhaust the region's territory, but there the resemblance between theory and reality is likely to end. It is most improbable that the real-world umlands will be hexagonal in shape or equal in size. (For an example, see Figure 6.1.) However, observed variations should be explicable in terms of the fact that the real world does not match the homogeneous character of the isotropic plain assumed in central place theory. Where cities of a specified hierarchical rank – including places that also function above that level – are relatively closely spaced, umlands should be comparatively small, and vice versa. The shape of each city's umland, in turn, should reflect the location and strength of competing centers. In effect, the shapes of the umlands are controlled by the geometrical arrangement of the points representing the members of the hierarchical order under consideration. To the extent that this geometrical arrangement deviates from the uniform triangular lattice of the theory, the shapes of the umlands will depart from the ideal hexagonal form.

The delimitation of umlands is also of interest to researchers concerned more generally with the nature of geographical regions. In brief, a geographical region may be defined as a contiguous section of the earth's surface exhibiting one or more distinctive characteristics that serve to distinguish it from neighboring areas. Two general categories of regions are recognized: uniform and nodal. A uniform region has a particular attribute (or a distinctive set of interrelated attributes) that is present more or less constantly throughout the region's areal extent – for example, a savanna region, as contrasted with a desert or a tropical forest. Nodal regions, in turn, possess a 'core,' or organizational focus, to which the remainder of the region is linked by recurring flows of traffic. In short, uniform regions are defined by static properties; nodal regions, by patterns of circulation (Whittlesey 1954, 36–7).

It is evident that a city together with its umland constitutes a nodal region, with the city itself as 'core.' Indeed, except possibly for political territories (which are very often city-centered), no other kind of nodal region exists. Recognition of nodal regions as a distinct taxonomic type testifies to the importance of cities and their umlands as major structural components of the cultural landscape. Without a study of umlands, it is doubtful that a geographic analysis of the spatial organization of society could be regarded as complete.

In applied geography, umlands can provide a useful areal framework for certain aspects of market research. Suppose, for example, that a manufacturer wishes to market a new product across a wide area. How should the promotional campaign be conducted in order to obtain the highest return for the advertising dollar? One satisfactory answer is to divide the target area into umlands and to allocate to each umland an advertising budget proportional to the umland's total population or total disposable income. The delimitation of umlands can also serve as a basis for analyzing sales figures and for assigning sales territories to agents in the field. For these kinds of purposes, umlands are generally superior to political territories such as states or counties because each umland forms a 'natural' sales region which can be serviced efficiently from the city at its center. With just this type of consideration in mind, Rand McNally has published for many years its *Commercial Atlas and Marketing Guide*, a frequently updated work that includes maps of the umlands of all major American cities (Rand McNally 1984, 52–3).

A network of umlands delimited for a set of high-order cities can be used as a framework for regional planning (Hall 1970, 14–18). This is particularly true for those aspects of regional planning related to the future spatial distribution of population, including the setting aside of land for various uses (residential, industrial, open space, and so forth) and the selection of routes for new transportation facilities. In addition, the close economic linkages between a city and its umland make the city-centered region a suitable unit of organization for implementing policies to reduce regional disparities. Because the umland is organically linked to its central city, benefits accruing to the latter may be expected to affect the entire umland. The city functions as a 'growth center' from which 'spread effects' flow outward to all parts of the surrounding territory (Hansen 1972; Moseley 1974). If any such policy can produce concrete results, the effects will naturally make themselves felt in the city's umland.

A city's umland, though primarily a 'central place' phenomenon – i.e. defined on the basis of retail trade and consumer services – is characterized normally by additional, non-retail town-country interactions. Umlands are connected to their central cities not only by the movements of shoppers but also by journey-to-work traffic, by social and recreational trips, and, if the central city is sufficiently large, by wholesale trade linkages. For high-order cities, moreover, there is often a close association between the retail umland and the region linked to the central city as a source of rural-to-urban migrants. In several important ways, therefore, the concept of an umland is broader in scope than might be inferred from its intimate relationship with the limited – though very important – realm of central place theory.

In some studies, the different types of linkages between a city and its surrounding territory are disaggregated and treated under separate headings. Depending on the criteria employed, each type of activity may give rise to a different sphere of influence on the map (Moisley 1958; Cazalis 1964). The external spatial relationships of individual cities are generally complex when analyzed in detail. Nevertheless, if the town-country linkages of different cities are considered simultaneously, each city on any given hierarchical level is seen to be surrounded by a zone within which its influence is stronger than that of any other comparable center. This zone of relative dominance is the region to which the term *umland* is properly applied, and experience suggests that relative dominance in the performance of central place activities is associated positively with relative dominance in other kinds of interactions. An umland should be conceptualized not as a single, isolated region, but as a component of a space-filling network of areas in which the boundaries of each cell are determined ultimately by the geographical arrangement of the cities of a particular hierarchical order.

Intercity flows and the urban 'pecking order'

A large country such as the United States or Canada may be viewed as a set of discrete nodal regions, each defined as the umland of a particular large city which offers a comprehensive array of consumer goods and services. The fully equipped cities that serve as the economic foci of these nodal regions may be termed 'regional capitals' – though with no necessary political connotation. (Vancouver, for example, is clearly the regional capital, though not the political capital, of British Columbia.) Within the umland of a regional capital, the organization of the urban system is essentially 'vertical': lower-order centers, together with their umlands, 'nest' within the umlands of higher-order centers and depend on the latter for goods and services they cannot provide for themselves. At the national scale, however, the organization of the set of regional capitals is essentially 'horizontal': flows of people, goods, and messages among these leading centers constitute the 'glue' that binds the various regions together and gives the nation its identity as a functioning economic unit. In short, interregional linkages are effected primarily by direct connections among the regional capitals.

Each regional capital represents the highest level of the urban hierarchy within its own region, but regional capitals may differ in population. Regions normally differ in resource endowment, population density, and wealth, and these differences will be reflected in variations in size among regional capitals. These variations, in turn, coupled with the relative locations of regional capitals, account for the fact that there are wide differences in the volumes of traffic connecting different pairs of regional capitals. Broadly speaking, there is a 'gravity' effect: large regional capitals that lie close together are connected by heavy traffic; small ones located far apart are linked by much smaller flows.

An illustration of this variation in the magnitudes of intercity flows is provided by Table 6.5, which shows the levels of domestic airline passenger traffic linking nine arbitrarily selected regional capitals across Canada in 1985 (Canada 1986). Although this table includes only nine cities, it accounts for 46 per cent of all domestic airline passenger trips in Canada in that year. The general effects of city size and distance on the volumes of traffic are readily apparent. Compare, for example, the amount of interaction between Toronto and Montreal – two very large cities situated fairly close together – and that between Halifax and Calgary – two smaller cities situated relatively far apart. As can be seen, the volume of traffic between Toronto and Montreal is many times greater than that between Halifax and Calgary. Other comparisons yield similar results.[6]

Implicit within the intercity flows recorded in Table 6.5 is a series of dominance-and-subordination relationships, analogous to the 'pecking order' found among chickens in a barnyard. In the ideal case, such a pecking order is completely transitive: each city is subordinate, in terms of the structure of intercity flows, to every city with higher ranking. By implication, the pecking order of regional capitals represents an evaluation of the relative importance of the various regions in sustaining the identity of the national economy.

In order to determine the pecking order for the cities of Table 6.5, we proceed as follows. First, we express each value in the table as a percentage of the total outflow from the city in question. These percentages appear in Table 6.6 and sum to 100 horizontally (because they are

6 It is possible to analyze a flow matrix of the type shown in Table 6.5 by means of a multiple regression model in which the dependent variable is the volume of pairwise interaction and the independent variables are city size and intercity distance. Such a model is actually a gravity model cast in a different form. For a discussion of some theoretical problems associated with the gravity model in this context, see Gould (1985, 115–23).

TABLE 6.5

Numbers of airline passengers traveling between selected Canadian cities (domestic flights, 1985)

From:	To: StJ	Hal	Que	Mon	Tor	Win	Edm	Cal	Van	Totals
St John's	–	48,820	820	18,660	51,300	2,550	4,340	4,910	3,870	135,270
Halifax	47,880	–	3,510	58,810	130,090	9,080	11,560	14,510	13,610	289,050
Quebec	900	4,050	–	42,350	48,680	3,290	2,720	3,290	4,530	109,810
Montreal	17,730	59,070	43,160	–	598,500	40,590	33,130	51,390	73,820	917,390
Toronto	48,960	129,850	47,770	599,090	–	161,020	133,760	198,010	226,030	1,544,490
Winnipeg	2,250	9,600	3,130	41,280	161,280	–	46,240	60,480	85,440	409,700
Edmonton	3,840	11,050	1,990	34,210	136,450	45,870	–	167,330	173,720	574,460
Calgary	4,420	14,790	2,920	52,050	201,510	59,180	183,850	–	215,650	734,370
Vancouver	3,170	12,460	3,590	72,980	232,170	87,490	175,200	214,310	–	801,370
Totals	129,150	289,690	106,890	919,430	1,559,980	409,070	590,800	714,230	796,670	5,515,910

SOURCE: *Air Passenger Origin and Destination: Domestic Report,* Catalogue No. 51–204 (Ottawa: Transportation Division, Statistics Canada 1986)

TABLE 6.6
Airline passenger movements as percentages of total outflow from each origin

From:	To: StJ	Hal	Que	Mon	Tor	Win	Edm	Cal	Van	Totals
St John's	–	36.09	0.61	13.79	37.92	1.89	3.21	3.63	2.86	100.0
Halifax	16.56	–	1.21	20.35	45.01	3.14	4.00	5.02	4.71	100.0
Quebec	0.82	3.69	–	38.57	44.33	3.00	2.48	3.00	4.13	100.0
Montreal	1.93	6.44	4.70	–	65.24	4.42	3.61	5.60	8.05	100.0
Toronto	3.17	8.41	3.09	38.79	–	10.43	8.66	12.82	14.63	100.0
Winnipeg	0.55	2.34	0.76	10.08	39.37	–	11.29	14.76	20.85	100.0
Edmonton	0.67	1.92	0.35	5.96	23.75	7.98	–	29.13	30.24	100.0
Calgary	0.60	2.01	0.40	7.09	27.44	8.06	25.04	–	29.37	100.0
Vancouver	0.40	1.55	0.45	9.11	28.97	10.92	21.86	26.74	–	100.0

SOURCE: Calculated from data in Table 6.5

percentages of outflow), but not vertically. Thus, for example, 38.57 per cent of all passengers flying from Quebec to the eight destinations included in the table terminate their journeys in Montreal. Similarly, 44.33 per cent of Quebec's outflow goes to Toronto – the largest of all outflows from Quebec. Table 6.6, in short, is designed to be read row-wise.

Next, we select a particular pair of cities and note the extent to which each member of the pair, viewed as an origin, relies on the other member as a destination. Let us compare Winnipeg and Edmonton in this way. Table 6.6 shows that 11.29 per cent of Winnipeg's outflow goes to Edmonton, whereas only 7.98 per cent of Edmonton's outflow goes to Winnipeg. Thus, Winnipeg relies on Edmonton more than Edmonton relies on Winnipeg. Following the convention introduced by Rouget (1972), we may say that Edmonton is dominant and Winnipeg is subordinate. In other words, Edmonton ranks above Winnipeg in the urban pecking order.

Table 6.7 summarizes the results obtained by thus comparing all possible pairs of cities in the analysis. (here, thirty-six different pairs). Like Table 6.6, Table 6.7 must be read row-wise. The number 1 in a cell signifies that the city named at the left-hand end of the row is dominant over the city named at the top of the corresponding column. Zeros indicate subordination. Thus, St John's is subordinate to Halifax, dominant over Quebec, subordinate to Montreal, and so forth.

The totals appearing at the right-hand side of Table 6.7 indicate the

TABLE 6.7
Dominance-and-subordination matrix derived from Table 6.6

	Stj	Hal	Que	Mon	Tor	Win	Edm	Cal	Van	Totals
St John's	–	0	1	0	0	0	0	0	0	1
Halifax	1	–	1	0	0	0	0	0	0	2
Quebec	0	0	–	0	0	0	0	0	0	0
Montreal	1	1	1	–	0	1	1	1	1	7
Toronto	1	1	1	1	–	1	1	1	1	8
Winnipeg	1	1	1	0	0	–	0	0	0	3
Edmonton	1	1	1	0	0	1	–	0	0	4
Calgary	1	1	1	0	0	1	1	–	0	5
Vancouver	1	1	1	0	0	1	1	1	–	6

SOURCE: Calculated from Table 6.6. See text for explanation.

number of centers dominated by each of the nine cities. Thus, Toronto dominates eight other places, Montreal seven, Vancouver six, and so forth. These totals, in effect, establish the pecking order for this particular set of cities: the higher the row total, the higher a city stands in the pecking order. The final ranked list is shown in Table 6.8.

As it happens, the pecking order for these nine Canadian regional capitals turns out to be completely transitive: each city dominates every city ranked lower and is subordinate to every city ranked higher. This is not an *inevitable* result. It is possible to encounter a situation in which city *A* dominates city *B* and city *B* dominates city *C*, but with city *A* subordinate to city *C*. Such a case of intransitivity is known as a circular triad. The presence of circular triads can be detected by inspecting the row totals in Table 6.7. If any circular triads are present, two or more of these totals will be identical. If many circular triads are present, the cities probably cannot validly be ranked: we would be dealing with equal partners rather than a pecking order. Probabilistic methods for evaluating the significance of circular triads have been developed by Kendall (1962, 144–54).

In the present case, not a single circular triad is found: hence the pecking order presented in Table 6.8 validly describes the interrelationships of the nine cities. But the pecking order is not identical to a ranking of the cities by population. There are two cases in point (Table 6.8). First, Calgary outranks Edmonton in the pecking order, though it is smaller than Edmonton in size. Second, Quebec ranks seventh in size but ninth in the pecking order, being outranked by both Halifax and St John's. Clearly the relative national importance of individual regional capitals is not neces-

TABLE 6.8
Relative dominance ('pecking order') of selected Canadian cities according to structure of intercity flows

Rank in pecking order	City	Population (1986)	Rank by population within this set of cities
1	Toronto	3,427,168	1
2	Montreal	2,921,357	2
3	Vancouver	1,380,729	3
4	Calgary	671,326	5
5	Edmonton	785,465	4
6	Winnipeg	625,304	6
7	Halifax	295,990	8
8	St John's	161,901	9
9	Quebec	603,267	7

SOURCES: Pecking order from Table 6.7; metropolitan populations from *1986 Census of Canada*

sarily a simple function of city size. Other pertinent factors include presumably relative location, regional resource endowments, and the pattern of corporate linkages.

In conclusion, the concept of the urban pecking order is by no means incompatible with that of hierarchical organization. In fact, the two concepts are complementary, the principal distinction being one of scale. Hierarchical organization refers to the structure of the urban system within individual regions of limited extent; the pecking order refers to interrelationships among the regional capitals across the nation as a whole. Each regional capital is simultaneously the summit of a local hierarchy of towns and the point of articulation through which this local structure becomes integrated into the national urban system.

Summary

The term *centrality* is used to denote the extent to which a town acts as a central place. A distinction is made between centrality and nodality. The former is a measure of the town's importance to external customers; the latter, of the total importance of the town as a shopping and service center.

A crude measure of centrality is provided by the size of a town's population. However, both centrality and nodality are measured most effectively by collecting data that show the number of functional units of each central function present in each central place. These data can be converted to a composite measure known as the Davies functional index. This index measures the functional complexity of central places more realistically than counts of either functions or functional units alone.

In its original form, the Davies functional index is a measure of nodality rather than centrality. However, it can be transformed into a measure of centrality by a simple modification based on the ratio between the total nodality and the total population of the region under investigation.

The umland of a central place, on a specified hierarchical level, is the area within which the influence of the central place is stronger than that of any other center operating on the same hierarchical level. The delimitation of umlands requires at least partial classification of towns into hierarchical orders, because each central place above the hamlet level is surrounded by a nested series of umlands: there is one umland for each hierarchical level on which the center performs. Therefore, in delimiting an umland for a center, it is necessary first to specify the hierarchical level to which this umland belongs. The relevant competing centers are then identified as places that function on, or above, the hierarchical level in question.

Three techniques for the delimitation of umlands were reviewed: the field survey approach, analysis of newspaper circulation, and the use of a formula derived from the Reilly gravity model. The field survey gives the most trustworthy results; the Reilly method is the easiest to apply. The delimitation of umlands has practical value for certain aspects of market research and regional planning.

The final section of the chapter demonstrated how data on airline passenger traffic can be used to investigate relationships of dominance and subordination among a country's leading cities. For Canada, it was shown that regional capitals are characterized by a clearly defined pecking order, which is not perfectly correlated with city size.

7

The problem of hierarchical structuring

The fundamental objective of central place research is to explain the size and spacing of real towns and cities by constructing a hierarchical classification of these settlements and by relating this classification to the spatial patterns displayed in the Christallerian theoretical models. The approach to explanation involves two conceptually distinct stages. First, one constructs a hierarchical classification. Second, one relates this classification to central place theory. This chapter is concerned with demonstrating how these two tasks may be accomplished.

Central place theory is not a general theory of the size and relative location of towns. It is a partial theory, concerned only with the distribution of consumer goods and services at the retail level. It explicitly excludes, by the nature of its assumptions, the urbanizing effects of long-distance trade, manufacturing, and the various activities identified earlier as 'exceptional' functions (i.e. mineral extraction, resort activity, and so forth – see chapter 3). Accordingly, the value of central place theory as a conceptual framework for understanding the size and spacing of towns is revealed most clearly in areas where nonretail activities (i.e. mining, manufacturing, and so forth) are absent or only weakly represented. Central place theory is most fruitful for rural areas relatively remote from major centers of industrial activity. The ideas discussed in this chapter should be construed as applying primarily to areas of this kind.

Case studies of central place networks comprise the largest single subarea within the entire urban systems literature. The earliest significant study of this type – a paper on East Anglia by Dickinson (1932) – was published the year before Christaller's work appeared. Christaller's own contribution on Bavarian towns must also be mentioned (Christaller 1933), and the theme was later taken up by Smailes, Bracey, and Carter in

England and by Brush in the United States (Smailes 1944, 1947; Bracey 1952, 1953, 1956, 1962; Brush 1953; Brush and Bracey 1955; Carter 1955, 1956). In the late 1950s, central place research entered its 'golden age,' a period of high productivity that lasted approximately from 1958 until 1970. Highlights included a series of works by Berry and his associates (Berry and Garrison 1958a, 1958b; Berry, Barnum, and Tennant 1962; Berry et al 1965; Berry 1967), together with other important studies by Coppolani (1959), Borchert and Adams (1963), Carter and Davies (1963), Scott (1964), Hodge (1965), Prost (1965), and Barnum (1966).

After 1970 the pace of research slackened, but valuable contributions have since been made by, among others, Lewis (1970, 1975), Preston (1971a, 1975, 1979, 1983), Rowley (1971), Stabler and Williams (1973), Dahms (1975, 1981), Garner (1975), Ackerman (1978), Evenden (1980), and Sarbit and Greer-Wootten (1980). An interesting recent development has been the use of central place concepts in anthropology and archeology. The original stimulus for anthropologists was a very important series of papers published at the height of the 'golden age' by Skinner (1964, 1965a, 1965b); relevant later works include those of Crissman (1976), Kelley (1976), C.A. Smith (1976), and Skinner himself (1976, 1977). In archeology the contributions of G.A. Johnson (1972), Marcus (1973), and Collis (1981) are noteworthy.

In employing central place concepts to analyze the urban pattern of any rural area, two fundamental points must be kept in mind. First, no matter how bucolic the area may appear, it is unlikely that noncentral functions will be completely absent from the local urban economy. Here and there a few small manufacturing plants may have sprung up, or perhaps the region contains a military base or scattered mines and quarries, or an occasional resort. Although these elements are not included in central place theory, they cannot simply be ignored, and their presence must be taken into account in a comprehensive analysis of the observed urban system. In short, even in areas where central place activity is overwhelmingly the principal reason for the existence of urban centers, certain aspects of the observed pattern may reflect the presence of other kinds of urban functions.

Second, the real world is never perfectly isotropic, except perhaps over a few very small areas. Even regions normally described as 'plains' are not utterly devoid of physiographic variation. Diversity is provided, for example, by river valleys and by the effects of differential erosion and deposition. (Fluvial geomorphologists are just as interested in plains as in hills and mountains.) In northern regions, moreover, continental glaciation has left its mark in the form of ground moraine, spillways, drumlins, and a variety of other features. From a farmer's point of view the quality of the land is never completely uniform over large areas, and differences in land capability are naturally reflected in place-to-place variation in the density of the rural population. Consequently, the spatial pattern of purchasing power faced by central place entrepreneurs in the real world is at best only a rough approximation of the absolutely uniform demand surface assumed in theory.

Taken together, the above two points imply that we should not expect the real world to display the mathematical regularity of theoretical models. One possible criticism of central place theory is that the theory must be false because it does not conform exactly to our observations of reality. Although this criticism raises a number of philosophical issues that are both important and profound, it misses the point. A match between theory and reality should be expected only to the extent that reality fulfills the assumptions of the theory. Moreover, in situations where the real world does not fulfill the theoretical assumptions, the observed pattern should differ from the theoretical models in predictable ways. For example, the presence of a manufacturing plant in a particular town within an otherwise 'pure' central place system should make that town larger, not smaller, than other centers of the same hierarchical rank. If predictions of this sort are consistently borne out by the facts, the discrepancies between theory and reality actually *support* the theory rather than contradict it. Failure to appreciate this important principle has led to misunderstanding, and, as a result, the explanatory power of central place theory has sometimes been underestimated.

Hierarchy or continuum? – a problem of interpretation

In the simplified world portrayed in central place theory, all central places belonging to a particular hierarchical order are identical in the number and variety of central goods and services they provide. It is assumed, moreover, that the amount of revenue accruing to each central place over a given period of time is directly proportional to the physical size of the corresponding market area (assumption 5 in the list discussed at the beginning of chapter 5, above). It is generally held to follow from this assumption that all central places belonging to a particular hierarchical order should be identical in population size, or very nearly so. Because all market areas on a given level of the hierarchy are identical in size and in population density, they generate equal amounts of revenue, which can be construed as supporting equal numbers of storekeepers and their dependents within the central places. Accordingly, the criterion of discrete stratification, described earlier in terms of the bundles of goods and services provided by the various centers, should apply with equal force to counts of population. In short, the hierarchical structure of the urban system should reveal itself as a series of distinct 'plateaus' of population size.

Between one plateau and the next, what sort of increase in population size have we a right to expect? We can give a partial answer by relating the increases in population to the value of the parameter K in the Christallerian models: the ratio between the sizes of market areas on any two adjacent levels in the hierarchy. If populations are proportional to revenues, and revenues, in turn, to the sizes of market areas, K also identifies the relative population sizes of central places on successive hierarchical levels. For example, suppose we are dealing with a system structured in accordance with the $K = 4$ model. The market areas in this model increase in size by a factor of 4 as we ascend the hierarchy, and we may therefore expect the same factor to apply to the populations of the central places. If centers on the lowest level contain 100 inhabitants, those on successively higher levels should have populations of 400, 1,600, 6,400, and so forth.

Actually the situation is not quite that simple. Strictly speaking, the population of a central place is governed not by the physical size of its market area but by the number of consumers that the market area contains. Central places on the very lowest level of a hierarchy serve only rural consumers. However, central places belonging to all hierarchical orders except the lowest serve not only rural consumers but also the populations of lower-order central places. Consider, for example, a $K = 3$ hierarchy with three levels of centers, termed hamlets, villages, and towns, respectively. The market area of a town-level center is exactly nine times as large as the market area of a hamlet in terms of area, but the town serves more than nine times as many customers as the hamlet because the town's market area includes not only farmers but also the populations of some villages and hamlets. (To be specific, each town-level center in the $K = 3$ model serves the equivalent of two complete villages and six hamlets.) Thus, in order to calculate the expected population size of a town-level center, we need to know several things: namely, (a) the physical size of hamlet-level market areas; (b) the density of the rural population; (c) the population of the hamlets; (d) the population of the villages; and (e) the ratio of the population of a town-level center to the total number of customers served by each town. This latter ratio, moreover – as shown by the algebraic model discussed in chapter 5 – is not necessarily the same for every level of the hierarchy, so it cannot simply be derived from the data pertaining to the hamlets. As can be seen, the calculation of the expected population sizes of central places can be quite a complicated business! For our present purposes, however, we may ignore the above complexities and accept the straightforward assumption that central place populations are directly proportional to the sizes of the relevant market areas.

In central place models based on hexagonal market areas, the smallest possible value of K is 3. In models in which the market areas are squares, the smallest possible value of K is 2 (see chapter 5). Accordingly, we should expect the populations of central places to display at least a doubling in size as we ascend from one level of the hierarchy to the next. Moreover, because the ideal market area is a perfect circle, and because hexagons are closer in shape to circles than are squares, patterns involving hexagonal market areas might be more likely than patterns involving squares. In this case, the populations of central places should increase at least threefold from level to level. Generalizing, we may say that we expect places on any given hierarchical level to be 'at least two or three times as large' as places on the level immediately below. This will be our rule of thumb in the discussion that follows.

It is usually difficult to discern any clearly marked 'plateaus' of size in lists of real-world urban populations. A general consideration of population data for any sizable area such as an American state or a Canadian province normally suggests a continuum of population size, with no major gaps or discontinuities. Thus the real world does not seem to exhibit the 'staircase effect' that hierarchical structuring would lead us to expect. In order to bring this issue into sharper focus, let us consider a specific example in some detail.

The type of evidence usually advanced to support the idea of a continuum of urban size is known as a rank-size graph. The preparation of a rank-size graph involves two steps. First, all the urban centers within a selected region are ranked in descending order according to population. Second, the variable 'population size' is plotted against the variable 'rank' on special graph paper in which both axes are scaled logarithmically. The use of logarithmic scales should not be a cause for alarm; this is merely a standard way of ensuring that cities of widely different sizes can be shown on a single graph without congestion. By convention, population size is plotted on the vertical axis and rank on the horizontal. The

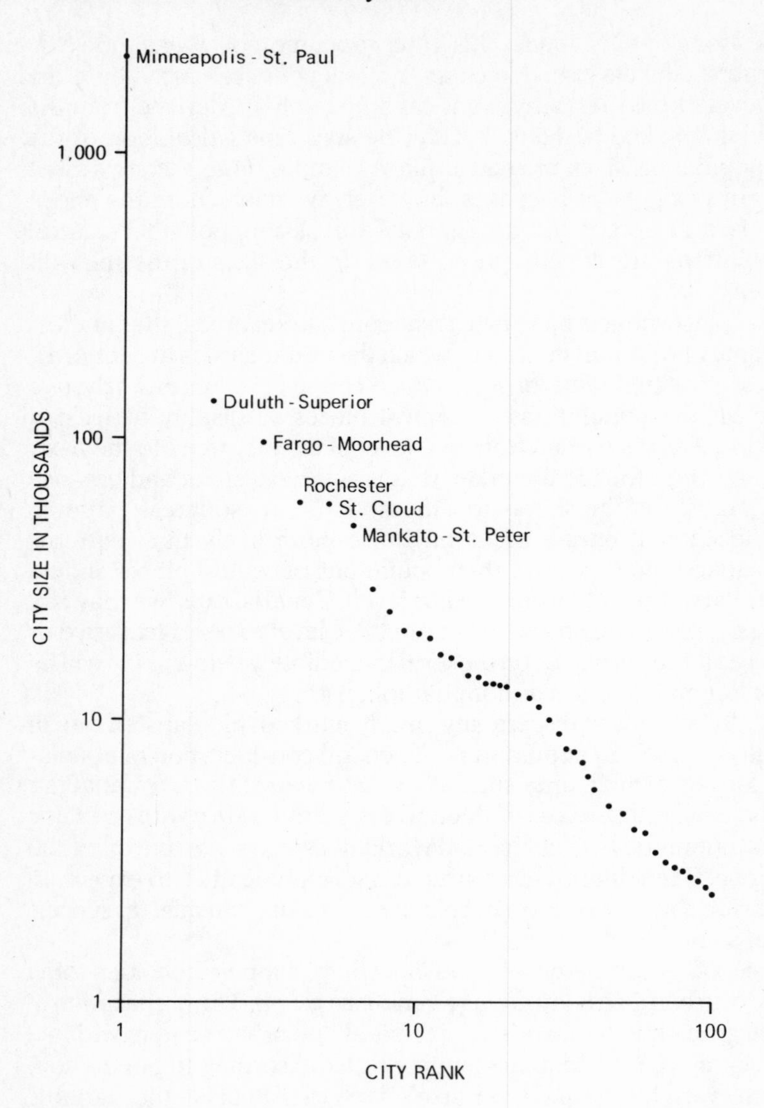

Figure 7.1
The rank-size continuum of city size in southern and central Minnesota, 1980 (logarithmic scales). Individual cities are plotted from number 1 to 20; then every second city from 20 to 40; then every fifth city down to 100. Population data are taken from the 1980 US Census.

plotted points trace out a line that slopes downward to the right, usually at an angle of about forty-five degrees. An example appears in Figure 7.1, which is based on the 100 largest urban centers (using 1980 populations) in the state of Minnesota south of a line extending from Fargo-Moorhead to Duluth-Superior.

The Minnesota rank-size graph does display one rather definite break in size: namely, the break that separates the Minneapolis–St Paul metropolitan area from all other centers. Within the context of southern and central Minnesota, the Twin Cities metropolitan area is a 'primate city' – outstandingly dominant in terms of both size and economic influence within the region that it serves. In its role as a central place, Minneapolis–St Paul is unquestionably superior to all other places included on the graph, and we may reasonably conclude that the gap between Minneapolis–St Paul and the second-ranked center (Duluth-Superior) represents a break between two distinct hierarchical orders.

Below this point, however, further evidence of hierarchical structuring appears to be absent. The practiced eye might detect a hint of a break between center number 6 (Mankato–St Peter) and center number 7 (Winona), but the overall impression is one of a smoothly descending continuum rather than a series of well-defined plateaus. As a matter of fact, almost all rank-size graphs convey this general impression of continuity, and this has led one writer to declare categorically that hierarchical orders do not exist: 'Like pool, pond, and lake, the terms hamlet, village, and town are convenient modes of expression; but they do not refer to structurally distinct natural entities' (Vining 1955, 169). On the surface, at least, the smooth gradations of size displayed by rank-size graphs appear incompatible with the idea of a discretely stratified hierarchy of towns.[1]

We can begin to move toward a resolution of this dilemma by examining the actual geographical situations occupied by specific urban centers. As a point of departure, let us consider the pattern of urban centers in the territory surrounding the Minneapolis–St Paul metropolitan area. By carefully inspecting suitable maps, and by referring constantly to data on the populations of towns, we find that the Twin Cities metropolis is surrounded by the following five middle-sized cities:

1 Rank-size graphs are important in studies of interregional variation in the frequency distribution of city size. This latter topic is discussed in detail in chapter 11, which also looks at the concept of the primate city. Almost all rank-size graphs appear to support the idea of a continuum rather than that of a hierarchy.

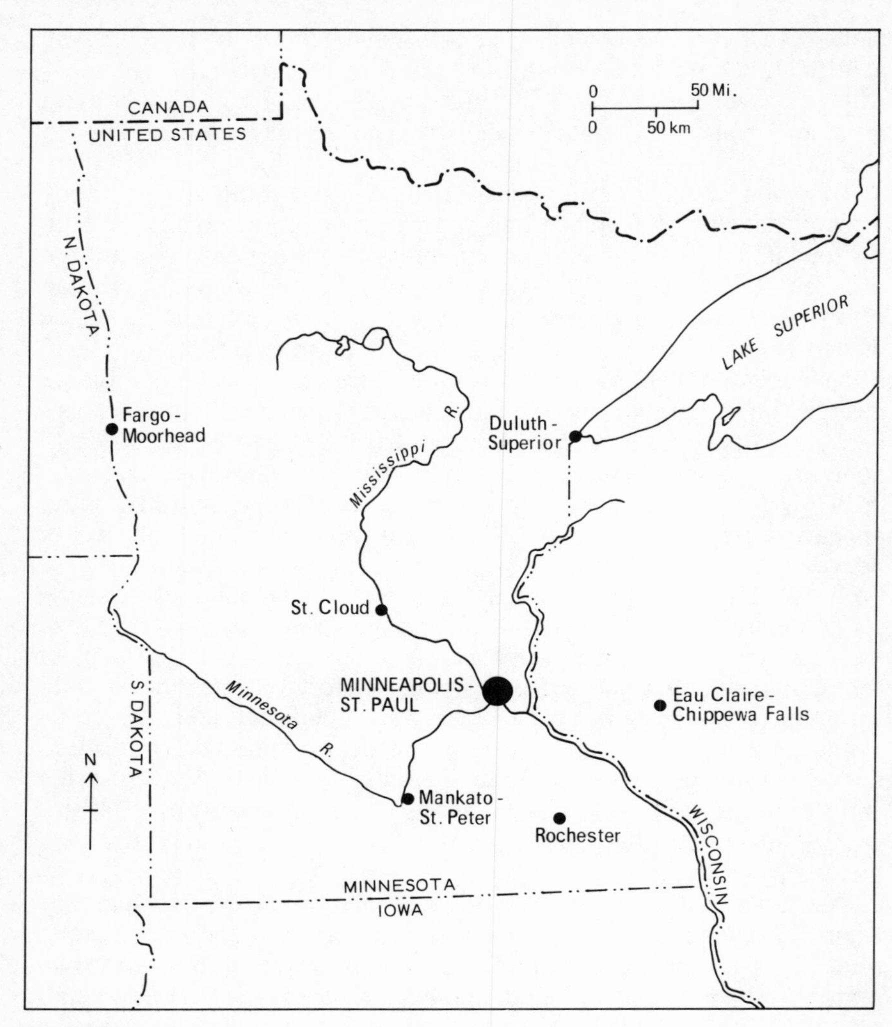

Figure 7.2
Large central places near Minneapolis–St Paul, Minnesota

(a) to the east, in Wisconsin, the twin center of Eau Claire–Chippewa Falls; (b) to the southeast, Rochester; (c) to the southwest, Mankato, together with its nearby satellite of St Peter; (d) to the northwest, St Cloud; and (e) to the north, Duluth-Superior (Figure 7.2). These five centers possess four important characteristics in common.

TABLE 7.1
Some hierarchical relationships in the region surrounding Minneapolis–St Paul

Urban centers	Population (1980)	Distance from higher-order center (miles)
Dominant regional metropolis		
Minneapolis–St Paul	2,114,000	
Centers surrounding Minneapolis–St Paul		
Duluth–Superior	132,321	135
Eau Claire–Chippewa Falls	67,747	88
Rochester	57,855	76
St Cloud	55,282	59
Mankato–St Peter	46,852	67
Average (5 centers)	72,011	85
Centers surrounding Mankato–St Peter		
Albert Lea	19,190	48
Owatonna	18,632	39
Faribault	16,241	37
New Ulm	13,755	24
Fairmont	11,506	42
Hutchinson	9,244	53
Average (6 centers)	14,761	41
Centers surrounding New Ulm		
St James	4,346	24
Sleepy Eye	3,581	13
Madelia	2,130	18
Lake Crystal	2,078	19
Gaylord	1,933	21
Fairfax	1,405	20
Winthrop	1,376	16
Average (7 centers)	2,407	19

SOURCES: Population counts from 1980 US Census; distances are straight-line mileages between central business districts of the cities in question.

First, they are broadly similar in size. As Table 7.1 shows, the southernmost four all have between 46,000 and 68,000 inhabitants (1980 data). Duluth-Superior is noticeably larger, at 132,000, but still in the same class as the others when compared with Minneapolis–St Paul, at 2.1 million.

Second, they are located at similar distances from the Twin Cities metropolitan area. Measured in terms of straight-line distances, four of the five lie between 59 and 88 miles from downtown Minneapolis (Table 7.1). Duluth-Superior, at 135 miles, is once again the exception.

Third, they are quite evenly spaced around Minneapolis–St Paul in terms of compass directions. The smallest angle subtended at Minneapolis by any pair of neighboring centers is 54 degrees (by Rochester and Eau Claire); the largest, 99 degrees (by Mankato and St Cloud). The 'expected' value, based on the assumption of equal angles around the common focus at Minneapolis, is 72 degrees (i.e. 360 divided by 5).

Finally, each of the five cities is the largest center within its own local area. The 'local area' of each center need not be precisely defined; but, in rough terms, it is the area within a circle having its radius equal to one-half the distance to the nearest higher-order city – that is, one-half the distance to Minneapolis–St Paul. Rochester, for example, lies 76 miles from the Twin Cities, and its 'local area' is therefore defined as the area within 38 miles of Rochester. The largest town within this radius is Austin, which, with only 23,020 inhabitants, is not even half as large as Rochester. The local preeminence of the other four cities under discussion is even more clearly marked than that of Rochester.

The four properties of size and spacing just described are strongly reminiscent of the patterns displayed in the classic Christallerian models – see especially Figures 5.3 and 5.5 in chapter 5, above. Theoretically, of course, Minneapolis–St Paul should be surrounded by a ring of six lower-order centers rather than five. However, the lack of a major city to the northeast, between Duluth-Superior and Eau Claire, can be interpreted as a logical response to relatively low consumer demand in this area, which is part of Wisconsin's thinly populated 'cut-over country.' Theoretically, too, the Twin Cities metropolitan area should not be so overwhelmingly dominant in size; but non-retail functions such as manufacturing, wholesaling, and state government play vastly greater roles in Minneapolis–St Paul than in any of the other centers. Theoretically, again, Duluth-Superior should not be conspicuously larger than the other lower-order centers; indeed, if anything, it should be smaller, because northern Minnesota in general is even more thinly populated than northern Wisconsin. But again, Duluth has significant noncentral functions as an ore-exporting port and steel-producing center, functions not found in the more southerly centers. And finally, theory commands that the lower-order centers should be equidistant from the metropolis; but reality responds by permitting each city to exploit the unique advantages of particular locations: Mankato at the natural focus of routes created by the 'elbow' of the Minnesota River; Duluth at the head of Lake Superior; St Cloud at the point where the old trail to the northwest left the valley of the Mississippi River, and so forth. The more one ponders the possible effects of all conceivably relevant details, the more one becomes convinced that the observed urban structure, though obviously adapted to real-world conditions, is fundamentally Christallerian and, accordingly, hierarchical.

But this is only the beginning. We can extend our analysis by shifting our attention outward from Minneapolis–St Paul to one of the five surrounding middle-sized cities in order to examine the size and placement of smaller towns within this center's area of influence. For example, let us consider the area surrounding Mankato–St Peter. Now if Mankato, like Minneapolis, is surrounded by a ring of smaller towns that form a distinct hierarchical order, these towns should be closer to Mankato than the distance from Mankato to Minneapolis. In all theoretical models of hierarchical structuring, the distance between adjacent members of any particular hierarchical order is significantly greater than the distance between adjacent members of the order immediately below. If we wish, we can be mathematically precise about these distances. For example, the distances increase in the $K = 3$ model by the square root of 3 as we ascend the hierarchy, and in the $K = 4$ model, by a factor of 2. We should not expect real urban patterns to be mathematically precise; but at least we have a good theoretical basis for assuming that lower-order centers surrounding Mankato should be separated from Mankato by distances appreciably shorter than the sixty-seven miles separating Mankato from the Twin Cities metropolitan area.

The question of population size can also be brought into play. Our earlier remarks on this theme suggest that towns belonging to the order immediately below that of Mankato should contain approximately 15,000 to 23,000 inhabitants – that is, one-third to one-half of the population of Mankato itself (including North Mankato and the satellite community of St Peter). These figures, of course, represent nothing more than a general guide. The enormous size of Minneapolis–St Paul in relation to the five cities on the Mankato level shows very clearly how the presence of non-retail functions can alter the expected ratios. However, we are now concerned with relatively small towns in which massive concentrations of manufacturing and other noncentral activities are unlikely. The smaller the towns, the greater the probability that they serve almost entirely as

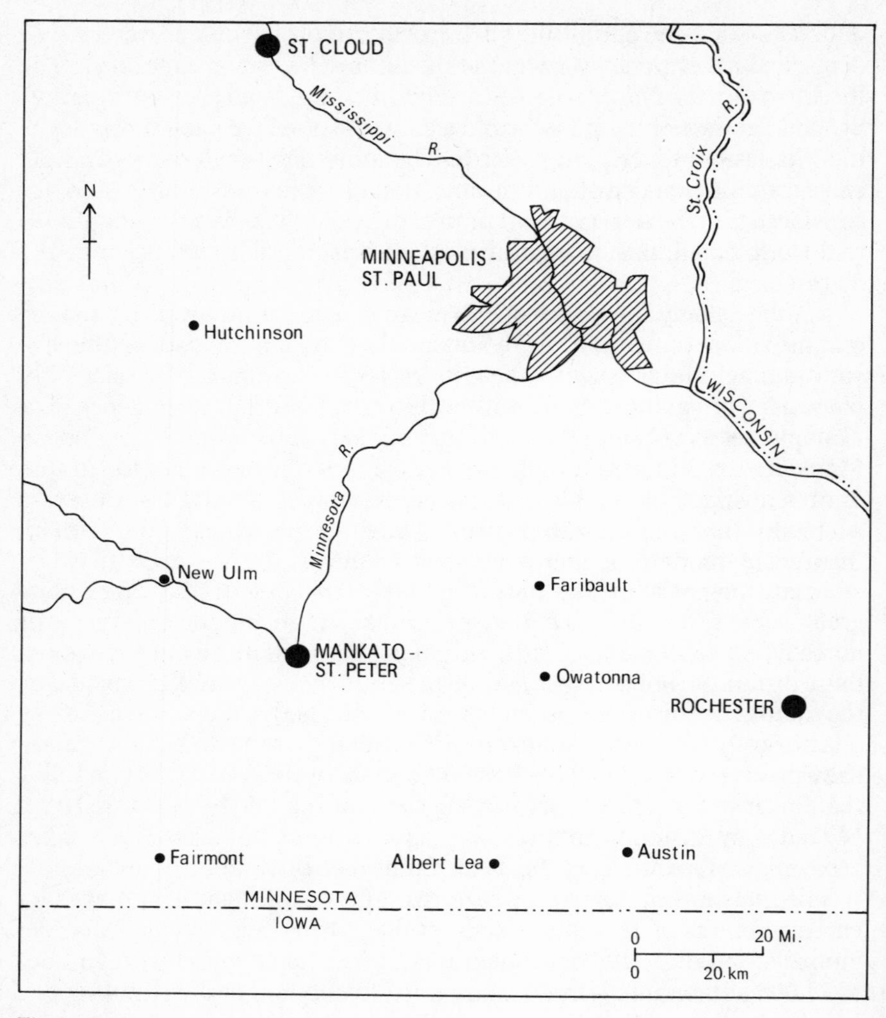

Figure 7.3
Intermediate-level central places near Mankato–St Peter, Minnesota

central places, and hence the greater the likelihood that their population sizes follow the theoretically expected pattern.

As it happens, Mankato is surrounded by six towns for which the relevant statistics on size and distance correspond remarkably well to the anticipated values. The towns in question are Faribault, Owatonna,

Albert Lea, Fairmont, New Ulm, and Hutchinson. Figure 7.3 and Table 7.1 provide information concerning the location, size, and spacing of these six towns. These centers collectively display the same set of general properties as the ring of cities surrounding Minneapolis–St Paul.

First, they are similar to one another in size (Table 7.1). Moreover, the largest three are situated east of Mankato in a region characterized by higher rural population density than is found in areas farther west.

Second, they are located at similar distances from their 'mother city,' Mankato. The distance to New Ulm is unusually short, perhaps because New Ulm is situated at the confluence of the Minnesota River with one of its largest tributaries, the Cottonwood. As Table 7.1 shows, the average distance from Mankato to the six surrounding towns is almost exactly one-half of the average distance from Minneapolis to the five surrounding cities on the Mankato level.

Third, they are generally well spaced around Mankato in terms of compass directions. An exception occurs in the case of the small angle subtended at Mankato by Faribault and Owatonna; but the angular distribution of the remaining centers is satisfactory.

Fourth, with two exceptions, each of these six towns is the largest center within its own 'local area,' the latter now defined as the area within a radius equal to one-half of the distance from the town in question to Mankato. The two exceptions deserve special attention for the light they cast on the limitations of our working definition of the term *local area*.

The first exception is Albert Lea. This town lies 48 miles from Mankato, and its local area is therefore assumed to have a radius of 24 miles. Within this distance, 20 miles to the east of Albert Lea, lies the town of Austin. And Austin, with a population of 23,020, contains some 3,800 more inhabitants than Albert Lea. In a similar fashion, Faribault's local area encompasses the town of Owatonna, which contains 2,400 more inhabitants than Faribault itself (Table 7.1). Thus Albert Lea and Faribault are not the largest centers within their own 'local areas.' However, it is apparent from the data on population size that Albert Lea, Austin, Faribault, and Owatonna all belong to the same hierarchical level. By definition, therefore, these places stand on an equal competitive footing with one another and may be assumed to serve 'local areas' that are mutually exclusive, even if this means that these areas cannot be circular in shape. Thus the concept of 'local area' is not entirely lacking in subtlety, and our working definition must always be used with caution.

Often in the real world a town does not lie at the geometrical center of the area it dominates; and this area, in any case, is most unlikely to be a perfect circle. Owatonna, for example, may be expected to serve an area that extends relatively far toward the south, in the direction of Albert Lea, but only a short distance toward the north, where it is truncated by competition from Faribault (Figure 7.3). In short, the shape of each 'local area' is determined in practice by the locations and commercial strength of nearby competing towns. Perfect circularity should not be expected; skewness of shape is the norm.[2]

Further extensions of this analysis appear to be justified. Around New Ulm, for example, is found a distinctive ring of seven smaller towns that lie between 13 and 24 miles from New Ulm and contain an average of 2,400 inhabitants. Included in this group are Sleepy Eye, Winthrop, and Madelia (Table 7.1). Below this level there appear to be two additional orders, though it would be necessary to go beyond population size and consider data on consumer shopping habits and on the variety of goods offered in each center before a firm conclusion could be drawn.

Similar results can be obtained not only for other major centers in the present study area, such as St Cloud, Rochester, and Eau Claire, but for other study areas as well. The message is clear: whenever urban centers are analyzed in their actual spatial setting, hierarchical relationships of the kind postulated in central place theory are readily identifiable.

The nature of these hierarchical relationships can be clarified by considering the set of shopping opportunities presented by the urban system to a representative consumer. For example, consider a resident of New Ulm. Many of the goods and services required by this consumer can be obtained within New Ulm itself, but for certain items a journey to a larger center will be necessary. In the normal course of events the destination of this journey will be the nearby higher-order center of Mankato. Other cities comparable to Mankato, such as Rochester and St Cloud, are simply too far away to be patronized with great frequency, though they may be visited occasionally for special reasons. If a center larger than Mankato is required, the resident of New Ulm will normally make a trip to Minneapolis. Minneapolis, Mankato, and New Ulm are thus linked into an integrated system by a recurring pattern of consumer travel that affirms their relative status as members of three different hierarchical orders. Many examples of such sequences could be cited:

2 Research has shown, in fact, that both Owatonna and Faribault have precisely the kind of skewed trade area that we have postulated. Owatonna's trade area is skewed toward the south and west, and Faribault's toward the north (Borchert and Adams 1963, p. 8).

Minneapolis-Rochester-Winona, Minneapolis–St Cloud–Willmar, and so forth; each list could obviously be lengthened by the addition of smaller centers. These sequences are examples of what Brookfield and Hart, in their memorable study of the geography of Melanesia, aptly call 'town chains' (Brookfield and Hart 1971, 406–8). By definition, each link in the chain represents a different hierarchical order. In essence, it is by tracing such chains downward from an obviously prominent city (such as Minneapolis–St Paul) that the type of hierarchical structuring illustrated in Table 7.1 can be shown to exist.

We are now ready to return to the problem posed by the rank-size graph shown in Figure 7.1. If we base our characterization of the urban system on the evidence presented in this graph, we are inclined to favor a continuum of city size; but if we emphasize the analysis of specific town chains, we find hierarchical structuring. Can both descriptions be valid? Indeed they can. In order to justify this assertion, two important points need to be made.

First, repeating a point made earlier, the real world does not conform to the assumptions on which central place theory is based. The real world is not perfectly isotropic, and real towns, except for very small ones, rarely function solely as central places. The sharp distinctions that exist in theory between the successive orders of the central place hierarchy become blurred under real-world conditions. For example, suppose that some towns belonging to a particular hierarchical order become 'contaminated' by the addition of varying amounts of manufacturing activity. The 'pure' towns will retain their original population size, whereas the 'contaminated' towns will experience some expansion. The towns receiving the most manufacturing may approach, in size, the cities on the next higher hierarchical level, thereby obscuring the original clear separation of the orders. Another important type of example arises from regional variations in rural population density. Towns of a given order may be relatively small where rural density is low, but distinctly larger – and able to offer a wider variety of goods and services – where rural density is high. Once again the result is a blurring of the steps in the hierarchy.

Second, the preparation of a rank-size graph is an aggregative procedure. Once a study area has been selected, the urban centers are ranked according to size regardless of their relative locations on the ground. The study of town chains, in contrast, is a locationally specific or elemental form of analysis – the elements (i.e. towns and cities) of an urban system are examined one at a time instead of simultaneously (Berry and Barnum 1962). Given the blurring of orders described above, the aggregative procedure followed in preparing a rank-size graph is virtually guaranteed to conceal any hierarchical relationships, particularly because the areas selected are usually large – often entire countries. The larger the area chosen, the greater the probability of encountering subregions that differ in rural population density and the degree to which the towns contain noncentral functions. When all the urban centers of such an area are aggregated into a single ranked list, distinctions among different subregions are automatically suppressed, creating an impression of overall continuity.

Rank-size graphs can play a major role in studies of the frequency distribution of city size (see also chapter 11). In the present context, however, they are 'inadmissible evidence.' They cannot reveal or explain hierarchical structuring because they do not take into account the relative positions of urban centers on the ground (or on the map). With the data thus abstracted from location, questions concerning hierarchical relationships simply cannot be addressed.

We conclude, therefore, that the concept of an urban continuum and the concept of hierarchical structuring are by no means mutually incompatible. They are simply two different things. The former is concerned exclusively with population size; the latter, with size and with spatial relations. An aggregative ranking of towns by size, especially for a large region, will usually produce a more or less smooth continuum. An elemental investigation, in contrast, by taking account of the relative location of each village and town, will normally reveal the steps of a hierarchy.

A Canadian case study

The above discussion of urban centers in the region of Minneapolis–St Paul has been somewhat informal, not providing an exhaustive analysis, but showing that the concept of an urban hierarchy is inherently a concept of spatial organization. Before we can assess the degree of correspondence between a real urban system and the theoretical models, we must have a more detailed empirical example in view. The following case study is designed to fulfill this need.

CHARACTERISTICS OF THE STUDY AREA

The area selected for intensive study lies in a rural section of southern Ontario. Specifically, it consists of Grey and Bruce counties, except for the

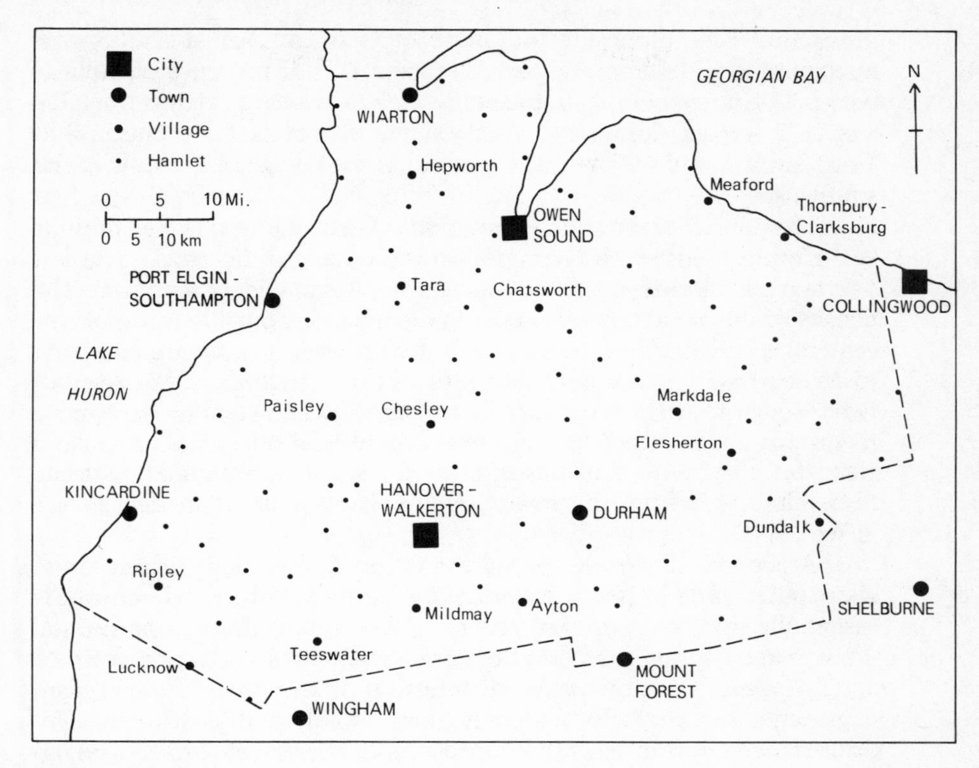

Figure 7.4
The hierarchy of central places in Bruce and Grey counties, Ontario. The method of classification is explained in the text.

four northernmost townships of Bruce County. These latter townships form the Bruce Peninsula, a region of thin soils and sparse population that supports negligible urban development. The study area forms a compact block of territory measuring approximately sixty miles (ninety-six kilometers) from east to west and forty-five miles (seventy-two kilometers) from north to south (Figure 7.4).

The study area possesses four characteristics that make it a suitable region in which to use central place theory as the basic conceptual framework for analyzing the size and spacing of towns.

1. It is an area of subdued relief. The bulk of the region is covered by morainic deposits and exhibits the gently undulating terrain typical of areas that have undergone continental glaciation. The only prominent topographic feature is the Niagara Escarpment, a north-facing erosional

break that runs alongside the shore of Georgian Bay and attains a maximum local relief of 800 feet (244 meters) near the town of Collingwood. The range of relief in the study area as a whole extends from 580 feet (177 meters) above sea level on the shores of Lake Huron and Georgian Bay to 1,700 feet (518 meters) at the village of Dundalk in the southeast.

2. The density of the rural population is fairly uniform. Mixed farming is the norm, with some emphasis on the raising of livestock. The few swampy and stony sections of land are small and widely scattered. The density of the 'rural' population (all persons living outside incorporated centers) is between eleven and twenty-two persons per square mile (four to eight persons per square kilometer) in twenty-three of the region's twenty-eight townships (1971 data). Moreover, all five of the remaining townships are adjacent to large towns, and field observations make it clear that the 'rural' densities of these townships are inflated by urban overspill. Away from the towns, sharp discontinuities in density are absent.

3. A network of closely spaced roads covers the entire region. As in many other parts of North America, the pattern of the road network is essentially that of a gridiron, reflecting the rectangular system of land division used during the period of early settlement. The roads in Bruce and Grey counties are spaced at intervals of 1.25 to 1.50 miles, with alignments running approximately north–south and east–west. Unopened road allowances are rare. Most roads are gravel-surfaced, but no part of the area is more than nine miles (14.5 kilometers) from a paved highway. In general, movement is equally unrestricted in all directions from any location.

4. All the urban centers function primarily, if not exclusively, as central places. Manufacturing is poorly represented and is almost entirely confined to the six or seven largest towns. Mineral extraction is limited to a few local gravel pits and one major site: a dolomite quarry near Wiarton. Some of the towns situated along the shores of Lake Huron and Georgian Bay receive trade from summer cottagers, and there are several ski resorts along the Niagara Escarpment in the northeast, but fulltime employment in the tourist industry is small. There is one significant military base: a tank range north of Meaford. Although these various noncentral activities must not be overlooked, the region's towns and villages are first and foremost service centers for the rural population: urban places containing 1,000 or more inhabitants account for only 46 per cent of the area's total population.

In general, therefore, the study area comes reasonably close to fulfilling the main assumptions on which central place theory is built. The area is by no means perfectly isotropic, but it is broadly homogeneous, both physically and economically. In short, we should expect to find a substantial degree of regularity in the size and spacing of the region's urban centers.

The area as a whole contains 104 central places. These range in size from Owen Sound, a city of almost 20,000 inhabitants, to hamlets containing no more than ten people. Three of the 104 central places are twin centers: namely, Hanover-Walkerton, Port Elgin–Southampton, and Thornbury-Clarksburg (Figure 7.4). The case of Hanover-Walkerton requires a word of explanation. These two towns, containing 5,100 and 4,500 inhabitants, respectively, are separated by four miles of open country; ribbon development along the connecting highway has not yet brought the two built-up areas into physical contact. Moreover, they are situated on opposite sides of the Grey-Bruce county boundary and have long carried on a friendly rivalry based partly on the traditional ethnic distinction between the Scottish settlers of southern Bruce County and the Germans of southern Grey. Nevertheless, they function as a single urban complex in the eyes of the surrounding population. Depending on one's tribal loyalties, each is regarded as a sort of annex of the other. This being the case, it was decided to treat them as a single central place for the purpose of this analysis.

Most of the study area falls within the sphere of influence of Owen Sound. However, the southwestern portion regards Hanover-Walkerton as its natural focus, and the extreme northeast looks to Collingwood (Figure 7.4). At this point it is appropriate to introduce the concept of 'mutual independence.' Two central places are mutually independent if neither center's residents habitually buy goods or services in the other. In the present case, inquiries in the field confirm that Owen Sound, Hanover-Walkerton, and Collingwood, taken in pairs, are mutually independent in this sense. Rather than shopping in Owen Sound or Collingwood, residents of Hanover-Walkerton direct their steps southward to the large cities of Kitchener-Waterloo and (to a lesser extent) London. Shoppers from Collingwood generally go to Barrie; those from Owen Sound go directly to metropolitan Toronto. Also, Owen Sound, Hanover-Walkerton, and Collingwood resemble one another in functional complexity – they each offer essentially the same variety of consumer goods and services. Finally, they are easily the largest centers in or near the study area in terms of population. These findings indicate that Owen

Sound, Hanover-Walkerton, and Collingwood are members of the same hierarchical order and that the study area, which is defined by county and township boundaries, overlaps the tributary regions of all three centers.[3]

THE IDENTIFICATION OF HIERARCHICAL ORDERS

Using both fieldwork and published directories as sources, an inventory of all functional units in the study area's 104 central places was prepared. (Data were collected from 1972 to 1974.) The area was found to contain a grand total of 2,478 functional units representing eighty-one different central functions. On average, therefore, there are twenty-four functional units per central place; but of course, the actual numbers vary widely. At one extreme, there are thirteen centers with only a single functional unit each; at the other, there are 454 functional units in Owen Sound. The frequency distribution of functional units occurring in the central places displays marked positive skewness, with only twenty-one of the 104 centers possessing more than the 'expected' number of twenty-four functional units each. In other words, there are many small centers and only a few large ones. This in itself constitutes prima facie evidence for hierarchical structuring.

The data in the functional inventory were converted to centrality scores by the method described in chapter 6 (modified Davies index). The results are summarized in Table 7.2 for the twenty-five places having the largest centrality scores. All the centers not shown in Table 7.2, as well as some that do appear, belong to the very lowest hierarchical order. The seventy-nine centers not shown have centrality values that range downward from +5 to −17. Approximately forty-five of the 104 central places in the study area have negative centrality scores.[4]

3 At one time I felt that the boundary of a study area in central place research should not be drawn, as here, along arbitrary political lines but should follow as closely as possible the limits of the tributary area of a single, preselected city (Marshall 1969, 68–79). Now I prefer a different view. No matter how the study area is defined, centers that lie outside the area may be significant, as witness Collingwood in the present study. On balance, I now favor a 'seamless web' over a series of self-contained 'functional wholes.'

4 Because calculation of centrality scores requires data on population size, accurate values cannot be obtained if population figures are not available. The Canadian census provides population counts for all incorporated centers and for unincorporated centers of fifty or more inhabitants. The study area contains thirty-six centers for which no population data are given. For the purpose of estimating centrality scores, each of these places has arbitrarily been assigned a population of twenty-five persons.

TABLE 7.2
Population and functional data for the twenty-five leading central places in Bruce and Grey counties, Ontario

Central place[a]	Population[b]	Number of central functions	Number of functional units	Nodality[c]	Centrality[d]
Owen Sound	18,469	76	454	2,318	969
Hanover-Walkerton	9,542	69	309	1,263	566
Wiarton	2,222	50	105	412	250
Durham	2,448	56	103	397	219
Kincardine	3,239	56	138	452	215
Port Elgin–Southampton	4,891	56	152	559	202
Lucknow	1,047	42	63	212	136
Dundalk	1,022	41	66	191	117
Chesley	1,693	49	72	229	105
Markdale	1,236	43	69	192	102
Ripley	448	29	40	121	89
Flesherton	524	31	48	115	77
Thornbury-Clarksburg	1,609	41	79	182	65
Tara	643	29	42	107	60
Teeswater	983	35	54	116	45
Chatsworth	399	25	36	74	45
Meaford	4,045	51	107	339	44
Paisley	793	32	45	94	36
Ayton	423	25	30	63	32
Hepworth	372	17	28	49	22
Mildmay	963	30	42	90	20
Cargill	233	15	17	29	12
Holstein	176	10	12	24	11
Ceylon	66	6	6	16	11
Feversham	161	12	14	21	9

a Places are ranked according to their centrality scores, which appear in the extreme right-hand column. Of the 79 centers not shown, none has a centrality score greater than 5.
b For 1971, the census year closest to the field seasons when the functional data were compiled.
c The 25 places listed in this table account for 95 per cent of the total nodality within the study area. The remaining 5 per cent is accounted for by the 79 small centers not listed.
d Values for both centrality and nodality are rounded to the nearest integer.

TABLE 7.3
Coefficients of rank correlation among measures of central place importance

	Population	Number of central functions	Number of functional units	Nodality
Centrality	0.84	0.88	0.87	0.93
Nodality	0.97	0.98	0.97	
Number of functional units	0.99	0.99		
Number of central functions	0.99			

NOTE: These results refer to the twenty-five central places listed in Table 7.2. All correlation coefficients are significant at the 99 per cent confidence level.

As shown in Table 7.3, the centrality scores are positively correlated with other measures of central place importance such as population size and numbers of central functions. Accordingly, one of these other measures – especially population size, which is usually the easiest to obtain – can be used as a first approximation or as a general guide in preliminary work. However, the centrality scores are the only values that incorporate an attempt to separate external and internal demands and thus should be given priority.

The only center whose centrality score is poorly predicted by the other indicators is Meaford. This town ranks seventeenth in centrality, but between fourth and seventh according to all the other measures. The discrepancy appears to have several causes. First, Meaford lies close to the study area's only significant military base and houses many of the base's civilian workers. Second, Meaford also serves to some extent as a dormitory for Owen Sound, twenty miles distant. There is more commuting to Owen Sound from Meaford than from either Wiarton or Port Elgin–Southampton, though all three of these towns are about equally distant from Owen Sound and it is not clear why Meaford should be preferred. Third, the density of the rural population within Meaford's tributary area is lower than average, partly because soil conditions are relatively poor and partly because the nearby tank range occupies a large area that would otherwise be farmed. As a result of all these factors, the ratio between internal and external demands for goods and services is biased in favor of the internal component. Accordingly, Meaford's population and total nodality are exceptionally large in relation to the size of its centrality score (Table 7.2).

We have now arrived at the question that lies at the very heart of central place research. Given our data on centrality, how should we identify the functional discontinuities or 'breaks' that separate the central places into distinct hierarchical orders? The type of evidence presented in Table 7.2, taken by itself, does not provide enough information for satisfactory classification of centers. The tabulated figures, of course, are abstracted from location. In order to produce a functionally meaningful classification we need to know not only each center's centrality score but also its geographical position and, most important, its relationships with other central places in terms of habitual patterns of consumer movement.

The collection and interpretation of information on patterns of consumer movement should be guided by the principle of hierarchical dependence. It is this principle, in fact, not the list of centrality scores, that is the basis for the identification of significant discontinuities in functional complexity.

The principle of hierarchical dependence may be stated formally as follows. Let *A* and *B* be two central places such that *A* has greater centrality than *B* and *B* contains not less than six different central functions. Then, if *A* is used regularly by the residents of *B* as a major source of supply for goods and services offered in *A* but not offered in *B*, center *B* belongs to a lower hierarchical order than center *A*.

Some explanatory comments are in order. The first of the two preliminary conditions – namely, that *A* has greater centrality than *B* – is merely a matter of identification. We need to keep track of which center is the larger of the two (as measured by centrality). The second condition – namely, that *B* contains at least six functions – may seem somewhat arbitrary but is quite acceptable within the North American context, and probably also within the European. In essence, this second condition implies that all centers containing fewer than six different central functions belong to the lowest hierarchical order, normally termed the hamlets. However, although all places with fewer than six functions are certainly hamlets, it does not follow that all places in the hamlet class have fewer than six functions. The number six is sufficiently small that we need not carry out any detailed investigation of consumer behavior below this level. The upper boundary of the hamlet class lies somewhere above this level – its precise location to be determined by systematic application of the principle that we are now discussing.

The main part of the principle exploits the fact that a hierarchical relationship is a form of dependence in which a smaller central place relies on one or more larger centers for goods and services that it does not provide for itself. The existence of such dependence is a sure sign that the central places in question belong to different hierarchical orders. However, if center *B* is dependent on center *A*, it does not follow that *B* belongs to the order immediately below that of *A*. One or more intermediate orders may intervene. In practice this problem is solved by commencing the analysis with the highest-ranking center and working downward, comparing each central place with all higher-ranking centers in turn. The relative status of any two places then follows naturally from the number of functional discontinuities that lie between them in the ranked list of centers.

In order to illustrate the general character of this stage of the analysis, we now outline the application of the principle of hierarchical dependence to the first eight centers listed in Table 7.2. The following remarks should be read with constant reference to Figure 7.4.

Owen Sound. This is the highest-ranking center, and therefore no 'upward' comparison with any other place within the study area is possible. When residents of Owen Sound shop out of town they leave the study area entirely, usually going to Toronto and sometimes to Kitchener-Waterloo.

Hanover-Walkerton. As indicated earlier, there is no relationship of dependence between Hanover-Walkerton and Owen Sound. This does not mean, of course, that there is no interaction whatsoever; it means that the residents of Hanover-Walkerton do not *rely* on Owen Sound. Both consumers and retailers regard these two centers as competing with each other on an approximately equal footing. Accordingly, there is no basis for placing these centers in different hierarchical orders.

Wiarton. For Wiarton two upward comparisons must be carried out – one with Hanover-Walkerton and one with Owen Sound. Interaction with Hanover-Walkerton is rare and would be even rarer except that residents of Wiarton are obliged to travel to Walkerton whenever they need to visit their county seat. However, Wiarton depends heavily on Owen Sound, and we therefore conclude that Owen Sound and Wiarton belong to different hierarchical orders. Further, Owen Sound's centrality score is almost four times as great as that of Wiarton (Table 7.2). Also, Wiarton is much closer to Owen Sound than to Hanover-Walkerton (Figure 7.4). It is not surprising that Owen Sound is the center on which the residents of Wiarton depend.

Durham. Residents of Durham do not shop in Wiarton, because the direct route from Durham to Wiarton passes through the larger and better equipped center of Owen Sound (Figure 7.4). Durham, in fact, depends on both Owen Sound and Hanover-Walkerton, with a preference for the latter. Keeping the centrality scores in mind (Table 7.2), we infer that Durham belongs to the same order as Wiarton – that is, one step below the hierarchical level represented by Hanover-Walkerton and Owen Sound.

Kincardine. Kincardine has no significant interaction with Owen Sound, Wiarton, or Durham – note the long distances involved – but is strongly dependent on Hanover-Walkerton. Kincardine thus belongs to the same order as Durham and Wiarton.

Port Elgin–Southampton. This center has no significant interaction with Wiarton, Durham, or Kincardine. Like Durham, it depends on both Owen Sound and Hanover-Walkerton, though with a preference for Owen Sound. We conclude that Port Elgin–Southampton belongs to the same hierarchical level as Wiarton, Durham, and Kincardine.

Lucknow. A new factor now enters the picture, for Lucknow depends mainly on Wingham, just outside the southern boundary of our study area (Figure 7.4). Accordingly, the dependency relationships of Wingham must also be investigated. Wingham depends partly on Hanover-Walkerton and partly on Goderich, another center outside our area. Wingham's dependence on Hanover-Walkerton, coupled with Lucknow's dependence on Wingham, signifies that Lucknow belongs two levels below Hanover-Walkerton. This is corroborated by Lucknow's partial dependence on Kincardine. There thus appears to be a significant functional discontinuity in Table 7.2 between Lucknow and Port Elgin–Southampton. The centrality scores are consistent with this conclusion.

Dundalk. Dundalk, like Lucknow, involves centers outside our study area (Shelburne and Collingwood). It belongs two levels below Owen Sound, in the same hierarchical order as Lucknow.

And so the process continues. The above sketches of dependency relationships are fragmentary but show that the task of classification must be approached in an orderly manner. By definition, a central place cannot be hierarchically dependent on another place that has less centrality than itself. It is logical, therefore, to start at the top of the list and to proceed by comparing each center successively with every place possessing greater centrality. In practice, most comparisons produce negative results, with no significant interaction between the centers. Few central places have close and regular ties with more than three or four places of greater centrality. But the cases in which dependence does exist are crucial, because they, and they alone, make possible the identification of the breaks between the hierarchical orders.

In Bruce and Grey counties, the procedure described above results in the identification of four hierarchical orders. For convenience, we will refer to these orders, in ascending sequence, as hamlets, villages, towns, and cities. (These labels, as used here, carry no implications with regard to the legal rank of incorporated centers.) The two counties have two cities, four towns, fifteen villages, and eighty-three hamlets. One additional city-level center and three additional town-level centers lie just outside the study area's boundary. The complete classification is shown in Figure 7.4.

We have accomplished the first of the two tasks to which this chapter is addressed: namely, the construction of a hierarchical classification. We now seek to relate this classification to central place theory.

Theory and reality: tests of correspondence

Evaluating the degree of correspondence between real-world observations and theoretical models, in any field of inquiry, reveals the extent to which a theory helps us understand reality. If a theory is internally consistent, and if both its assumptions and its conclusions correspond reasonably well with the facts of the case, then we can justifiably claim that the theory provides a satisfactory explanation of our observations.

There can be no doubt about the internal consistency of central place theory: the logical coherence of the Christallerian argument is beyond reproach. Further, our study area comes tolerably close to meeting Christaller's assumptions – isotropic plain, general absence of nonretail activities in the towns, and so forth. Do the actual size and spacing of central places in Bruce and Grey counties agree with the geometric patterns displayed in the models that are the end products of Christaller's theory?

It is appropriate to proceed by considering, in turn, each of the four diagnostic criteria of hierarchical structuring identified in chapter 5: (1) discrete stratification of centrality; (2) incremental baskets of goods; (3) a numerical pyramid in the frequency of centers in successive orders; and (4) interstitial placement of the orders.

DISCRETE STRATIFICATION OF CENTRALITY

It is clear that there is variation among the centrality scores within each of our four hierarchical orders, particularly on the village level; but perfect uniformity within each order cannot be expected under real-world conditions. More important, the functional discontinuities between orders correspond to distinct gaps in the ranked list of centrality values (Table 7.2). It is significant, moreover, that the functional discontinuities were identified by means of the principle of hierarchical dependence, not simply by inspection of the centrality scores.

There is very little overlap between adjacent orders in terms of the centers' populations. The largest town is barely half the size of the smallest city. Only one of the fifteen villages – Meaford – is larger than the smallest town; only three of the eighty-three hamlets are larger than the smallest village, and they are locally well known as dormitory settlements for nearby higher-order centers. In general, we conclude that the four orders of central places are satisfactorily discrete.

There is a great temptation, in this type of work, to introduce numerical grouping procedures that can be applied directly to the centrality scores, to the populations, or to the original data on functions and functional units. This approach, however, ignores locational interdependencies within the urban system, and therefore it inevitably begs the question. We stress once again the necessity of basing the hierarchical classification primarily on knowledge of the travel patterns of consumers.[5]

INCREMENTAL BASKETS OF GOODS

The question of incremental baskets of goods may be investigated in the following way. As a first step, we prepare a table (not reproduced here) that shows, for each central function, the percentage of the central places in each order that possess the function in question. In Bruce and Grey counties, for example, furniture stores are found in 6 per cent of the hamlets, in 93 per cent of the villages, and in 100 per cent of the places on the town and city levels. Next, we define a function as being 'available' in any hierarchical order in which it is present in at least 50 per cent of all centers. Furniture stores, therefore, are 'available' in villages, towns, and cities, but not in hamlets. Now, in order that the criterion of incremental baskets of goods may be fulfilled, each function should remain 'available' in every order that is higher than the lowest order in which it attains the 50

5 If a study area is very large (say, the entire American Midwest), and if rural population density changes markedly from one side of the area to the other, then the populations and centrality scores of adjacent orders of central places might overlap considerably. This overlap, however, would be caused by adaptation of centers of all orders to the geographic variation in rural density, not by the absence of hierarchical structuring.

TABLE 7.4
Incremental functions on each level of central place hierarchy, Bruce and Grey

Level	(Additional) functions present	
Hamlet	Gas station	Post office
	General store	
Village	Automobile dealership	Grain and feed dealer
	Automobile repairs	Grocery store
	Bakery store	Hardware
	Bank	Home furnishings
	Barber shop	Hotel/motel
	Beauty parlor	Insurance agent
	Cartage agent	Jewelry store
	Dairy store	Lawyer
	Doctor	Lumber, building supplies
	Drugstore	Plumbing, heating
	Fairground (annual fair)	Pool hall
	Family shoe store	Public library
	Farm machinery	Restaurant, cafeteria
	Fire department	Variety store/smoke shop
	Fresh meats, delicatessen	Veterinarian
	Funeral home	Weekly newspaper
Town	Automobile accessories	General hospital
	Beer store	Junior department store
	Bowling lanes	Liquor store
	Camera store	Men's and boy's clothing
	Chiropractor	Movie theater
	Cold storage lockers	Optician
	Dentist	Paint and wallpaper
	Dry cleaning, laundry	Real estate agent
	Eaton's order office	Sears order office
	Family clothing store	Tavern/bar
	Flower shop	Women's clothing
City	Antique shop	Men's shoe store
	Auctioneer	Millinery shop
	Bookstore	Photography studio
	Chartered accountant	Piano tuner
	Children's wear	Provincial savings office
	Chinaware, gifts	Records and tapes
	County seat	Riding school
	Custom tailor	Sewing shop
	Daily newspaper	Sporting goods
	Dead animal removal	Stationery, office supplies
	Furrier	Trust company office
	Land registry office	

NOTE: This table includes eighty of the eighty-one functions covered by the study. One function – the specialized candy store – does not achieve the 50 per cent level of occurrence in any of the four hierarchical orders.

per cent level. In actual fact, every function in the present study area succeeds in satisfying this condition. Accordingly, the goods and services available on any given level of the hierarchy are also available on all higher levels.

It is also logical to expect that the percentage level of occurrence of any function within a particular hierarchical order will not be lower than its level of occurrence in any lower order. In other words, the percentages should increase, or at least not decline, as we ascend the levels of the hierarchy. This condition is violated by only seven of our eighty-one functions, and never drastically. In the vast majority of cases, therefore, we observe the expected trend of increasing 'availability' on successively higher hierarchical levels.

We now define each function as being 'incremental' in the lowest order in which it is 'available.' This makes it possible to prepare lists of the functions that comprise the incremental baskets of goods on the various hierarchical levels. Table 7.4 shows the functions that are incremental on each level of the hierarchy in Bruce and Grey counties. These lists, of course, are based on one specific study area, and it should not be assumed that exactly the same pattern of incremental functions will occur in other regions. However, Table 7.4 closely resembles the results obtained by other researchers in various parts of the American Midwest (Brush 1953; Hassinger 1957; Berry, Barnum, and Tennant 1962; Borchert and Adams 1963).

It is useful to carry this portion of the analysis a little further by considering the size of the 'step' taken by each function as it becomes incremental. Returning to the example of the furniture stores, the level of occurrence rises from 6 per cent in the hamlets to 93 per cent in the villages, and the difference of 87 percentage points is the 'step' taken by this function as it becomes a member of the village-level basket of goods. How large should we expect such steps to be? At one extreme, the maximum possible step is a full 100 percentage points: that is, from complete absence in one order to universal presence in the order immediately above. (This actually does occur with six of the functions in the present study.) At the other extreme, the minimum possible step is a single percentage point: that is, from 49 per cent to 50 per cent.

If we assume, as a null hypothesis, that the size of the step is a random variable possessing a normal distribution, we can infer that the average size of the steps will be fifty percentage points. To put this another way, we can infer that forty of our eighty-one central functions should exhibit steps of greater than fifty points. How do these theoretical expectations compare with our actual results? In fact, the observed average step turns out to be sixty-eight percentage points, which is significantly larger than the expected value of fifty under the hypothesis of randomness. Also, sixty-two of the eighty-one functions have step-values greater than fifty points. Thus the observed steps are much too large to have resulted from a purely random allocation of central functions to central places. This finding demonstrates that our hierarchical classification based on patterns of consumer movement is statistically significant in terms of the overall distribution of central functions among the four orders. In effect, the analysis of the step-values of the functions provides independent corroboration of the validity of the hierarchical classification.

THE FREQUENCY OF CENTERS IN SUCCESSIVE ORDERS

The third criterion is that there should be a numerical pyramid in the frequency of central places occurring in successive orders. This criterion is by far the easiest of the four to apply, since one merely needs to inspect a single sequence of numbers. As noted earlier, the area covered by the present study contains two cities, four towns, fifteen villages, and eighty-three hamlets. Clearly the 'pyramid' requirement is not violated.

At the same time, however, the observed sequence of frequencies does not correspond precisely to any sequence that can be derived from the basic Christallerian models. We will return to this point later in this chapter under the heading 'The Value of *K*.'

INTERSTITIAL PLACEMENT OF THE ORDERS

We have already stressed that we should not expect a system of central places to display perfect Christallerian geometry under real-world conditions. But just what should we expect as far as the network's geometrical properties are concerned? How much distortion of the uniform theoretical pattern is permissible? What sort of spatial regularities should we be attempting to detect? How shall we decide whether, or to what extent, the observed arrangement of central places follows the spirit, if not the letter, of the Christallerian models?

An appropriate overall strategy for addressing these questions is to use central place theory as a basis for making empirically testable predictions that do not demand rigid adherence to the theory's geometrical perfection. Several testable predictions of this kind are outlined in the paragraphs that follow.

Median centers of settlement

The first set of predictions involves the concept of the median center, which was introduced in chapter 3 in connection with our centrographic analysis of the spatial distribution of functional classes of cities at the continental scale. Consider a large region that contains a perfect Christallerian arrangement of central places: the median center of each order of central places falls on, or very close to, the median center of the complete network. Our first test, then, considers the extent to which the median centers in Bruce and Grey counties are spatially coincident.

Figure 7.5 shows the locations of the median centers for (i) all 104 central places considered simultaneously, (ii) the eighty-three hamlets, (iii) the fifteen villages, and (iv) the towns and cities taken as a single class. The towns and cities are combined because they are only ten in number even when we include the four that lie just outside the study area (Figure 7.4).

The median centers for the hamlets and villages fall within five miles (eight kilometers) of the median center for all places. These three median centers, moreover, are centrally placed within the study area as a whole, which is fully in accord with theoretical expectations. The median center for the towns and cities is significantly displaced toward the southern boundary of the study area, because three of the four towns and cities lying outside the study area are adjacent to the area's southern edge. Inspection of Figure 7.4 reveals that the median center for the towns and cities would fall much closer to the expected position if the four external centers were simply ignored.

In summary, although the four median centers do not coincide, none is far removed from the location where central place theory predicts that it should be found.

Mean distance deviations

The second set of predictions involves another centrographic measure: mean distance deviation (MDD). In chapter 3 we saw that the MDD of a set of points is the average distance of members of the set from their median center.

We can arrive at an expected value for the MDD of each order of central places in the following way. First, in order to simplify the calculations, we assume that our study area is a rectangle. This rectangle must satisfy two conditions: its area must be identical to that of the study area, and the ratio between the lengths of its short and long sides must be equal to the ratio between the average width and the average length of the study area.

These two conditions lead immediately to two simultaneous equations that completely determine the dimensions of the rectangle. In the present case the required total area is 2,895 square miles (7,498 square kilometers) and the ratio of the sides is taken to be 3:4. Solving the simultaneous equations shows that the required rectangle has sides 46.6 and 62.1 miles (75.0 and 100.0 kilometers) in length, respectively.

If we now assume, in accordance with theory, that our rectangle contains a uniform distribution of central places and that the median center of these places lies at the center of the rectangle, a good estimate of the mean distance deviation is given by the following formula:

$$v = \frac{1}{4\sqrt{2}}\left[\frac{2ab}{a+b} + \sqrt{a^2+b^2}\right], \tag{7.1}$$

where v is the predicted MDD value and a and b are the lengths of the sides of the rectangle. (The derivation of this formula is somewhat tedious and is accordingly omitted; also, the formula should not be used if the ratio of the lengths is greater than 10:1.)

Given the values of 46.6 and 62.1 miles for a and b, the expected value of the MDD, v, is 23.1 miles (37.2 kilometers). The observed MDD values, obtained by actual measurement, are as follows: hamlets, 21.3 miles (34.3 kilometers); villages, 22.1 miles (35.6 kilometers); towns and cities, 26.9 miles (43.3 kilometers).

The observed values for the hamlets and the villages are very close to the expected value of 23.1 miles. The observed value for the ten towns and cities is larger than expected, but there are two good reasons for this departure. First, the fact that four of the ten towns and cities lie outside the study area inevitably inflates the average distance to the median location. Second, the six towns and cities lying inside the study area occupy locations that are notably peripheral. Four are ports, situated on the coasts that serve as the area's boundaries on the west and north (Figure 7.4). In contrast, only two of the fifteen village-level centers occupy coastal sites. The region's largest centers show an above-average affinity for coastal locations, which can be explained by the importance of water transport during the period of early settlement. The large MDD value for the towns and cities may therefore be viewed, at least in part, as a legacy of the mid-nineteenth century, when the region's external linkages depended more heavily on lake-borne shipping than on transport by land. Although two early 'colonization roads' penetrated the region from the

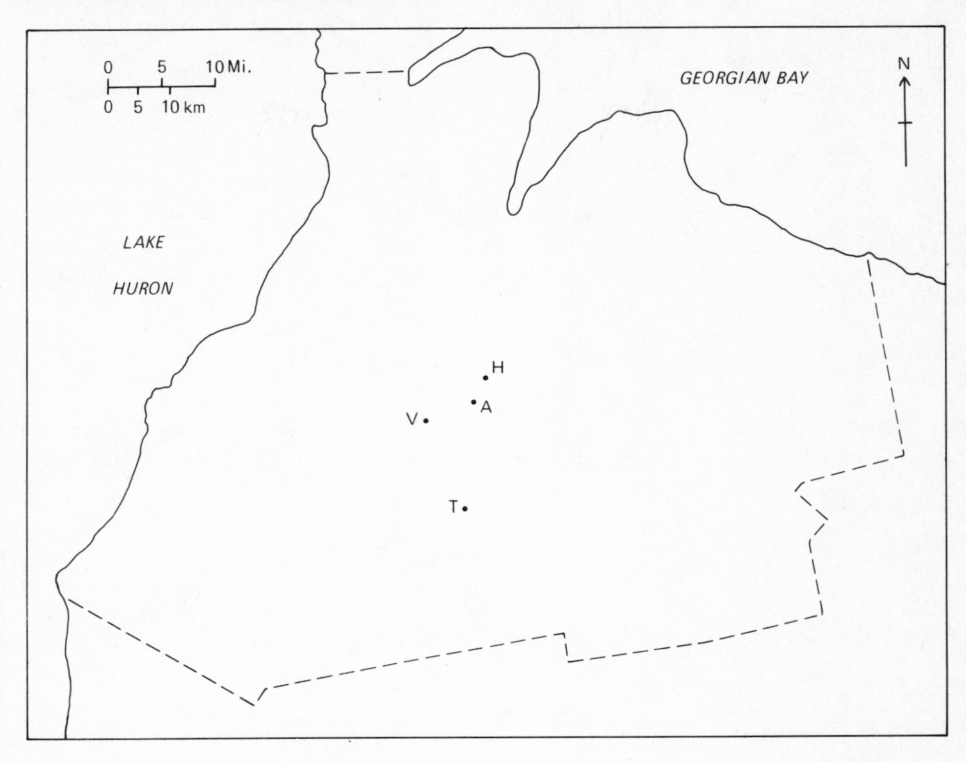

Figure 7.5
The locations of selected median centers within the regional urban system shown in Figure 7.4. Types of median centers: *A* = all central places, regardless of hierarchical order; *H* = hamlets; *V* = villages; *T* = towns and cities combined.

south, many settlers made the journey from older areas by way of Lake Huron or Georgian Bay, traveling by boat and entering the study area at the ports of Kincardine, Port Elgin–Southampton, and, above all, Owen Sound. As the leading points of entry to the region, these centers gained an early advantage that they have retained to the present day.

The median centers and their associated MDD values provide objective evidence that the members of each order of central places are distributed uniformly throughout all parts of the study area. This general homogeneity of pattern, in turn, implies that members of the lower orders occupy the interstices between members of the higher orders, as in the models.

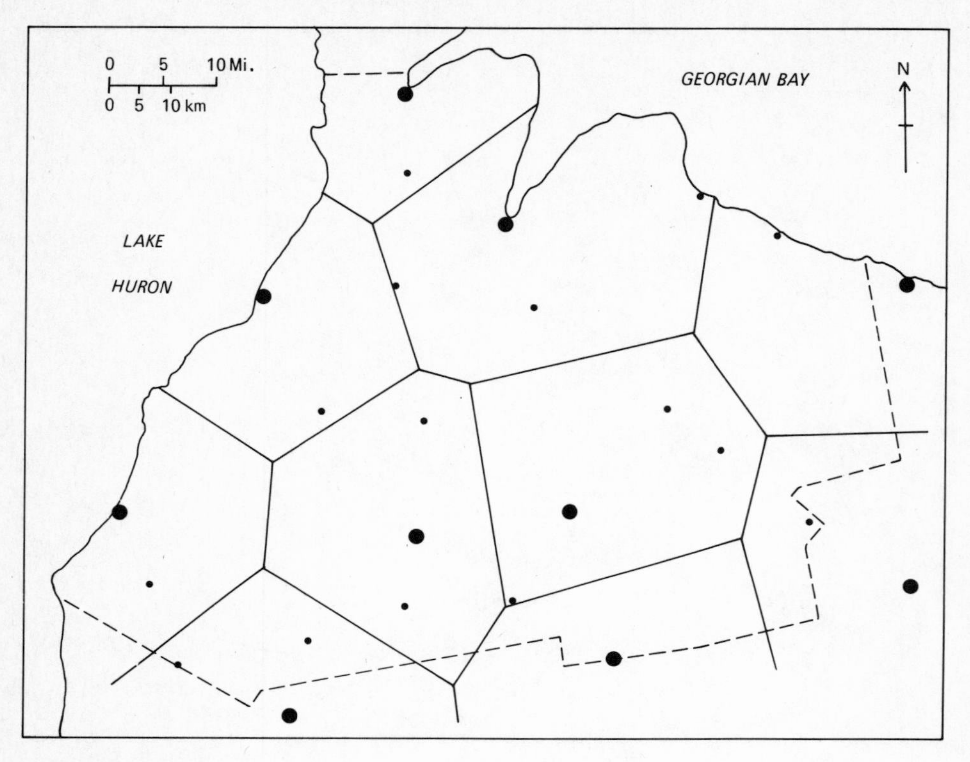

Figure 7.6
Thiessen polygons around central places of town and city rank. Large dots represent town-level and city-level centers; small dots represent villages. For names of centers, see Figure 7.4.

Thiessen polygons

A surface containing *n* fixed control centers can be uniquely partitioned into *n* subregions by assigning every point on the surface to its nearest control center. This can be accomplished by constructing the perpendicular bisectors of lines joining adjacent control centers and by extending these bisectors until they intersect one another. The result is that each control center is surrounded by a polygon that contains all points lying closer to the enclosed control center than to any control center outside the polygon. The polygons thus created have several names. Mathematicians refer to them as Voronoi polygons or Dirichlet regions, but geographers usually call them Thiessen polygons. The name honors Alfred Thiessen, a meteorologist who used this approach as the basis of a method for estimating regional precipitation averages from data collected at unevenly spaced recording stations (Thiessen 1911; Kopec 1963; Getis and Boots 1978, 126–37).

In Figure 7.6, Thiessen polygons are drawn around the ten central places of town rank or higher in our present study area. Most of the polygons are 'incomplete' – truncated by the boundary of the study area. This does not pose a problem in the present analysis, although it does preclude detailed study of the properties of the polygons themselves. For example, central place theory leads us to expect that complete Thiessen polygons will have an average of six sides, but this prediction cannot be given a fair test unless the number of complete polygons is reasonably large. (In Figure 7.6, the Hanover-Walkerton polygon has six sides and the Durham polygon has five.) Notice also that the polygons meet in threes at each vertex. In practice this is almost invariably the case, but a meeting of four or more edges at a single vertex is not a geometrical impossibility and therefore great care must always be exercised when the polygons are being constructed.

Our specific purpose in this section is to examine the relationship between the tessellation of town-level Thiessen polygons and the locations of village-level central places. Central place theory predicts that the villages will be situated as far as possible from the towns – not forgetting that city-level centers are also 'towns' – in order to minimize competition. Accordingly, the villages should be located close to the edges of the town-level Thiessen polygons, which define the set of locations maximally remote from the towns. The fifteen village-level centers are included in Figure 7.6. To what extent is our theoretical expectation borne out by the facts?

Table 7.5 records the distances from each village to its nearest town-level center and to the closest point on the perimeter of that center's Thiessen polygon. Three pertinent observations may be made. First, every village lies closer to the relevant perimeter than to its nearest town. In every case but three (Ripley, Chatsworth, and Mildmay), the distance to the nearest town is more than twice as great as the distance to the perimeter. Second, the largest distance from village to perimeter (Markdale, 6.2 miles) is smaller than the smallest distance from village to town (Mildmay, 6.6 miles). Third, on average, the distance from village to town is more than three times as great as the distance from village to perimeter.

Given these findings, it is fair to describe the villages as being 'remote' relative to the towns and cities. No village lies exactly on the edge of any

TABLE 7.5
Positions of village-level central places in relation to town-level Thiessen polygons

Village	Nearest town or city	Distance to nearest town or city (miles)	Distance to edge of Thiessen polygon (miles)
Lucknow	Wingham	11.3	2.3
Dundalk	Shelburne	11.2	5.4
Chesley	Hanover-Walkerton	11.0	4.1
Markdale	Durham	12.9	6.2
Ripley	Kincardine	7.5	5.6
Flesherton	Durham	14.8	4.1
Thornbury-Clarksburg	Collingwood	12.8	5.8
Tara	Owen Sound	11.5	0.2
Teeswater	Wingham	7.5	3.3
Chatsworth	Owen Sound	7.9	5.8
Meaford	Owen Sound	17.6	1.4
Paisley	Port Elgin–Southampton	11.9	1.6
Ayton	Durham	10.4	0.6
Hepworth	Wiarton	7.3	1.9
Mildmay	Hanover-Walkerton	6.6	4.1
Average value		10.8	3.5
Standard deviation		3.1	2.0

polygon, but such precision is too much to hope for under real-world conditions. The villages do show a distinct preference for locations where competition from higher-order centers is relatively weak. Setting perfect hexagonal geometry aside, this is precisely what the theory leads us to expect.[6]

The Thiessen polygons in Figure 7.6 are reasonably good approximations of the town-level tributary areas of the centers on the town and city levels. However, city-level centers have city-level as well as town-level tributary areas. Consequently the polygons surrounding Owen Sound, Hanover-Walkerton, and Collingwood in Figure 7.6 do not show the maximum extent of the spheres of influence of these three centers. In addition, the use of Thiessen polygons involves the implicit assumption

6 A parallel analysis was carried out for hamlets, using villages, towns, and cities as control centers for construction of village-level polygons. The results are not as striking as those for villages, but nevertheless seventy-six of the eighty-three hamlets lie closer to the nearest polygon edge than to the nearest village, town, or city.

that the boundaries of the town-level tributary areas fall exactly halfway between neighboring town-level centers. In reality, each boundary tends to be slightly displaced toward the smaller of the two centers, reflecting the greater power of attraction exerted by the larger town. (See also the discussion of the Reilly gravity model in chapter 6.)

Inter-level distances

A further aspect of the spatial pattern of a system of central places involves the distances from the members of one particular order to the nearest places of higher rank. To take a concrete example, we can measure the distances from each of the eighty-three hamlets in Bruce and Grey counties to the nearest center of village rank or above. (Towns and cities also function as villages and must therefore be included as potential village-level destinations for customers originating in hamlets.) The eighty-three measurements in question vary from a low of 1.6 miles to a high of 10.5 miles, with an average value of 6.2 miles. The last value is the average distance – measured as the crow flies – that hamlet residents must cover to reach the village-level basket of goods and services. Average inter-level distances of this type form the subject of the present section.

Pursuing our usual strategy, we begin by attempting to produce a plausible set of 'expected' values, using central place theory as a guide. For convenience, let us retain the hamlet-to-village example as a working illustration throughout the following discussion. The method now to be described is entirely general, but its logic is most readily clarified if we keep a concrete example in view.

As an opening gambit we arbitrarily postulate that centers of village rank and higher are arranged in the form of a perfect triangular lattice, just as in the Christallerian models. How far apart will neighboring centers in this lattice be situated? Assuming that each center is surrounded by an identical hexagonal tributary area and that the sum of these tributary areas is equal to the total area of the study area, it can be shown that

$$d = [2A/N\sqrt{3}]^{\frac{1}{2}}, \tag{7.2}$$

where d is the distance between adjacent lattice points, A is the total area of the study area, and N is the number of points in the lattice.

As noted earlier, the total area of our study area, A, is 2,895 square miles. Determining the appropriate value for N, however, involves an element of judgment. We know, of course, that there are fifteen villages, four towns, and two cities inside Bruce and Grey counties, but it can be

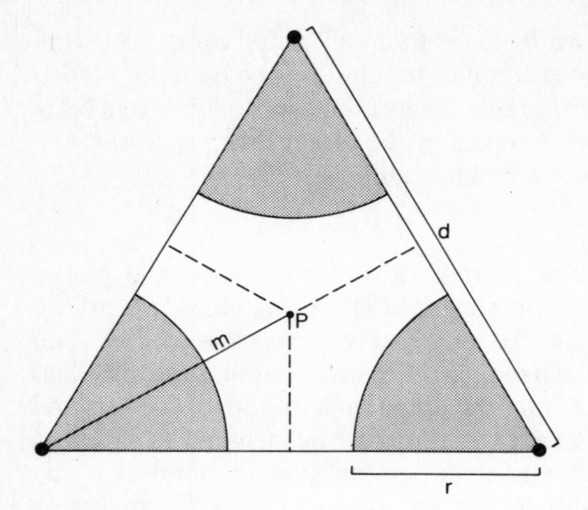

Figure 7.7
Selected geometrical elements of a single triangle within a triangular lattice of central places. See text for explanation of symbols.

argued that some allowance should be made for the four additional centers of town and city rank that lie just outside the study area. After all, the tributary areas of these external centers extend into our study area and thereby diminish the total area served by the twenty-one internal centers. However, such losses of area are partly offset by the fact that two of the internal centers – Lucknow and Dundalk – lie close to the study area's boundary; their tributary areas are therefore not wholly contained within our defined territory. On balance it seems appropriate to regard the study area as containing the equivalent of twenty-two village-level tributary areas rather than twenty-one. Accordingly, we adopt 22 as the value of N.

Taking the values of A and N to be 2,895 and 22, respectively, the value of d obtained from the above formula is 12.3 miles. This, then, is the distance separating adjacent centers of village rank and above in a hypothetical triangular lattice that matches the study area in terms of total area and total number of village-level tributary areas.

Figure 7.7 represents a single triangle in this hypothetical lattice. The triangle is equilateral, each vertex being the site of a central place of village rank or above. For reasons that will shortly become apparent, we wish to subdivide this triangle in such a way that three identical corner portions, lying within radius r of each vertex, contain, when added together, exactly one-half of the total area of the triangle. In Figure 7.7 the required corner portions are indicated by shading. Using standard theorems of Euclidean geometry, it can be shown that

$$r = \frac{d}{2}\left[\frac{\sqrt{3}}{\pi}\right]^{\frac{1}{2}}, \tag{7.3}$$

where r is the required radius and d, as before, is the length of each side of the triangle. Here, since d is 12.3 miles, the radius r is 4.6 miles.

Let us assume that every triangle in our hypothetical village-level lattice is subdivided in the manner shown in Figure 7.7. This means, in effect, that every village, town, and city is enclosed within a circle of radius r and that the sum of the areas of these circles is equal to one-half of the total area of the study area. Let us further assume that centers of hamlet rank are now placed randomly within the study area – the location selected for each hamlet is completely unaffected by the locations of other hamlets and of higher-order centers. Given this randomness, we may infer that, on average, 50 per cent of all hamlets will fall within the 50 per cent of the total area that lies within radius r of the village-level centers. In other words, the distance to the nearest village will be less than r for one-half of the hamlets and greater than r for the other half. Accordingly, the distance r may be taken as a reasonable estimate of the average distance to the nearest village when the hamlets are randomly distributed.

But, of course, a random distribution of hamlets is not what central place theory leads us to expect. Indeed, we should expect that the location of each hamlet is influenced by the locations of other centers; and in particular, that the hamlets will tend to be situated far from the village-level centers rather than close to them. We should be sorely disappointed, in fact, if the observed average distance from the hamlets to the nearest village is as small as r; for this would imply that the hamlets are not 'avoiding' the villages as successfully as theory demands. Accordingly, let us take r as our *minimum* estimate of the expected average value. In the present case, as calculated above, this minimum estimate is 4.6 miles.

The obvious next step is to devise a *maximum* estimate of the expected average distance from the hamlets to the nearest village. Referring again to figure 7.7, we see that no hamlet can be farther from its nearest village than the distance from any vertex of the triangle to P, the triangle's midpoint. It can easily be shown that this distance is given by

$$m = d/\sqrt{3}, \tag{7.4}$$

TABLE 7.6
Average distances (in miles) to nearest center of higher order

	Theoretical values[a]			Index of remoteness:
	Probable minimum, r	Probable maximum, m	Observed value, x	$100\left(\frac{x-r}{m-r}\right)$
From hamlets to village or higher	4.6	7.1	6.2	64
From villages to town or higher	7.6	11.8	10.8	76
From towns to city	13.6	21.1	19.8	83

a Theoretical values based on the assumption that the study area contains 22 village-level tributary areas, 8 town-level tributary areas, and 2.5 city-level tributary areas. The method of calculation is explained in the text.

where m is the distance from P to any vertex and d is the length of each side of the triangle. If we assume that every hamlet lies as far as possible from three surrounding villages, it follows that the average distance from the hamlets to the nearest village is simply m. Accordingly, we adopt m as our maximum estimate of the expected average value. In the present case, with d being 12.3 miles, the value of m is 7.1 miles.

In reality, of course, the lattice of central places of village rank and higher is not perfectly triangular. Generally speaking, the lattice tends to be attenuated (i.e. the centers are relatively far apart) where the density of the rural population is below the regional average. Conversely, the lattice tends to be compressed where rural density is comparatively high. Real-world triangles, moreover, tend to be scalene rather than equilateral. As a result, the minimum and maximum estimates described above should be regarded as subject to variation. This variation, however, is sharply constrained by the fact that the estimates depend on the total area of the study area, which is a fixed quantity. Thus, if some of the real-world triangles are larger than those in the hypothetical lattice, others must necessarily be smaller, and the effects of these opposing deviations will tend to cancel each other out. Therefore, the values calculated above can be accepted as plausible estimates of the limits between which the true average distance to the nearest higher-order center should be found to lie.

We can go one step further and predict that the true average value should lie closer to the maximum estimate than to the minimum: the maximum estimate involves the notion of minimizing competition, whereas the minimum estimate involves the decidedly un-Christallerian notion of randomness in the spatial pattern. By the same token, an observed value smaller than the minimum estimate would imply that the hamlets huddle close to the higher-order centers instead of giving them a wide berth. This, too, is inconsistent with basic Christallerian principles.

For Bruce and Grey counties the above approach yields the results shown in Table 7.6. It is encouraging to find that the observed average distances fall within the expected limits in every case. Moreover, as shown by the simple 'remoteness' index in the right-hand column of the table, each observed value lies closer to the relevant maximum estimate than to the corresponding minimum estimate. The results for the villages and for the towns are especially satisfactory. For the hamlets the anticipated tendency toward remoteness is less strongly marked, at least partly because they are very numerous in relation to higher-order places. In fact, each village-level triangle of the type shown in Figure 7.7 must accommodate an average of two hamlets rather than one or less. Since these two hamlets cannot both occupy the midpoint of the triangle, and since they will attempt – according to the theory – to avoid not only the higher-order centers but also each other, some reduction in the remoteness index of the hamlets is likely. The same constraint does not apply for other levels of the hierarchy.

The observed average distances correspond remarkably well to the values that would be obtained from a perfect $K = 3$ arrangement. In the $K = 3$ pattern the distances increase by the square root of three as we ascend from level to level. Thus, setting the hamlet-to-village distance at the observed value of 6.2 miles, the distances on successively higher levels would be 10.7 miles and 18.6 miles, respectively. Of course, the close correspondence between these theoretical values and the observed distances does not imply that our study area conforms precisely to the theoretical pattern. Indeed, a glance at Figure 7.4 makes it clear that the study area does not exhibit the classical $K = 3$ geometry. The correspondence of the values does suggest, however, that both the $K = 3$ model and the real urban system are constructed on the same general principles of spatial organization. In particular, the similarity of the distances suggests that real-world conditions are likely to favor the development of structures in which the values of K are small.

Given the observed values reported in Table 7.6, we can make some general predictions concerning other kinds of inter-level distances. Consider, for example, the average distance from the hamlets to the nearest center of *town* rank or above. For some hamlets, the nearest village-level center is also a town-level (or even city-level) center, and in these cases the distance from hamlet to town is identical to that from hamlet to village. For other hamlets, however, the nearest village-level center is only a village, and the distance to the nearest town is necessarily greater than to the nearest village. The latter situation, in fact, is true for fifty-three (64 per cent) of the eighty-three hamlets in the present study. It follows that the average distance from hamlets to the nearest town must be greater than the average distance from hamlets to the nearest village. By a parallel argument, the average distance from hamlets to cities must be greater than that from hamlets to towns. The actual values are as follows: from hamlets to the nearest village or higher, 6.2 miles; from hamlets to the nearest town or higher, 9.8 miles; from hamlets to the nearest city, 14.7 miles.

In the same way, the average distance from villages to the nearest city must be greater than the average distance from villages to the nearest town. The observed averages are as follows: from villages to the nearest town or higher, 10.8 miles; from villages to the nearest city, 15.7 miles.

In every case, therefore, the anticipated increase in average distance is in fact observed.

Finally, the distance relationships discussed in this section are not inevitable properties of the spatial arrangement of urban centers. It is possible to design a system in which, for example, the average distance from hamlets to the nearest village is greater than from villages to the nearest town. This would be the case if hamlets and villages occurred as segregated groups at opposite sides of the study area, with towns occupying the middle ground. One is likely to be told, of course, that an arrangement of this kind is utterly unrealistic and that spatial segregation on the basis of levels of centrality simply does not occur. However, this objection is actually a declaration of faith in the reality of interstitial placement of the orders. And interstitial placement – geometrically precise in the models but flexible in the real world – is the hallmark of Christallerian spatial organization.

OVERVIEW OF THE LOCATIONAL ANALYSIS

The four techniques described above – from the investigation of median centers to the analysis of inter-level distances – are united by their common concern with aspects of spatial arrangement. Taken together, they provide a general method for evaluating the extent to which an observed system of central places displays – in flexible form – the geometrical properties of the Christallerian models.

Each technique takes the form of a test. The expected outcome of the test is specified in advance by assuming that the size and spacing of urban centers can be explained, largely if not entirely, by central place theory. If a real urban system passes the test, we have no reason to reject the hypothesis that central place principles are at work. If it fails, we infer that other factors must be involved, and that central place theory, alone and unaided, does not furnish an adequate explanation. As it happens, the central place pattern of Bruce and Grey counties passes all the tests with flying colors. However, despite the satisfaction one may obtain from this specific local result, understanding the strategy of investigation is more fundamental than knowing the characteristics of any particular study area.

The value of *K*

Given the prominent role played by the index K in the Christallerian models, it is natural to inquire what value or values of K can best be employed to represent the structure of a real-world urban system. Because the regular triangular geometry of the theoretical models is not encountered in the real world, this question cannot be satisfactorily answered simply by examining a map. The visual comparison of map and theory usually suggests that the observed pattern of settlement is a compound structure: some centers conform to one model, some to another, and some to no regular locational rule whatsoever. Accordingly, the search for appropriate values of K is carried out not in terms of geometry but in terms of the frequencies of central places in the various hierarchical orders. By focusing on these frequencies we can derive values of K that describe the urban pattern in terms of average conditions but do not necessarily describe the locational peculiarities of any individual central place.

The standard approach to the identification of relevant K-values is to construct a geometrically permissible model that reproduces, as closely as possible, the observed numbers of central places belonging to successive orders (Woldenberg 1968; Parr 1978, 1980). This approach may be illustrated with reference to the results obtained in Bruce and Grey counties. First, we write down the numbers of central places belonging to the various levels, commencing with the highest order – thus: 2, 4, 15, 83. Second, we sum these values cumulatively in order to generate a second series of numbers in which the nth member is the sum of the first n numbers in the original series – thus: 2, 6, 21, 104. This second series of numbers describes the number of tributary areas on each hierarchical level, taking into account the fact that each central place possesses a distinct tributary area for each level on which it functions. Third and finally, starting with the second number, we divide each number in the second series by the number that precedes it – thus: 6/2 = 3.0; 21/6 = 3.5; 104/21 = 5.0.

The quotients thus obtained may be termed mean real-world *K*-values. In the present case, for example, the mean size of a town-level tributary area is 3.5 times as large as that of a village-level tributary area. Similarly, the mean size of a village-level tributary area is 5.0 times as large as that of a hamlet-level tributary area.

Mean real-world *K*-values differ from the *K*-values found in the classic Christallerian models in two respects. First, they are not necessarily integers. Second, those that are integers do not necessarily belong to the limited set of whole numbers that arise from perfect hexagonal geometry (see chapter 5). These differences, of course, simply reflect the fact that real central place systems do not display the flawless geometrical precision of theoretical models.

Nevertheless, by selecting a set of theoretically permissible *K*-values that approximately reproduces the observed series, we can construct a model that expresses the hierarchical structure of the observed urban system in a theoretically feasible form. We ask, in other words, what pattern in the realm of theory comes closest to matching the structure we actually observe. By carrying out a series of trial-and-error experiments involving both square and hexagonal market areas, we discover that the best-fitting model for Bruce and Grey counties is a hexagon-based, variable-*K* system in which the values of *K* are 3, 4, and 4, respectively. Given two city-level centers, this set of *K*-values produces four towns, eighteen villages, and seventy-two hamlets; the real system contains four towns, fifteen villages, and eighty-three hamlets. Measured in percentage terms, the discrepancies between the observed and theoretical values are not large.[7]

These results suggest three general conclusions. First, even in regions where the original system of land division was essentially rectangular,

7 In reworking Christaller's Bavarian data, Parr (1980) found that he could reduce discrepancies between observed and theoretical values by constructing models in which some market areas are hexagonal and others are rectangular. The rectangle is, of course, an unorthodox shape in this context, but Parr speculates that it might result from historical changes within the urban network. The approach is provocative and deserves further study.

hexagonal geometry may be relevant to an understanding of the central place pattern. The road network of Bruce and Grey counties is strongly rectangular, yet the best-fitting theoretical model incorporates hexagonal market areas. Second, the small mean real-world *K*-values support the hypothesis that large values of *K* offer too few spatial opportunities to entrepreneurs and are therefore unlikely to occur under real-world conditions. Third, the variable-*K* character of the best-fitting model suggests that it is not unusual for the marketing ($K = 3$) and traffic ($K = 4$) principles to be at work simultaneously in determining the overall spatial structure of real central place systems.

Concluding remarks

No study area can match the assumptions of central place theory in every detail. Although a geographer may speak of his or her study area as a 'laboratory,' the real world is not like the laboratory of the physical scientist. In the latter, factors that influence the outcome of an experiment can be largely controlled by the investigator, and hence a close correspondence may be anticipated between the empirical results and the predictions derived from a relevant theory. In geographical research, however, we must 'take things as they come.' The factors influencing an observable spatial pattern, even assuming all of them to be known, cannot be controlled by the investigator and will normally vary in strength from one part of the 'laboratory' to another. Consequently, a perfect match between theory and observation is unlikely.

Features of real urban systems most likely to be predicted by central place theory with tolerable accuracy are aggregate or average characteristics – for example, the positions of median centers of settlement and the general proximity of lower-order central places to the edges of higher-order Thiessen polygons. The Christallerian models are fundamentally 'statistical' models: they predict certain aspects of the behavior of the urban system as a whole, not the behavior of individual central places. Yet the destiny of each individual center is inextricably intertwined with the development of the system to which it belongs. Accordingly, full understanding of any individual center requires appreciation of the character of the larger systemic context. The ability of Christallerian models to provide insight into the structure of this larger context has earned central place theory its place of honor in the pantheon of geographical ideas.

Summary

The overall objective of this chapter has been to demonstrate the value of central place theory in explaining the size and geographical arrangement of real towns. Two regions – one in Minnesota and one in Ontario – are used as illustrations, but the primary emphasis is placed on concepts and methods rather than empirical details.

The chapter begins with two important reminders. First, central place theory is a partial, not a general, theory of the size and spacing of towns. Second, we should not expect real urban systems to exhibit the mathematical perfection displayed by central place models. Real urban systems are not completely regular, because real towns do not function solely as central places and the real world is not perfectly isotropic. However, there is general agreement that central place theory is the best conceptual framework available for analyzing the size and spacing of towns in agricultural regions where manufacturing and other nonretail urban activities are absent or weakly represented.

In the past, a major source of doubt concerning the value of central place theory has been the apparent incompatibility between the concept of hierarchical structuring and the smooth continuum of town size revealed by rank-size graphs. A careful analysis of this problem shows that no incompatibility exists. Rank-size graphs are concerned exclusively with town size, whereas the concept of hierarchical structuring involves not only the sizes of urban centers but also their locations. If cities and towns are studied in their actual geographical settings, it is possible to identify 'town chains' in which smaller centers are linked to larger by means of established patterns of consumer movement. Each link represents a different hierarchical order. However, because the real world is not homogeneous and real towns do not function solely as central places, the gaps that separate the different levels of the hierarchy within individual town chains are generally lost to view when many towns are placed simultaneously on a rank-size graph. Therefore, depending on the approach selected, it is usually possible to demonstrate the existence of both a continuum and a hierarchy within the same region. This dualism is illustrated above by means of a brief study of town chains in southern Minnesota.

In classifying central places into hierarchical orders, the identification of valid break-points cannot be accomplished exclusively by inspection of centrality scores. A defensible classification requires investigation of patterns of consumer movement. The guiding principle here is that two central places belong to different hierarchical orders if the residents of the smaller center habitually depend on the larger for goods and services that the smaller center does not provide. By applying this principle of hierarchical dependence systematically to all possible pairs of central places in the study area, the investigator can reliably identify the functional discontinuities that separate the various orders. (In practice, not every possible pair of centers needs to be considered: places containing fewer than six central functions may safely be assumed to be hamlets.)

The application of the principle of hierarchical dependence to a rural section of southern Ontario yields a classification consisting of four hierarchical orders. To complete the analysis, this classification is then evaluated against the four standard diagnostic criteria of hierarchical structuring derived from central place theory: (1) discrete stratification of centrality; (2) incremental baskets of goods; (3) a numerical pyramid in the frequency of central places in successive orders; and (4) interstitial placement of the orders. The application of the last criterion calls for special techniques that take account of the spatial arrangement of urban centers. Four appropriate methods of locational analysis are described. These methods are based on the following spatial concepts: (1) median centers of the settlement pattern; (2) mean distance deviations; (3) Thiessen polygons; and (4) inter-level distances.

When the standard diagnostic criteria are applied to the Ontario study area, the results indicate that the urban network conforms quite closely to Christallerian principles of spatial organization. Complete mathematical regularity is understandably absent; but, in all essential respects, the size, frequency, and spatial arrangement of the centers on each hierarchical level are similar to what would occur under the uniform conditions assumed in the theoretical models. In general, the results support the view that central place theory provides a satisfactory explanation of the geographical arrangement of urban centers in agricultural regions.

The final portion of the Ontario case study deals with the relative sizes of market areas on the different levels of the hierarchy. Comparisons are drawn between the *K*-values occurring in the theoretical models and the *K*-values obtained from the empirical data. As expected, the empirical *K*-values are found to be small, none being larger than 5. However, the empirical *K*-values are not necessarily integers, and they do not necessarily correspond to the *K*-values permissible in the realm of theory.

A key to the general character of central place research lies in the nature of the relationship between theories and observations. Theories, in a word, are tentative explanations. Like all theories except those in the physical sciences, central place theory is intended not to present an exact replica of reality but to function as a framework of relevant ideas within which to pursue satisfactory explanations. Reduced to its essentials, the methodology of central place research involves two distinct stages. First, using centrality scores and the principle of hierarchical dependence, the urban centers of a selected region are classified into hierarchical orders. Second, using central place theory as a basis for the design of appropriate tests, the empirically established hierarchy is evaluated by a series of comparisons between 'observed' and 'expected' characteristics. If the urban system passes the tests, we have no grounds for rejecting the view that the spatial organization of the system results primarily from the operation of central place principles. If the system fails the tests, we infer that additional factors beyond those encompassed in central place theory must be at work. Such factors – for example, a notably irregular pattern of rural population density or the presence of a cluster of manufacturing towns – can then be made the objects of further investigation.

From an epistemological perspective, the purpose of central place research is not to prove that central place theory, viewed as a self-contained corpus of abstract ideas, is either true or false. Indeed, truth is not a concept by which theories, as such, may usefully be judged. The ultimate purpose is to discover, separately for each region, the extent to which central place theory helps us to explain the size and spacing of towns. In short, it is not the *truth* of central place theory that is at issue, but its value as an instrument of understanding.

8

A theory that failed: central places according to Lösch

This chapter, as its title implies, deals with a theory that has not led to any important new insights. Why should we consider this theory at all? Why waste paper on a theory that failed?

The answer is that the weaknesses in Lösch's version of central place theory are not widely appreciated or understood. Lösch is still generally held in high esteem – fully comparable with that accorded to Christaller. Indeed, Christaller and Lösch are often regarded as an inseparable pair, like Sears and Roebuck, or Rodgers and Hammerstein. No doubt this view stems largely from the fact that both men saw the hexagon as the ideal theoretical shape for a city's umland. As we shall see, however, that is about as far as the resemblance goes.

The Löschian economic landscape

When August Lösch (1906–45) devised the theoretical economic landscape that bears his name, he viewed it primarily as a model for the location of manufacturing (Lösch 1954, 101–37). It has long been recognized, however, that the Löschian landscape is a tenable construct only for those types of productive activity in which raw materials exert no significant locational attraction (Isard 1956, 274). Accordingly, Lösch's approach is considered inadequate with respect to manufacturing. His model is universally regarded as a contribution to the theory of central places.

In terms of fundamental assumptions, two crucial differences between Lösch and Christaller can be identified. First, with regard to the isotropic plain, Lösch moved a step beyond Christaller and postulated that the farms covering the plain are explicitly hexagonal in shape. The farmsteads (that is, the actual farm buildings) are situated at the centers of the farms and thus form a regular triangular lattice in their own right. As the urban system develops within this precisely formulated rural environment, only farmstead locations are permitted to become central places. Lösch briefly considered the possibility of locating central places at the midpoints of the triangles formed by the farmsteads, but he rejected this alternative in favor of locating them solely at farmstead sites (Lösch 1954, 121–2).

Second, Lösch rejected the Christallerian assumption that a central place containing a function with threshold *t* will also contain all other functions having thresholds equal to or smaller than *t*. Lösch replaced this requirement with the assumption that, whenever possible, all excess profits must be eliminated, or at least minimized. In the Christallerian models, suppliers of functions with progressively lower thresholds locate, so far as possible, in extant central places, and thus excess profits are eliminated only for the suppliers of hierarchical marginal functions. The suppliers of all other functions, in order to ensure the elimination of their excess profits, would have to ignore the locations occupied by existing centers and position themselves as close to one another as their thresholds permit, thus minimizing the sizes of their market areas. This is precisely how suppliers are assumed to behave in Lösch's model: the locational choice made by each entrepreneur takes into account the locations of all suppliers of the same type but ignores the locations of suppliers of other types.

Lösch's assumption that the farmsteads on the plain form a perfect triangular lattice creates a sort of 'playing board' on which the development of the urban system can proceed. Lösch begins the game by asking what sizes of market areas are possible in such a setting. Once this question is answered, an urban system is created by superimposing all possible sizes of market areas (up to an arbitrary maximum) simultaneously on the plain. It is assumed that each possible size of market area is suitable for one particular function. In addition, the minimization of excess profit is achieved by assuming that each function is assigned to the smallest possible market area large enough to provide the necessary threshold.

The question of what sizes of market areas are possible is approached as follows. First, it is assumed that all market areas are hexagonal, so that space-filling tessellations can be formed. This automatically ensures that every consumer has access to every central function. Second, the center of every market area must fall on a farmstead site, because only farmstead locations are allowed to become central places. For a permissible size of

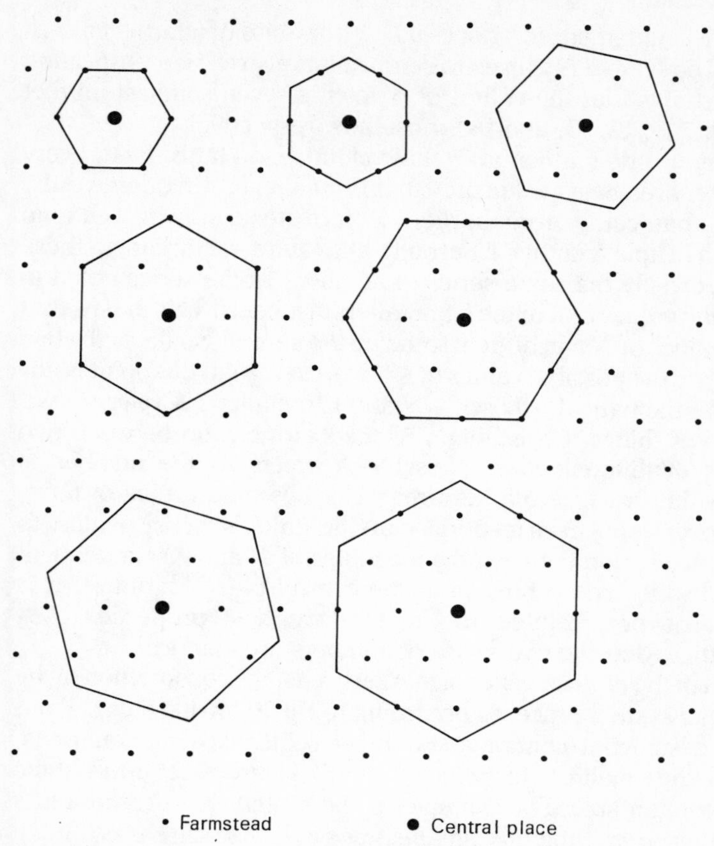

Figure 8.1
The seven smallest Löschian market areas, drawn to the same scale. The numbers of farmsteads served are 3, 4, 7, 9, 12, 13, and 16, respectively. See text for detailed explanation.

market area, therefore, an indefinitely extended tessellation can be laid down on the plain with the center of every hexagon falling precisely on a farmstead.

The seven smallest possible market areas are shown in Figure 8.1. The 'size' of any market area in the Löschian landscape is measured by the effective total demand within its borders. By convention, this demand is quantified by counting farmsteads, including the farmstead that occupies the central place's position at the center of the market area. Borderline farmsteads are counted as either one-half or one-third of a farmstead, as appropriate. The symbol *N* is used to denote market area size. Inspection of Figure 8.1 shows that the values of *N* for the seven smallest market areas are 3, 4, 7, 9, 12, 13, and 16 farmsteads, respectively.

At this point a little mathematics must be taken on faith. First, every possible market area has a value of *N* that is an integer. (In other words, the fractions that come from borderline farmsteads always end up summing to a whole number.) Second, and quite remarkably, these integers form exactly the same series of numbers as the series noted in chapter 5 for the values of *K* in the Christallerian models! This means that the possible values of *N* can be generated by the same formula as the one used to generate the possible values of *K*. However – and this point is not particularly mathematical – Lösch's *N* and Christaller's *K* refer to two entirely different things. Christaller's *K* denotes the ratio between two market areas of different sizes; Lösch's *N* refers to the number of farmsteads within a single market area. The Löschian lattice of farmsteads, moreover, is not even mentioned in Christaller's theory. Failure to make a clear distinction between the meanings of *N* and *K* is a frequent source of ambiguity and confusion in the central place literature. It is purely a coincidence, related to the mathematical properties of a triangular lattice, that the two series of numbers are identical.

Given a list of the possible sizes of market areas, the construction of the Löschian urban system proceeds according to the following steps. First, the system is assumed to contain an unbroken sequence of market areas, running from the smallest, for which $N = 3$ (Figure 8.1), up to some arbitrary maximum size. For example, if the system is to contain fifty different market areas, the fifty smallest possible sizes are used. Each market area, of course, is the basis of a space-filling tessellation or 'market net' which can be extended indefinitely in all directions across the isotropic plain. Second, as mentioned above, a one-to-one correspondence is assumed between market nets and central functions: each different market area used in the model is taken to be the appropriate size for one particular function. This assumption is somewhat restrictive, since one can conceive of various situations in which certain market areas are suitable for several functions, whereas others are not suitable for any functions at all. Lösch was aware of these possibilities, but he nevertheless produced a model in which each market net is used only once.

A minor problem of concordance could arise if the threshold of some particular function, expressed as a number of farmsteads, did not correspond exactly to one of the permissible values of *N*. In this case the function is assumed to make use of the smallest market area that exceeds the required threshold size. For example, N cannot take the value of 10, and a function having a threshold of ten farmsteads would have to be matched with the market area for which $N = 12$. Such a function would inevitably earn excess profits. This possible problem of concordance explains why Lösch postulated that excess profits would be 'minimized' rather than 'eliminated' in his model. The crucial assumption, however, is that the model incorporates each permissible net of market areas once, and once only. For all practical purposes, Lösch assumed that the possible values of N were identical to the threshold requirements of the central functions provided by the system.

The next step is the arbitrary selection of one farmstead location to serve as a 'metropolis' – a central place that provides every function available in the economy. The metropolis, in other words, is a center for every size of market area contained in the model. The market nets are now superimposed on the plain, subject to the constraint that the metropolis supplies all functions. Imagine the different market nets drawn on separate sheets of tracing paper, and visualize the metropolis as a thin, sharply pointed spindle sticking up out of the plain. The nets of market areas are impaled on this spindle, like bills of sale in an old-time country store. Care is taken, however, to ensure that the spindle passes exactly through the center of one hexagon in each net. In this way the metropolis can provide every central place function the economy contains.

If the market nets are suitably rotated about the central spindle (metropolis), the various lattices of supply points will coincide at certain locations on the plain, thereby creating multifunctional central places. Will the plain produce more than one metropolis: that is, will there be additional locations, other than the site of our original spindle, at which suppliers of *every* type of function will coincide? The answer – always assuming the plain to be sufficiently large – is in the affirmative and can be reached by two quite different routes. We must, in short, make a distinction between two varieties of the Löschian landscape: namely, the unrestricted and the truncated cases.

In the unrestricted or single-origin case, one metropolis (i.e. the original spindle) is located arbitrarily, and all other metropolises are derived by geometrical construction. The market nets are rotated so as to maximize the spatial coincidence of the different kinds of suppliers, and a metropolis springs into existence wherever all the different lattices of suppliers happen to coincide. As might be expected, in view of the inherent geometrical symmetry of the model, the resulting metropolises form a perfect triangular lattice in their own right. Moreover, the distance between any two adjacent metropolises bears a simple relationship to the sizes of market areas used in the construction. Specifically, if the farmsteads are taken to be one unit of distance apart, the distance between neighboring metropolises is equal to the square root of the least common multiple of all the values of *N* included in the model. For example, let the farmsteads be one mile apart, and let a simple system be constructed containing only the four smallest sizes of market areas. The relevant values of *N* are 3, 4, 7, and 9. The least common multiple of these four integers is 252, and the distance between adjacent metropolises is the square root of this number, or 15.9 miles.[1]

In the truncated or multiple-origin case, the market net having the largest size of market area is the first to be placed on the plain. It is then assumed that a metropolis will arise at the center of every hexagon in this net. Therefore, not one, but every metropolitan location is determined arbitrarily. Also, every farmstead lies within the market area of a metropolis, which is not the case in the unrestricted models. Systems of market area nets are then constructed around each metropolitan center independently, with the spatial extension of each net being abruptly terminated when the boundary of the metropolitan tributary area, defined by the size of the largest market area, is reached.

The structure of a truncated Löschian landscape based on the first 267 market area sizes is shown schematically in Figure 8.2. (The number 267 has no theoretical significance; it is based on considerations of graphic clarity relating to the large working diagram from which Figure 8.2 is derived.) The 267th Löschian number – excluding $N = 1$, which is taken to represent self-sufficiency on each farm – is $N = 972$. Hence the metropolitan center located at point *M* in the figure serves a tributary region containing 972 farmsteads. Note, however, that Figure 8.2 represents only one-sixth of the complete tributary area of the metropolis at *M*. Since the landscape surrounding each metropolis is radially symmetrical, the pattern shown in the diagram is repeated in every sixty-degree sector around every metropolis. Accordingly, the kite-shaped quadrilateral *MPQR* is the fundamental building-block of the entire urban system.

1 A diagram of this simple four-function system is provided by Parr (1973, 187). The rules governing the rotation of market area nets within the Löschian landscape are moderately complicated and need not concern us here. In addition to Lösch's own discussion (Lösch 1954, 124–30), see Parr (1973), Tarrant (1973), Beavon (1977), Marshall (1977b, 1978), and Sohns (1978) for details.

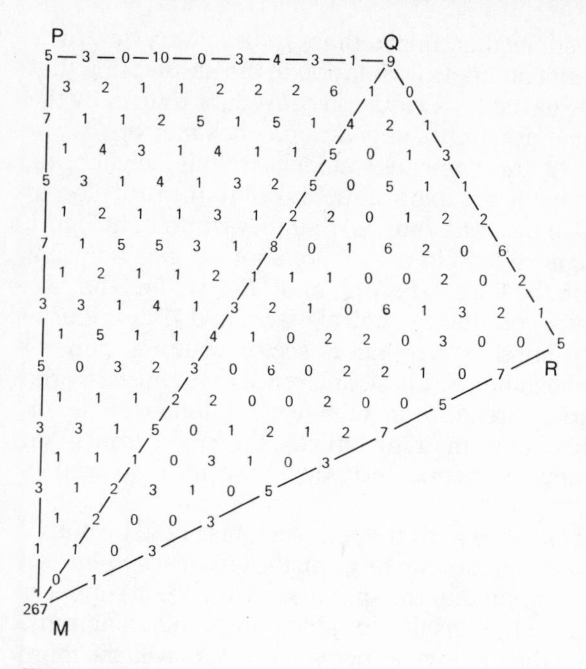

Figure 8.2
A sixty-degree sector of a truncated Löschian landscape based on the 267 smallest market areas. Each number occupies the site of a farmstead and indicates the number of different central functions offered for sale at that location. The metropolis at point *M* provides all 267 functions and serves as the origin for the construction of the figure.

The metropolis at point *M* provides, by definition, all of the 267 functions included in this example. For 89 of these functions – the 89 having the smallest market areas – at least one additional point of supply occurs elsewhere within the area covered by the diagram. The remaining 178 functions, however, have market areas sufficiently large that the nearest possible points of supply to *M* lie outside the boundary defined by the line *PQR*. These 178 functions, therefore, can be supplied only from the metropolis and not from any other central place. The suppliers of all but one of these 178 functions are not as close together as their threshold market areas dictate. (The exception is function number 267, the highest-ranking function of all; the threshold market area of this function defines the spacing of the metropolises in the first place.) The fact that these suppliers are spaced farther apart than is necessary for their economic survival represents an implicit violation of the assumption that no excess profits should be earned – a violation caused, of course, by the arbitrary truncation of the market area nets around each metropolis.

An interesting feature of the truncated models is that neighboring thirty-degree sectors differ in the numbers of supply points (central places) and establishments (retail and service outlets) that they contain. In Figure 8.2, for example, if we ignore supply points common to two adjacent sectors (i.e. points lying on the lines *MP*, *MQ*, and *MR*), we find that the left-hand or 'city-rich' sector contains 69 central places and 152 establishments, whereas the right-hand or 'city-poor' sector contains only 44 central places and 95 establishments. These differences are caused by the fact that the nets of market areas lie at numerous different angles in relation to the lines of truncation. In many lattices, therefore, truncation removes more points of supply from one thirty-degree sector than from its neighbors on either side.

Two points concerning city-rich and city-poor sectors should be noted. First, the differences between sectors can be given the greatest emphasis if we discard the earlier convention that the spatial coincidence of suppliers should be maximized. Lösch himself, to judge from his diagrams, evidently felt that stressing the differences between sectors was the more important objective, and it has become customary not to insist on the maximization of spatial coincidence in the truncated models. Second, city-rich and city-poor sectors, being purely a product of truncation, do not exist in the unrestricted models. Indeed, it can be shown that all sectors are alike in the unrestricted models regardless of whether or not the spatial coincidence of suppliers is maximized (Marshall 1977b, 10–11). Lösch was apparently unaware that city-rich and city-poor sectors cannot occur in the unrestricted case.

The presence of alternating city-rich and city-poor sectors in the truncated models has attracted much attention, because this pattern replicates, in a general way, sectoral differences in the density of population and the level of urban development that have been observed in the surroundings of real metropolises (Lösch 1954, 438–9; Nicholls 1970). It seems most unlikely, however, that such real-world differences can be attributed to the same factors as those that produce city-rich and city-poor sectors in the models. As one writer has noted, sectoral differences in development around real cities are almost certainly the result of 'the long-term advantages conferred upon locations along or near inter-metropolitan arteries'; but these differences in accessibility are 'excluded by the nature of the assumptions' in Lösch's theory (Parr 1973, 206–7). A particular spatial pattern, in other words, may have more than one possible set of causes, and a model that 'fits' does not necessarily explain.

Geometric relations between Lösch and Christaller

The general character of the Löschian landscape may be further clarified by a brief examination of the purely geometric relationships between this landscape and the models proposed by Christaller. Christaller's models are embedded within the Löschian landscape (Marshall 1977a). Each Christallerian model can be extracted from the complete Löschian landscape by selecting an appropriate subset of the many different sizes of market areas that the Löschian landscape contains. For example, in order to create the $K = 3$ model, we select a set of market areas having Löschian numbers, N, such that each value of N is exactly three times as large as its predecessor. There is an infinite number of sets possessing this property, commencing with the following examples:

$N = 3, 9, 27, 81, \ldots$

$N = 4, 12, 36, 108, \ldots$

$N = 7, 21, 63, 189, \ldots$

$N = 13, 39, 117, 351, \ldots$

$N = 16, 48, 144, 432, \ldots$

$N = 19, 57, 171, 513, \ldots$

The spatial arrangement of central places of various orders is identical in all models belonging to this series and corresponds to the pattern shown in Figure 5.3 (chapter 5). The difference between any two models in this series lies solely in the proportion of farmstead sites that fail to become central places: the larger the initial value of N, the greater this proportion. Note that the word 'arrangement' refers only to the positions of the central places relative to one another. The absolute spacing between centers is a separate issue, depending on the distance between neighboring farmsteads in each particular instance.

In a similar manner, Christaller's $K = 4$ model can be derived by selecting a set of Löschian market areas in which each value of N is *four* times as large as its predecessor. Again there is an infinite number of sets having this property, and again the arrangement of central places is the

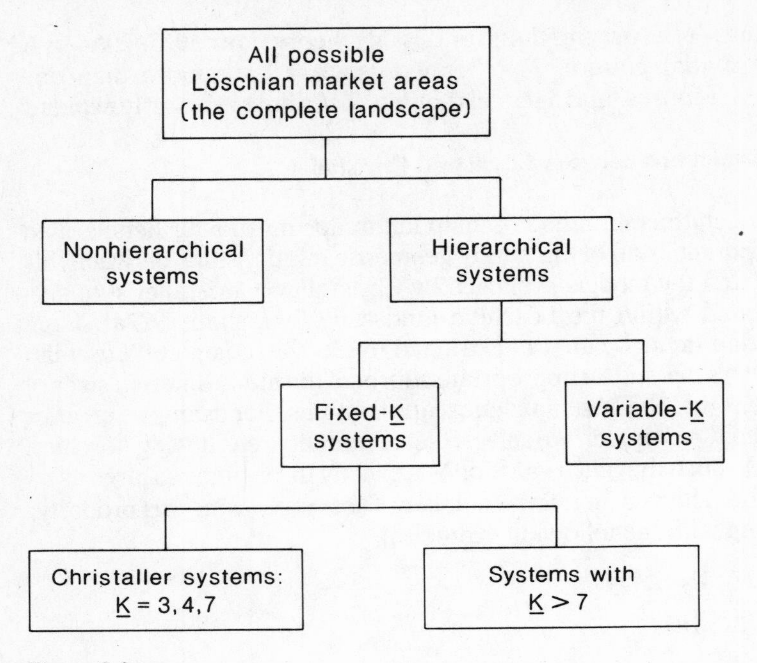

Figure 8.3
The interrelationships among Christaller's central place systems, other hierarchically structured systems, and the complete Löschian landscape

same in all cases, corresponding to the pattern shown in Figure 5.5 (chapter 5). The first few sets in this series are as follows:

$N = 3, 12, 48, 192, \ldots$

$N = 4, 16, 64, 256, \ldots$

$N = 7, 28, 112, 448, \ldots$

$N = 9, 36, 144, 576, \ldots$

$N = 13, 52, 208, 832, \ldots$

$N = 19, 76, 304, 1{,}216, \ldots$

We have already noted that the possible values of Lösch's N and the possible values of Christaller's K form exactly the same series of numbers. The product of any two members of this series is itself a member of the series (Marshall 1975c, 425). As a result, any permissible value of N may be selected as the size of the smallest market area in extracting any Christallerian model from the Löschian landscape. It follows that each of the Christallerian models is geometrically contained within the Löschian landscape in literally an infinite number of ways. (The extraction of the $K = 7$ model is left as an exercise for the reader.)

It is possible, of course, to select sets of Löschian market areas that do not correspond to any of the Christallerian models. The following summary, which should be read in conjunction with Figure 8.3, describes the various categories into which all possible sets of Löschian market areas fall.

Hierarchical systems. If, in a set of Löschian numbers (i.e. market area sizes), each number is an integer multiple of the next smaller number in the set, the associated system of central places is hierarchical in form. This 'integer multiple' property is extremely important. It can be used, in fact, as the fundamental defining characteristic of 'hierarchy.' In the language of logic, this property is a sufficient condition for the existence of hierarchical structuring. So long as this basic condition is fulfilled, the associated system of central places will possess all four diagnostic properties of an urban hierarchy: namely, discrete stratification of centrality, incremental baskets of goods, a numerical pyramid in order membership, and interstitial placement of orders. The integer multipliers in such a system are themselves always Löschian numbers.

Non-hierarchical systems. If the members of a set of Löschian numbers are related to one another by multipliers not all of which are integers, the associated central place system is nonhierarchical. An example – again, one of an infinite number of possibilities – is the system represented by $N = 9, 19, 57, 97$. Although 57 is an integer multiple of 19, not all the multipliers are integers. It is generally agreed that nonhierarchical systems cannot help us to explain the structure of real central place networks.

Fixed-K systems. If, within a hierarchically structured system, the multipliers linking the successive sizes of market areas have a constant value, we have a fixed-K pattern. If the value of K is 3, 4, or 7, we have one of the three classic Christaller models; if K is greater than 7, we have a fixed-K pattern, but one not proposed by Christaller himself.

Variable-K systems. Finally, if two or more different integers are present as market area multipliers within a hierarchical system, we have a variable-K pattern (Marshall 1977a). An example is shown in Figure 5.7 (chapter 5).

Parr (1978) has suggested the name 'general hierarchical model,' or simply 'GH model,' for the family of hierarchical patterns. Because this family includes both fixed-*K* and variable-*K* members, the GH model is considerably more flexible than the classic Christaller models as a basis for analyzing real central place systems. The three basic Christallerian patterns, in fact, can be regarded as special cases of the GH model.

An evaluation of the Löschian models

Both the unrestricted and the truncated forms of the Löschian landscape give rise to a number of difficulties not encountered in the Christallerian version of central place theory. Some of these difficulties are specific to either the unrestricted or the truncated case; some are common to both.

The unrestricted models involve unrealistically large distances between metropolitan centers. Consider, for example, a landscape with farmsteads exactly one mile apart. This implies that each hexagonal farm is 554 acres, a not unreasonable figure for general farming in the temperate middle latitudes. Now suppose that only the ten smallest market areas are included in the model. With one function apiece, this is obviously an unrealistically small number of market area sizes. But even with only these ten small market areas, the distance between neighboring metropolises turns out to be 2,495 miles! (Recall that the intermetropolitan distance is the square root of the least common multiple of all the values of *N* included in the model – the least common multiple of the first ten Löschian numbers is 6,224,400.) If more market areas are included, even larger distances are obtained. Real-world distances between fully equipped metropolitan centers are nothing like so great. For this reason alone, the unrestricted models are not particularly helpful for understanding real urban patterns.

The truncated models avoid this difficulty by permitting us to assume a reasonable distance between neighboring metropolises. For example, we may postulate a distance of 150 miles, thus automatically determining the size of the largest market area in the system. However, serious new difficulties immediately appear. First, the truncation of market area tessellations at the boundary of each metropolitan sphere of influence undermines the economic viability of suppliers located near the boundary by depriving them of portions of their natural market areas. Since these market areas, before truncation, are either equal to, or just barely larger than, the necessary threshold size, many of the affected suppliers will be forced out of business. In consequence, either there will be some portions of the plain not provided with certain functions, or else suppliers located farther from the boundary and closer to each metropolis will earn significant excess profits by expanding their market areas into the affected zones. Both outcomes are inconsistent with Lösch's initial assumptions.

Second, as indicated earlier, many functions can be supplied only from a metropolis, because the necessary distance between adjacent supply points in the relevant market net is greater than the radius of the metropolitan sphere of influence – that is, greater than the radius of the largest market area in the model. Once again, significant excess profits sneak in by the back door, because entrepreneurs who supply these functions from each metropolis will presumably attempt to expand their market areas in order to serve, and benefit from, the entire metropolitan sphere of influence. Moreover, the number of functions affected in this way is not trivial. In the model illustrated in Figure 8.2, for example, two-thirds of all functions in the economy are in this category. It can be shown, in fact, that two-thirds is the expected proportion; Figure 8.2 is not an exceptional case. Accordingly, this problem cannot be brushed aside on the grounds that only a few functions are affected.

Third, the truncated models incorporate unrealistically large numbers of different sizes of market areas. In other words, they make provision for too many central functions of different types. As an example, let us retain the convention that farmsteads are one mile apart, and let us assume a distance of 150 miles between neighboring metropolises. Without dwelling on the necessary mathematics, we can state that a truncated model built to these specifications includes 4,906 different sizes of market areas (Marshall 1975c, 1978). By no stretch of the imagination does the real world contain 4,906 different types of retail and service outlets. The most exhaustive studies suggest that the upper limit lies between 300 and 400, depending on the number of fine distinctions one is prepared to recognize (Berry, Barnum, and Tennant 1962; Borchert and Adams 1963; and especially Barnum 1966, 123–9). Further, 150 miles is not an excessively large distance between metropolitan centers in real-world terms. But if a larger distance is assumed, the number of central functions required for the completion of the model is even greater than 4,906.

The Löschian theorist is thus placed in a quandary. The unrestricted models succeed in minimizing excess profits, but they locate the metropolitan centers ridiculously far apart even if only a trivially small number of different functions is provided. The truncated models permit a reasonable spacing for the metropolises, but they also permit widespread earning of significant excess profits. In addition, the truncated models require the presence of many more types of retail and service businesses than are ever likely to be encountered in the real world.

Both types of models, moreover, pay almost no attention to the natural agglomerative tendency of real entrepreneurs. In the Löschian world, the spacing between suppliers of any given function is determined primarily by the principle that excess profits should be minimized, and each set of suppliers is therefore located on the plain with little regard for the locations occupied by suppliers of other types of functions. It is true that each metropolis serves as a point of reference for all functions, and that the market nets in the unrestricted case are rotated in order to maximize the average number of functions provided at each central place, but these controls are introduced after the size of market area adopted by each type of business has been determined in accordance with the fundamental assumption of minimum excess profits. Lösch's desire to minimize excess profits clearly took precedence over his interest in bringing about the spatial agglomeration of firms.

Lösch's preoccupation with excess profits leads to absurd results. Many central places in the Löschian landscape offer a combination of high-order and low-order functions, but with no middle-order functions. Quite frequently a single high-order function is provided in splendid isolation by a central place that performs no other kind of activity. In addition, the Löschian urban pattern is one of extremes, consisting of a few metropolises that provide every type of function and a large number of very small centers that provide no more than a handful of functions apiece. The system illustrated in Figure 8.2, for example, includes 267 different activities, but no center other than the metropolis offers more than ten functions! The system appears to be a clearly defined two-order hierarchy, since the gap in functional complexity between the metropolis and all other centers is extremely wide. The small centers, however, actually constitute a heterogeneous group of places that possess market areas of many different sizes and that repeatedly violate the criterion of incremental baskets of goods. The small centers do not form a homogeneous order of hamlets in the Christallerian sense. Thus Löschian urban systems cannot be said to be hierarchical in structure.

In view of all these difficulties, it is impossible to avoid the conclusion that Lösch's models are inferior to Christaller's as a theoretical basis for the investigation of real central place systems. Retail and service firms in the real world simply do not behave like the entrepreneurs postulated in Lösch's theory. Instead, they congregate at a limited number of locations, giving rise to hierarchically structured urban systems of the sort that Christaller's theory describes. Viewed as a problem in geometry, the Löschian landscape presents a considerable challenge, and it seems to have attracted a great deal of attention for that reason alone. However, we must not commit the fallacy of assuming that Lösch's theory, simply because it is more recent and more difficult to grasp than Christaller's, necessarily provides a more accurate description of reality. All the available evidence indicates that Christaller's models are much more realistic than those of Lösch. We must conclude that the Löschian landscape has little to offer as a basis for understanding the spatial structure of real urban networks.

Summary

This chapter has presented a description and evaluation of the family of theoretical models known collectively as the Löschian landscape. In the study of towns in their role as central places, the Löschian landscape represents the only fully articulated body of theory that might conceivably provide an alternative to Christaller's models.

The fundamental difference between the Löschian and Christallerian versions of central place theory lies in Lösch's desire to construct an urban system in which excess profits would be held to an absolute minimum. This means, in effect, that each different central function is matched with a market area of no more than threshold size. In Christaller's theory, excess profits are automatically minimized for the suppliers of hierarchical marginal functions, but all other suppliers are free to earn excess profits to the extent that this is permitted by the spacing of the hierarchical marginal suppliers. In each Christallerian model, the number of different sizes of market areas is equal to the number of hierarchical orders within the urban system. In each Löschian model, by contrast, the number of different sizes of market areas is equal to the number of central functions in the system, not the number of hierarchical orders. In fact, there are no hierarchical orders, in the Christallerian sense, within the Löschian models.

Two varieties of Löschian models may be distinguished. First, there are unrestricted models, in which the location of only a single metropolitan center is determined arbitrarily. Other metropolises emerge as a result of geometric interrelationships among the networks of different market areas. Second, there are truncated models, in which an entire triangular lattice of metropolitan centers is taken as given. In the truncated models, in other words, the lattice of metropolises is an assumption rather than a consequence of the construction of the model.

However, in both the unrestricted and the truncated cases, each metropolis is a center of supply for every type of central function present in the economy.

There is a simple geometric relationship between each of the classic Christallerian models and the fully developed Löschian landscape. Knowing the value of *K* in the required Christallerian model, one extracts from the Löschian landscape a series of market areas such that each area is exactly *K* times as large as its predecessor. In effect, the classic Christallerian models are incomplete versions of the full Löschian landscape.

Close examination of their structure reveals that the Löschian models invariably fail to match the general characteristics of real urban systems. In short, the Löschian models are empirically false. The unrestricted models are made internally consistent – in minimizing excess profits – only by placing metropolitan centers ludicrously far apart. The truncated models permit a plausible spacing of metropolises, but this allows widespread earning of excess profits (thereby violating Lösch's major assumption) and requires the presence of an unrealistically large number of different types of retail and service businesses.

The conclusion is that the Löschian landscape is distinctly inferior to the Christallerian models as a theoretical framework for explaining the size and spacing of real urban centers.

9

Cities and long-distance trade: the mercantile model

Central place theory, as has been stressed, is concerned with towns as centers of local trade. But trade, as a general category of economic activity, has a second major dimension: long-distance trade. The effects of long-distance trade on the spatial structure of urban systems are described by the mercantile model. This model was introduced by James Vance in 1970 in *The Merchant's World: The Geography of Wholesaling*, a book that ranks second only to Christaller's work as a contribution to our understanding of the size and geographical arrangement of cities (Vance 1970).

Vance's mercantile model and Christaller's central place theory represent complementary approaches to the study of the geography of urban systems. Neither approach, on its own, furnishes a complete account of the influence of trade on systemic structure. In combination, however, they cover this topic comprehensively. Each approach offers something that the other does not provide. In central place theory, the agent of trade is a retailer, and the customer is a final consumer of the product or service supplied. In the mercantile model, by contrast, the crucial agent of trade is a wholesaler; and the customer is not a final consumer, but another wholesaler, a manufacturer, or a retailer.

In order to place the mercantile model in perspective, we begin by considering two characteristics of central place theory that we have hitherto ignored.

Two limitations of central place theory

Quite apart from the fact that Christaller's approach addresses only one of the major functions that towns perform, central place theory is narrow in two other respects: it is static, and it is inward-looking.

Central place theory is static in that it lacks a sense of *development*. It provides a rational justification for a series of spatial patterns that make sense in terms of economic efficiency, but it does not incorporate any evolutionary process by which these plausible arrangements of local trade centers could have come into being (Houston 1963, 141). Central place theory paints a picture of a world that somehow springs into existence fully formed, with no historical antecedents. No matter how attractive we may find central place models in esthetic terms, they refer to a realm quite unlike the evolving world we actually inhabit.

It is not being argued here, incidentally, that Christaller was ignorant of history. Indeed, as Preston has emphasized, Christaller explicitly recognized the fact of change, and he discussed the dynamics of central places at considerable length (Christaller 1966, 84–132; Preston 1983). The Christallerian models, nevertheless, are static in the sense that they are intended as models of *pattern*. They cannot be regarded, in any historically meaningful sense, as models of *process*.

Central place theory is inward-looking in its exclusive concern with a self-contained system of consumer servicing that involves no explicit economic links with the outside world. No writer in the central place literature ever asks where the goods bought by the farmers on the isotropic plain originate. Some of these goods can doubtless be accounted for as products of local craftsmen; but even a superficial glance at reality tells us that many commodities, viewed in relation to the rather short distances covered by the average shopper, are exotic. In other words, the shelves of retail stores are stocked, at least in part, with items of nonlocal origin. But retailers do not pull their stocks of nonlocal goods out of thin air. They depend on wholesalers – which brings us back to the long-distance component of trade.

It is a truism, though one sometimes overlooked by central place enthusiasts, that every urban place is connected to the complex network of commodity movements that forms the worldwide system of long-distance trade. Local traffic created by this long-distance trading activity may be small in volume and quite unobstrusive, especially in tranquil rural districts where village gossip takes precedence over news of national importance. Total isolation, however, is impossible. Each local system of retail provision must therefore be seen as the terminus of an extended chain of production that involves numerous steps – from mining or harvesting, through processing, fabricating, and packaging, to storage, shipment, and final distribution. The goods offered for sale on the retail level in Iowa or Bavaria originate in every corner of the globe. And within the overall process of production and distribution an important role is played by the merchant wholesaler, a shadowy figure whose significance in the evolution of urban systems has only recently begun to be fully appreciated.

In a general way, the fact that the fortunes of towns are linked to long-distance trade has been known for some time, particularly through the works of writers such as Gras (1922), Van Cleef (1937), Bridenbaugh (1938), and, above all, Pirenne (1925, 1937). But the geographer's concern with spatial relationships, rather than corporate organization and political intrigue, received scant attention until the appearance of Vance's penetrating analysis in *The Merchant's World* (Vance 1970). Vance perceived that central place theory is static and inward-looking because it deals only with local or retail trade, and he presented a complementary analysis of long-distance or wholesale trade that is both outward-looking and dynamic. Like many important conceptual advances, Vance's mercantile model seems 'obvious' once stated, but this in no way detracts from its importance. Indeed, it only makes the rest of us seem somewhat slow and muddle-headed. After all, we were vexed by the very problem Vance managed to solve. One is reminded of Huxley's self-critical reaction upon learning of Darwin's principle of natural selection: 'How extremely stupid,' he said, 'not to have thought of that' (Howard 1982, 8).

Elements of the mercantile model

The mercantile model is a stage model; it attempts to capture the dynamics of a particular phenomenon by identifying a series of distinct evolutionary stages through which the phenomenon has passed. In essence, a stage model is a simplified and segmented form of history. Each stage is distinguished from its predecessor by the appearance of some significant new element in the overall process. Such models may, but need not, incorporate specific historical dates to which the beginning and end of each stage may be referred. In the general case, no specific dates are provided, and the model therefore stands as a conceptual abstraction rather than a concrete historical narrative. Normally the timing of the commencement of each stage will vary from place to place within the geographic area to which the model applies.[1]

The basic premise of the mercantile model is that the fundamental

1 The adoption of a stage model is motivated by a desire to facilitate understanding, and is not in any sense a denial of either the complexity or the continuity of the historical experience from which the model has been distilled. That experience, in any case, can never be *our* experience. As Oakeshott so perceptively observed, 'the facts of history are *present* facts' (Oakeshott 1933, 108; emphasis added).

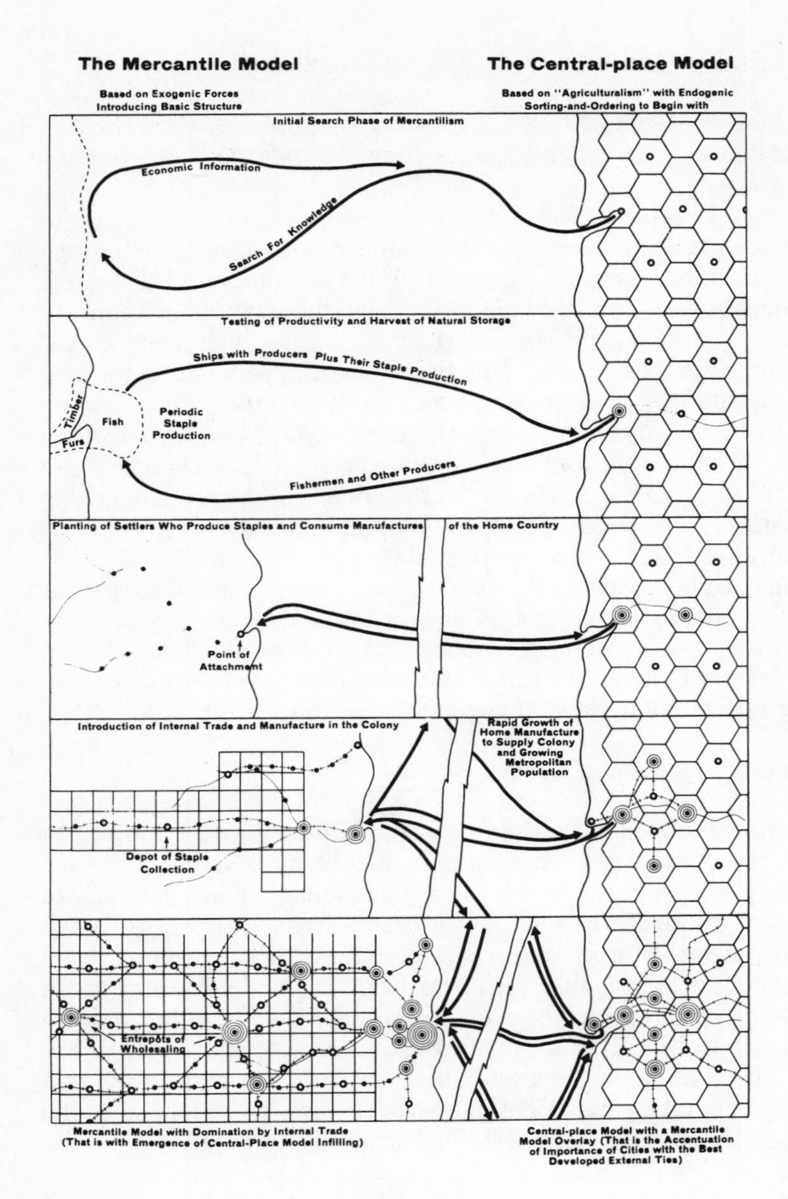

Figure 9.1
The Vance mercantile model depicting the evolution of the urban system of the New World. SOURCE: James E. Vance, Jr, *The Merchant's World: The Geography of Wholesaling* (Englewood Cliffs: Prentice-Hall 1970), 151, reproduced with permission

creative force molding the spatial structure of an urban system is the establishment of dependent outposts in new territories by distant commercial interests. As this basic premise implies, the mercantile model is first and foremost an account of European settlement in North America. Indeed, the model's applicability to other areas is at present an open question, especially for Europe itself, where the formative period of urban development prior to the era of overseas expansion was dominated by feudalism and thus by an emphasis on local rather than long-distance trade. It may well be that the mercantile model cannot be applied to European cities without significant modification. Given our current knowledge, the model is best regarded as a generalization tailored specifically for areas of European settlement in the New World.

The process of settlement encapsulated in the mercantile model consists of five stages.

Stage 1: Exploration. In this initial stage, the old land (Europe) sends out expeditions to explore the new land (America) and to bring back reports concerning the availability of natural resources and the suitability of the new land for settlement (Figure 9.1, topmost panel).

Stage 2: Harvesting of natural resources. In the second stage, the emphasis shifts from transmission of information to shipment of tangible products. At first, however, these products are limited to commodities available directly from nature. In the case of North America, the earliest significant product was the codfish of the Grand Banks of Newfoundland, followed in due course by beaver pelts and timber (Innis 1954, 1956b; Lower 1938; Wynn 1981). Settlement begins during this stage, but the earliest settlements are viewed by their old-country sponsors as principally the centers of control for exploiting natural resources, and only secondarily as being the nuclei of agricultural colonies.

Stage 3: Emergence of farm-based staple production. As time goes on, the number of permanent settlers in the new land steadily increases. More and more of the natural forest cover is cleared, viable farms and plantations are established, and it soon becomes possible for the new land to export not only natural products such as furs and timber but also agricultural commodities such as grain, salted meat, indigo, tobacco, and cotton. (Climate and other environmental factors obviously play important roles in the selection process: John Rolfe's successful experiments with tobacco in Virginia would have been doomed to failure if attempted in Massachusetts.) This stage is also characterized by the full flowering of the economic philosophy known as mercantilism. According to this view, it is the responsibility of the new land (i.e. the colonies) to provide the old land (i.e. the mother country) with foodstuffs and other raw or partly processed materials, whereas it is the prerogative of the mother country to supply the colonies with manufactured goods. Given the transoceanic character of the trade, the principal urban centers in the new land are the seaports – each of which derives much of its strength from a burgeoning community of wholesale merchants who control the flow of trade. Viewed from the perspective of mercantilists in the mother country, these seaports serve as the points of entry to the colonies and their resources. Such centers are termed 'points of attachment' in Vance's model (Figure 9.1, center panel).

The mercantile model is intended to apply to North America as a whole, not necessarily to each individual colony. At the regional, as distinct from the continental, scale, the second stage of the model may be omitted if no natural staple commodity exists in sufficient abundance. For example, as Earle (1977) has emphasized, no suitable natural staple was available to serve as a basis for transatlantic trade in the earliest days in Virginia, and the viability of the colony remained in serious doubt until the raising of tobacco was successfully established. Broadly speaking, the second stage is less applicable to the United States than to Canada, where its undoubted significance has given rise to the so-called staple theory of economic development (Innis 1956a; Watkins 1963; Wolfe 1968). The overwhelming importance of naturally occurring staples in Canada may be seen in the list of basic Canadian exports – fish, furs, timber, wheat, wood pulp, and minerals; only one commodity in this list is not obtained directly from nature.

Stage 4: Establishment of interior depot centers. The increasing population and the continuing demand for staple exports combine to draw settlement progressively farther into the interior. This process of penetration is marked by the selection of long-distance routes to facilitate the movement of staple products from the interior to the points of attachment on the coast. Towns are established at strategic locations along these routes to serve as 'depots of staple collection' (Figure 9.1, fourth panel). These inland towns are first and foremost instruments of long-distance trade rather than central places serving a local population. They are wholesale collection centers rather than centers of retail distribution. Most are established before the arrival of significant numbers of permanent rural settlers, a fact summed up in Wade's celebrated dictum that 'the towns were the spearheads of the frontier' (Wade 1959, 1). During this stage, also, the new land begins to develop its own manufacturing industries, most noticeably in the original seaboard cities. These centers not only offer the largest market and labor supply but also enjoy the readiest access to the necessary technical knowledge, most of which – at least during the initial phases of industrial development – emanates from cities in the mother country.

Stage 5: Economic maturity and central place infilling. The growth of the manufacturing sector ultimately leads to economic maturity: the new land no longer has to rely on the mother country for manufactured goods. In short, the new land attains economic independence. (Political independence is a separate issue; this is not explicitly considered in Vance's model but in the American experience was achieved quite early, before the end of stage 3.) Most important, however, is the fact that permanent agricultural settlement 'catches up' with the previously established network of depots of staple collection. The areas of land situated between the lines of long-distance movement now become dotted with systems of central places. But the depots of staple collection also take on a central place role; and usually, with their early start and superior accessibility, they emerge as the dominant central places of their respective districts. The same applies to the original points of attachment on the coast. Thus the wholesaling or mercantile cities are not a random subset of the complete central place network; they are the leading central places. Because of their dual role as the control points of long-distance trade and as the largest and most functionally complex central places, their pattern of location determines the basic character of the spatial structure of the entire national urban system (Figure 9.1, bottom panel).

As the above description of the mercantile model indicates, the geographical arrangement of the New World's major cities is basically exogenic – determined by forces originating outside the area. The spatial structure of the North American urban network is fundamentally a product of decisions made in the past by English merchants. By emphasizing the external origins of New World cities, the mercantile model takes on a distinctive outward-looking character that contrasts sharply with the inward-looking quality of central place theory. Given central place theory's traditional set of assumptions, changes in the spatial organization of a pure system of central places can come only from inside the system itself – for example, by means of a change in the density of purchasing power on the isotropic plain. Such a change is termed endogenic – it originates inside the very system where its effects come to be felt.

A contrast may now be drawn between the urban network of North America, which is fundamentally exogenic, and that of Europe, which is fundamentally endogenic. One implication is that an understanding of European urbanization should rest primarily on central place theory, whereas an understanding of the American pattern should rest primarily on the concept of mercantile settlement. The key word in the preceding sentence, however, is 'primarily.' On both sides of the Atlantic, both approaches are required – and even then, we must add an appreciation of the rise of modern manufacturing in order to complete the main features of the picture. Nevertheless, the best point of departure for a satisfactory explanation of the North American urban pattern is probably the mercantile model, whereas central place theory remains the most appropriate starting-point in the European case.

The five-stage model presented above goes beyond the purely 'mercantile' component of the process in question. Strictly speaking, the term *mercantile* refers only to wholesaling and to closely related supporting activities in fields such as banking, insurance, and transportation. Both manufacturing and retailing are specifically excluded. Thus the concept of a mercantile urban network should encompass neither the growth of manufacturing nor central place infilling. Accordingly, we are faced with an inconsistency: Vance began by identifying the mercantile element in urban growth as being distinct from the central place and manufacturing elements, yet he later extended his general description of the mercantile model to include both of these latter elements (Vance 1970, 138–67). This difficulty can be overcome if we simply apply the term *mercantile model* to a comprehensive view of settlement that is based on the historical primacy of mercantile (i.e. wholesaling) activity and that also acknowledges the subsequent importance of both the central place and manufacturing functions.

Given that the mercantile model recognizes the importance of central place activity, and given that it is process-oriented, we may ask in what sense the model provides the dynamic element that is missing from classical central place theory. The answer is not spelled out explicitly by Vance, but mercantile activity, with which the entire process of settlement in each region commences, 'fixes' the spatial pattern of the towns destined to become the leading central places of the fully developed system. These centers, of course, are small when first established – Pittsburgh in 1800 could boast only 1,565 residents – but each of them is nevertheless the largest place within its own local region. In the normal course of events, moreover, each maintains its dominant position as time progresses. As explained in chapter 1, the leading centers of an urban system exhibit a high degree of rank stability through time. In an important sense, therefore, the Christallerian view that the highest-order central places are the first to be established is quite correct! This view seems odd because we tend to think of these highest-order centers as they exist today, when many of them contain several million inhabitants. We naturally reject the notion that centers of metropolitan size could appear fully formed on a surface containing no other towns. But when we adopt an evolutionary point of view and think of these places as developing slowly from modest beginnings, always leading the pack but only recently attaining their present gigantic size, the idea that the highest-order centers are the earliest to appear becomes easy to understand.

Once established, the network of mercantile centers becomes the principal determinant of the geometry of the evolving central place system. As settlers take up land in the areas between the long-distance trade routes, central places are founded to cater to local needs. Because the real world is not isotropic, we should not expect the resulting pattern to display the geometrical regularity of the Christallerian models. In general, however, a hierarchical system consisting of numerous low-order centers and relatively few higher-order centers will emerge. Going a step further, the coming of the railroad and the appearance of localized concentrations of manufacturing may tend to obscure the basic central place pattern (Muller 1977). Nevertheless, in all but the most heavily industrialized districts, the contribution of central place activity to the overall character of the urban pattern is usually not difficult to discern. The urban system as a whole may be viewed as a compound structure incorporating linear, dispersed, and clustered elements that reflect the presence of mercantile, central place, and manufacturing activities, respectively.

The principal mercantile alignments

Mercantile activity in North America has produced an urban pattern dominated by seven geographically distinctive groups of cities. Each group is essentially linear in form, giving rise to the use of the term *alignments* in descriptions of the system's general structure (Vance 1970, 157–59). The seven mercantile alignments are shown in Figure 9.2.

The first linear group is the Seaboard alignment, consisting of the original points of attachment along the eastern seaboard. This group of centers is dominated by the 'big four' Atlantic ports of Boston, New York, Philadelphia, and Baltimore, but it also includes smaller cities such as Portland, Newport, Providence, Norfolk, Charleston, and Savannah. The

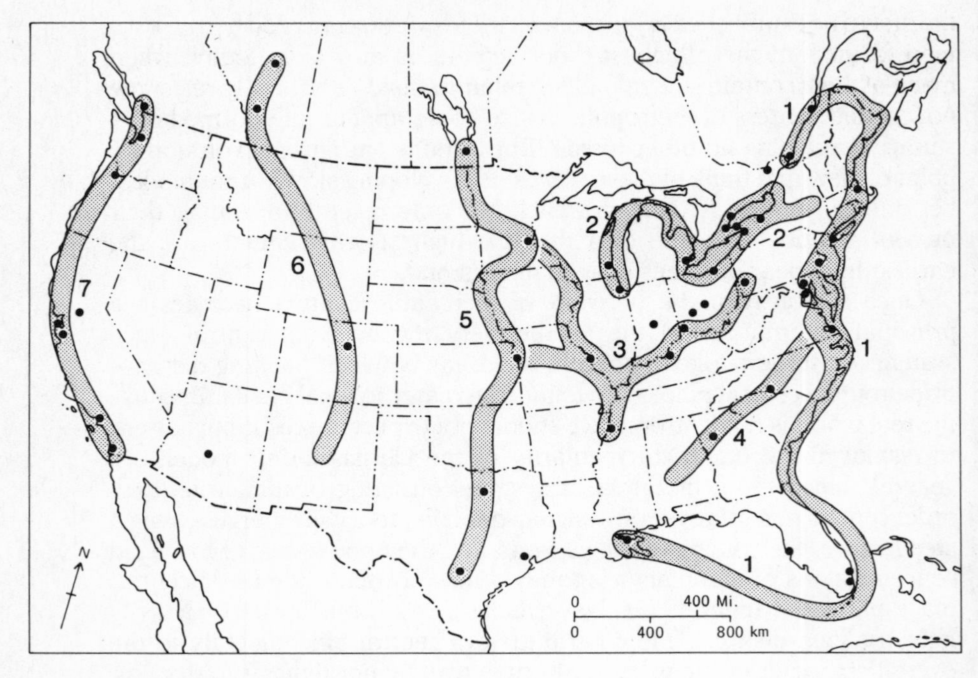

Figure 9.2
Vance's seven historic mercantile alignments: (1) Eastern Seaboard; (2) Great Lakes; (3) Ohio-Mississippi-Missouri, or River; (4) Appalachian Piedmont; (5) Eastern Plains; (6) Rocky Mountain Piedmont; and (7) Pacific Coast. The fifty cities shown are the forty largest metropolitan areas in the United States and the ten largest in Canada in 1980–1; only eleven of these cities lie outside the seven alignments.

city of New Orleans is also usually included – a special case because of its relative remoteness from the Atlantic but normally placed in the Seaboard alignment rather than the River alignment, described below.

The Seaboard alignment may be extended northward to include the Canadian ports of Saint John, Quebec, and Montreal – though perhaps not Halifax, the 'Warden of the North' (Raddall 1971), which has always been essentially a naval base rather than a commercial gateway. The inclusion of the St Lawrence River cities of Quebec and Montreal may seem odd, but these centers resemble the eastern seaboard cities in being bona fide points of entry to North America for Atlantic shipping, directly accessible to ocean-going vessels since the earliest days of settlement. In this respect the St Lawrence River below Montreal is functionally similar to inlets such as Delaware and Chesapeake bays, and even the Gulf of Mexico. Like New Orleans, Quebec and Montreal are 'Atlantic' ports despite their apparent seclusion in locations distant from the main body of the Atlantic Ocean.[2]

The second group is termed the Lakes alignment (Vance 1970, 158). It consists of the major Great Lakes ports together with the leading centers along the Erie Canal. It includes, on the US side, Albany, Syracuse, Rochester, Buffalo, Erie, Cleveland, Toledo, Detroit, Green Bay, Milwaukee, and Chicago, and, on the Canadian side, Kingston, Toronto, Hamilton, St Catharines, and – far to the northwest – Thunder Bay. Kingston and Albany stand at the Canadian and American entrances to this alignment, respectively, with the Lake Ontario–Welland Canal route joining the upstate New York route at Buffalo. From Buffalo westward the essentially linear arrangement of this set of towns is quite apparent, particularly from the perspective of lake-borne shipping. There is an interesting contrast in urban development between the American and Canadian sides of Lake Erie, with the major cities being situated exclusively on the American side. The main focus of settlement in the old British colony of Upper Canada was the north shore of Lake Ontario rather than that of Lake Erie. The concentration of urban growth on Lake Ontario has persisted to the present day: Toronto has recently overtaken Montreal in size to become the largest urban complex in Canada.

The third mercantile grouping runs roughly parallel to the second and is known as the River alignment. It encompasses the Ohio River, the lower Missouri, and the middle course of the Mississippi from La Crosse to Memphis. The alignment begins in the east at Pittsburgh, where the Allegheny and Monongahela rivers join to form the Ohio. Moving westward, it includes Cincinnati, Louisville, Evansville, Memphis, St Louis, and Kansas City (Figure 9.2). The Ohio-Mississippi-Missouri route was the principal thoroughfare followed both by traders and by settlers in the penetration of the trans-Appalachian west during the nineteenth century, and the leading cities of the area between the Great Lakes and the Gulf of Mexico are concentrated along this route.

The fourth alignment is found in the south, along the eastern and southern margins of the Appalachian Mountains. The chief cities of this

2 Unlike the other cities of the Seaboard alignment, Quebec and Montreal lack year-round ice-free conditions – a persistent problem for Montreal in its long-standing trade rivalry with New York. Also, Saint John, Portland, and Boston have all benefited from their ability to serve as winter outlets for Canadian goods.

Appalachian Piedmont alignment are Richmond, Greensboro, Charlotte, Greenville, Atlanta, and Montgomery. Except for Richmond, these centers are not the famous Fall Line cities; they are situated on the piedmont proper and not at the line of contact between the piedmont and the Atlantic coastal plain. (The Fall Line cities lie east of the mercantile alignment and include Richmond, Petersburg, Raleigh, Columbia, Augusta, Macon, and Columbus.) The Appalachian Piedmont alignment, as Vance has pointed out, differs from the Seaboard, Lakes, and River alignments in that it owes its importance more to the railroad than to navigable waterways (Vance 1970, 158). The leading staple exports were also unique – tobacco and cotton in the south, as compared with grain and other temperate products farther north.

A case might be made that Richmond belongs to the Seaboard alignment rather than the Appalachian Piedmont alignment. First, as noted above, Richmond is not situated on the piedmont proper but on the Fall Line. Second, and more crucial, the drowned lower course of the James River made Richmond during the eighteenth and nineteenth centuries accessible to ocean-going ships, an advantage denied to Greensboro, Charlotte, and the other cities of the piedmont alignment. Indeed, the navigability of the lower James was an important factor in the transfer of the capital of Virginia from Williamsburg to Richmond in 1780 (Stanard 1923, 37). Unlike all the other cities of the piedmont alignment, Richmond was a port engaged directly in transatlantic trade. And the city was by no means badly placed for this activity. In terms of sailing distance, Richmond is distinctly closer to the open Atlantic beyond the Virginia Capes than is Baltimore, a city whose membership in the Seaboard alignment is not in doubt. It therefore does not seem unreasonable to regard Richmond as a seaboard city. However, Vance rightly calls attention to Richmond's affinity with the more southerly piedmont centers, to which it was early linked by rail and with which it shared specific economic interests. Perhaps the most satisfactory solution is to rejoice in the uniqueness of Richmond's geographical situation and to regard it as belonging simultaneously to both the seaboard and piedmont alignments. After all, there is no compelling reason why membership in two or more alignments should be categorically forbidden.

The fifth major group of mercantile cities forms an alignment that runs north and south, approximately along the line of the ninety-fifth meridian (Figure 9.2). The common factor is the importance of this strip of territory as a staging area for the collection and processing of the agricultural products of the Great Plains and for the forwarding of these products to eastern markets. Grain and beef are the principal products. Grain elevators, flour mills, and meat-packing plants are conspicuous features of the urban landscape in all the major centers. Vance terms this zone the 'Collecting Alignment' – but perhaps this is inappropriate, because the general function of 'collecting' is not confined to this section of the continent. We will use the term Eastern Plains alignment, even though, in terms of physical geography, the towns in question actually lie within the American Interior Lowland, not within the Great Plains, as normally defined (Mather 1972).

The northernmost city of the Eastern Plains alignment is Winnipeg. Situated in the narrow gap between the international border and the rocky expanse of the Canadian Shield, Winnipeg is widely known for its historic role as the gateway to the Canadian Prairies. Several writers, including Careless (1970), Burghardt (1971), and Kerr (1977), have stressed the great importance of wholesaling in Winnipeg's early development. A similar history characterizes the other leading cities of the alignment, including Fargo-Moorhead, Minneapolis–St Paul, Sioux City, Omaha, St Joseph, Kansas City, Wichita, and Dallas–Forth Worth. Kansas City has already been listed above as a member of the River alignment. Like Richmond in the east, Kansas City can fairly be regarded as belonging to two distinct alignments.

The sixth mercantile alignment lies at the western margin of the Great Plains and is known as the Rocky Mountain Piedmont alignment. From north to south the principal cities are Edmonton, Calgary, Lethbridge, Great Falls, Billings, Cheyenne, Denver, Pueblo, and Albuquerque. These cities have acted not only as collection centers for the agricultural products of the piedmont zone but also as focal points for the exploitation of the mineral resources of the Rocky Mountains. In varying degree they are also gap towns, standing adjacent to the chief passes through the mountain barrier; and in several cases their early growth was greatly stimulated by the building of the transcontinental railroads. Calgary, situated near the narrow defile of the Kicking Horse Pass through the Canadian Rockies, provides a particularly clear example of this relationship.

The seventh and final group is the Pacific Coast alignment: Vancouver, Seattle, Tacoma, Portland, San Francisco, Monterey, Los Angeles, and San Diego. Victoria, the provincial capital of British Columbia, could be included in the early years, but its island location proved to be a liability. Vancouver, founded in 1886 as the western terminus of the Canadian Pacific Railway, rapidly eclipsed Victoria in size and commercial importance and went on to become Canada's third-largest metropolis (McCann 1978). Victoria's limited growth during the present century has been sustained less by mercantile activity than by its dual role as a political capital and naval base – like Halifax, situated at the opposite end of the country. Another Pacific center that might be added to the list is Prince Rupert, the town originally selected to be the terminus of the Grand Trunk Pacific rail line (later part of the Canadian National system). This line was completed as planned, but the greater prosperity of the Vancouver area soon led to second thoughts, and a new rail link was built that effectively reduced the Prince Rupert connection to branch-line status.

The Pacific coast of North America was claimed and settled from the sea in spite of its remoteness from Europe. In these days of direct over-the-pole air travel, an ocean voyage from England to California by way of Cape Horn seems daunting to say the least, yet during the nineteenth century such voyages were taken for granted. The cities of the Pacific Coast alignment were therefore no less exogenic in origin than their counterparts on the eastern seaboard. Their mercantile connections were maintained almost entirely by sea until the arrival of the first transcontinental railroad line in San Francisco in 1869. Even long after that date, ocean transportation remained paramount. The opening of the Panama Canal in 1914 did more than symbolize the United States's status as a two-ocean naval power; it also served the purely commercial interests of the cities of the Pacific Coast alignment.

Viewing the North American urban pattern as a whole, it is evident that the leading cities in both the United States and Canada are, to a remarkable degree, concentrated along the seven great mercantile alignments. Using present-day metropolitan population totals as a measure of overall importance, we find that thirty of the forty leading American cities and nine of the top ten Canadian cities lie on these major alignments (Figure 9.2). It is virtually inconceivable that this level of concentration could have occurred entirely by chance. In fact we can give our findings a crude but striking probabilistic interpretation. Suppose that the total area occupied by the strips of land comprising the seven alignments amounts to 20 per cent of the habitable territory of North America. This is quite a generous estimate, since it requires that each alignment be something like eighty miles in width. It then follows that, if the location for an individual city is selected randomly, the probability that the city will fall on one of the mercantile alignments is 0.20. We note that thirty-nine of the fifty leading cities actually do fall on one or another of the seven alignments. We now have a straightforward binomial problem: what is the likelihood of obtaining thirty-nine or more successes in fifty independent trials when the probability of success on a single trial is 0.20? The calculations show that this likelihood is less than one in a quadrillion (1 quadrillion = 10^{15}). Accordingly, it is futile to suppose that the affinity of the leading cities for the mercantile alignments is merely a random phenomenon.

In general, therefore, the geography of urbanism in both the United States and Canada is first and foremost a product of long-distance trading activity. The two other principal functions that cities perform – namely, central place activity and manufacturing – are adequately and even strongly represented in all the leading centers, but the spatial structure of the urban system is fundamentally an expression of mercantile needs. So far as North America's largest metropolises are concerned, the geographical arrangement that we observe today is the legacy of an era in which large-scale manufacturing was insignificant and in which long-distance trade was more important than central place activity in determining town locations and prospects for growth.

Given this general conclusion, it is worth considering the characteristics of the few leading cities of today that do not lie on the seven principal alignments. In Canada there is only one center among the top ten to consider: namely, Ottawa. And in this case, the explanation is readily found. Ottawa's role as capital was determined not by the natural operation of commercial forces but by a decision made by Queen Victoria. Her choice ensured that neither Montreal nor Toronto – then, as now, Canada's dominant mercantile cities – would gain an extra advantage by assuming the role of national capital. Ottawa today ranks fourth among Canadian cities in size, but it owes this position almost exclusively to the growth of the federal government. Its commercial importance is small, its role as a manufacturing center even smaller.

In the United States as well, the largest city not regarded as belonging to any mercantile alignment is the national capital. By position alone, Washington, DC, could belong to the Seaboard alignment: its tidewater location is similar to those of Baltimore and Philadelphia. Also, Georgetown, now part of the Washington metropolitan area, was once a thriving tobacco port engaged in transatlantic trade. But the circumstances of the city's origin, and its basic functional composition, suggest a different conclusion. Washington, like Ottawa, was a 'special creation,' chosen deliberately to prevent sectional discord; the city's growth has been almost entirely the result of the presence of the federal government, not of mercantile or industrial activities. Washington is not a true mercantile city and therefore should not be included in the Seaboard alignment despite the suggestive character of its location.

As Figure 9.2 shows, the remaining large cities outside the principal mercantile alignments are situated either in the Midwest (Columbus, Dayton, and Indianapolis) or in the sunbelt (Miami, Fort Lauderdale, Tampa–St Petersburg, Houston, Phoenix, and Sacramento).

The three Midwestern cities neatly illustrate the causal relationships that exist between locations and urban functions. Dayton, easily the most heavily committed to manufacturing among the three, can be viewed as an industrial satellite of Cincinnati, from which it is less than one hour's drive distant. In contrast, Columbus and Indianapolis are as much central places as manufacturing towns; and each of them occupies, in Christallerian terms, a $K = 4$ position: Columbus between Cincinnati and Cleveland, and Indianapolis between Cincinnati and Chicago. In short, both the locations and the functions of these three cities appear perfectly logical provided that the locations of the metropolises on the mercantile alignments are taken as given.

The cities of the sunbelt have generally experienced rapid growth during the past fifty years regardless of their positions in relation to the principal mercantile alignments. The causes of this growth are complex in detail, but two underlying factors can be identified: first, the attraction of an essentially winterless climate; second, the locationally footloose character of both the burgeoning tertiary sector of the economy and the newer kinds of manufacturing, notably the electronics and aerospace industries (Sternlieb and Hughes 1975; Perry and Watkins 1977). For some of the large sunbelt cities that do not belong to the mercantile alignments, we can point to specific factors that have encouraged rapid population growth. Examples include the influence of the National Aeronautics and Space Administration (NASA) in Houston, the impact of the Cuban refugee population in Miami, and Sacramento's status as the capital of the nation's most populous state. But strong urban growth is currently a general phenomenon across the southern tier of states. And, most important, the recent growth of the leading sunbelt cities is not primarily an exogenic mercantile phenomenon, but rather a response to endogenic structural changes in the ways in which Americans make their living. Perhaps in the very long run the south and west will emerge as the locus of most of North America's leading cities, and geographers of the future will look back on the late twentieth century as the closing years of an early exogenic phase of urbanism that lasted less than four centuries before giving way to an era dominated by endogenic change.

Metropolitan dominance

Closely related to the mercantile model of urban development is the concept of metropolitan dominance. This concept has evolved haltingly along several different paths and is not easy to define with precision, but there is general agreement that three basic ideas are involved. First, a true metropolis is not merely a city that is large in size and commercially important, but also a city that exercises significant financial control within the national economy (Gras 1922, 1926; Kerr 1968; Conzen 1977; Borchert 1978; Wheeler and Dillon 1985; Wheeler 1986). Second, a country's national territory can be divided into a series of distinct 'metropolitan regions' (sometimes called 'metropolitan communities'), each dominated economically by a city of metropolitan status (McKenzie et al 1933; Dickinson 1934, 1964; O.D. Duncan et al 1960; Duncan and Lieberson 1970). Third, the influence of a metropolis on its surrounding region makes itself felt in such a way that many social and economic variables – for example, personal income, level of education, and the age structure of the population – display fairly regular gradients across space, changing in value monotonically as distance from the metropolis increases (Bogue 1950; Friedmann and Miller 1965; Berry 1973).

In the present context, it is the first two of these ideas that are of interest. Given the natural tendency for closely related economic activities to agglomerate in space, major banks, stock exchanges, brokerage firms, and other financial institutions will tend to concentrate in cities that play leading roles in long-distance trade. Accordingly, the principal metropolitan centers – in the specific sense of financial control points – will normally be found along the main mercantile alignments and will, in general, be the largest cities on these alignments. This expected spatial correspondence between mercantile and financial activity certainly holds true in the United States and Canada. For example, all twelve of the centers designated as Federal Reserve Bank cities in the United States lie on the seven major mercantile alignments, and two-thirds of the thirty-six additional centers designated as Branch Cities in the Federal Reserve system are similarly located (B. Duncan and Lieberson 1970, 186–7). In Canada the financial dominance of Montreal and Toronto – more particularly the latter in recent years – is beyond dispute (Kerr 1968). Thus the geographies of long-distance trade and of finance may be seen as mutually reinforcing factors that have contributed substantially to the economic domination of the urban system by a relatively small number of favored metropolitan locations.[3]

However, not all the towns and cities engaged in mercantile activity along the principal alignments will achieve metropolitan status. Even allowing for lack of general agreement concerning the kind or amount of financial importance a city must possess before it deserves the label 'metropolis,' at any given time only a small minority of all urban places can be regarded as members of the metropolitan class. In this connection, Meyer (1980) has recently stressed the value of drawing a clear distinction between the control of exchange and the performance of physical movement. The former implies a 'head office' type of function likely to be found only in metropolitan centers; the latter is essentially a transportation function likely to be widespread in occurrence. Accordingly, while we may safely assume that all metropolitan centers are mercantile, a much larger number of urban places – many of them small – are mercantile without ever becoming metropolitan.

Meyer's distinction also has implications for the timing of metropolitan development, since metropolitanism can migrate from one city to others by means of the expansion or relocation of managerial responsibilities and the establishment of new financial institutions. A particular city on a mercantile alignment may pass through a strictly mercantile phase of early development as a wholesaling and transportation center before entering a truly metropolitan phase, in which it takes on decision-making powers formerly housed elsewhere. In general, metropolitanism in the United States and Canada has undergone westward diffusion from origins along the eastern seaboard and in Montreal, respectively. However, as Meyer himself points out, detailed studies of the dynamics of this process and of its geographical expression in the expanding North American space-economy have yet to be carried out.

The second aspect of metropolitan dominance that calls for comment is the idea that the national territory can be carved up into a set of mutually

3 The decisions – financial, commercial, and political – made in metropolitan centers affect not only the specific fortunes of other urban places but also the general character of society as a whole. This fact has attracted the attention of a number of historians, notably Schlesinger and Wade in the United States and Careless in Canada. These writers have regarded metropolitan dominance not only as an effective interpretive framework in its own right but also as a useful counterpoise to the 'frontierism' of Frederick Jackson Turner and his disciples (Turner 1920; Schlesinger 1933, 1940; Careless 1954, 1974; Wade 1959; Taylor 1972; Walsh 1981).

exclusive metropolitan regions. On one level, this idea raises interesting problems of definition and measurement: for example, the problem of deciding which cities should be classed as metropolises, and the task of selecting an appropriate method for delimiting the metropolitan umlands (see chapter 6). These definitional problems, however, are peripheral to our present concerns. On a deeper level, the idea of a space-filling tessellation of metropolitan regions confers on metropolitan dominance an areal dimension that contrasts sharply with the linear emphasis of the mercantile model. Indeed, given the general spatial concordance between metropolitanism and the mercantile system, the concept of metropolitan dominance perhaps represents an area-oriented version of the mercantile model, at least for leading cities. (Alternatively, the mercantile model represents a line-oriented, or route-oriented version of metropolitan dominance.) This interpretation is consistent with the fact that the major staple commodities on which the mercantile system depends are themselves areally based. The *trade* in furs or grain can be viewed as a network of points and lines, but the actual production of the raw material requires coverage of area. As a result of this dualism, we may conceptualize a mercantile system simultaneously in terms of points and lines *and* points and areas. And finally, with the mercantile model switched from the linear to the areal mode, the link between this model and central place theory is too obvious to require further comment.

We may close this chapter by stressing once again the creative role of mercantile activity in determining the overall spatial structure of the North American urban system. The early and continuing dominance shown by the leading mercantile cities, in size and in economic power, shows that the urban system has been constructed essentially 'from the top down' rather than 'from the bottom up.' The exogenically determined superstructure of major control points has been the given element in the pattern, serving as a skeletal master plan to which all other components of the evolving urban system have been forced to conform. Notwithstanding the importance of central place activity in areas lying outside the principal mercantile alignments, and notwithstanding the contribution of manufacturing to city growth in particular regions, the urban system of North America, viewed in the large, is fundamentally an expression of the spatial organization of long-distance trade.

Summary

Like central place theory, the mercantile model is concerned with towns as centers of trade. However, central place theory deals with local trade, and the mercantile model with long-distance trade. Central place theory describes the geography of retailing; the mercantile model, the geography of wholesaling.

The mercantile model, developed by Vance, applies specifically (though perhaps not exclusively) to the history of settlement by Europeans in North America. The model consists of a sequence of five stages: (1) exploration; (2) harvesting of natural resources; (3) emergence of farm-based staple production; (4) establishment of interior depot centers; and (5) economic maturity and central place infilling. As this process of settlement unfolds, the spatial structure of the evolving urban network is determined more by the strategies and requirements of long-distance trade than by the need to provide local rural populations with consumer goods and services.

With regard to pattern, as distinct from process, mercantile activity in North America has created seven major urban alignments, each consisting of a linear group of wholesaling cities that occupy geographically similar situations. The seven major alignments are as follows: (1) Seaboard, (2) Great Lakes, (3) River (or Ohio-Mississippi-Missouri), (4) Appalachian Piedmont, (5) Eastern Plains, (6) Rocky Mountain Piedmont, and (7) Pacific Coast.

The total area covered by these seven alignments is small; yet these narrow strips of territory contain thirty of the forty largest cities in the United States and nine of the ten largest in Canada. Clearly the overall spatial structure of the North American urban system is fundamentally an expression of the importance of long-distance trade. The only significant group of large cities not associated with the principal mercantile alignments is found in the sunbelt. These centers have grown large in response to recent endogenic changes within the American system itself, whereas the leading cities on the mercantile alignments owe their importance to exogenic influences that emanated from western Europe.

It was emphasized that central place theory and the mercantile model, viewed as distinct conceptual frameworks for the study of the size and spacing of towns, are complementary rather than mutually incompatible. Their complementarity hinges on the historical fact that the wholesaling cities located on the principal mercantile alignments became the leading central places in their respective districts The layout of the mercantile alignments consequently determined the gross geometry of the evolving central place system. As settlement proceeded, networks of lesser central places developed in the areas lying between the main wholesaling alignments. In due course, modern large-scale manufacturing appeared, further complicating the pattern. Today's mature urban system is a compound structure: the original linear arrangement of mercantile centers has been significantly modified by the addition of dispersed central places and clusters of manufacturing towns.

Finally, it was shown that the mercantile model is closely related to the concept of metropolitan dominance, in which the hallmark of metropolitan status is the presence of great financial power. This power tends to be concentrated in a few large cities which function, accordingly, as the major control points in the national economy. Metropolitan dominance is related to the mercantile model by virtue of the strong spatial association between financial activity and the principal wholesaling cities. All the leading American and Canadian metropolises are situated on the seven great mercantile alignments. In the final analysis, therefore, long-distance trade and metropolitanism have jointly determined the spatial arrangement of North America's leading cities, thereby creating a superstructure that has determined, in turn, the overall geographical character of both the American and Canadian urban systems.

10

The analysis of city growth rates

In this chapter, we turn to the manner in which cities undergo demographic change. Given that the population of the world as a whole continues to increase, and given the long-established drift of rural population into the towns (see chapter 2), the mere fact that cities grow may seem unworthy of serious consideration. However, urban growth is selective: during any given period of time, different cities grow at different rates. The selective character of city growth naturally stimulates interest in the mechanics of population change and in the causal factors at work in particular cities or regions.

The Metropolitan Statistical Area (MSA)

As in several earlier chapters, we will develop our understanding of general concepts and methods by focusing on a particular empirical example. For the present discussion, the selected example is that of rates of growth in the American metropolitan system during the period 1970–80.

We are now using the label 'metropolitan' in what might be termed the 'census bureau' sense, not in the more restricted sense of financial control point. The US census definition separates metropolitan centers from other places simply on the basis of size, with the dividing line at 50,000 inhabitants. Because of this rather low threshold of size, only a small proportion of the cities classified as 'metropolitan' in the census possess metropolitan characteristics in the financial sense. These two senses of 'metropolitan' are quite distinct, and care should be taken not to confuse them.

The official US census label for a metropolitan center is 'Metropolitan

Statistical Area,' usually abbreviated to MSA. ('Standard Metropolitan Statistical Area,' or SMSA, was in use from 1959 to 1983, but is now obsolete.) Each MSA consists of a county, or a group of contiguous counties, containing at least one incorporated center of 50,000 or more inhabitants. Counties that lie adjacent to the county (or counties) containing the principal urban center (or centers) are included in the MSA if they meet certain conditions pertaining to population density, proportion of population classified as urban rather than rural, and the strength of commuting links between the outlying counties and the county (or counties) containing the principal city (or cities). The use of data on commuting patterns reflects the basic intention that each MSA should represent a more or less complete labor market, excluding only a small number of comparatively long-distance commuters.[1]

Inclusion of outlying counties gives the typical MSA much more territory than the continuously built-up area at its center. Most MSAS contain considerably more rural territory than land covered by bricks and mortar! However, this combining of rural and urban territory reflects a concern with social relations rather than with land use in the narrow sense: the objective is to identify an areal unit that makes sense in terms of the economic and social integration of the population. For this purpose, the limits of the continuously built-up area are likely to be irrelevant.

When two or more MSAS are physically adjacent to one another, the census bureau may define a composite unit known as a 'Consolidated Metropolitan Statistical Area,' or CMSA. Just as MSAS are built up from counties, so CMSAS are built up from MSAS. Each MSA within a CMSA is officially termed a 'Primary Metropolitan Statistical Area,' or PMSA. This designation alerts users of census data to the fact that the MSA forms part of a more extensive CMSA unit. As might be expected, CMSAS are defined only for the nation's largest and most sprawling metropolitan complexes. The Los Angeles CMSA, for example, consists of the four adjoining (and nonoverlapping) PMSAS of Los Angeles–Long Beach, Anaheim–Santa

1 Some MSAS are not based on county units. In Louisiana, parishes are used instead of counties. In Virginia, incorporated cities are legally independent of the county structure, and each MSA therefore includes one or more cities (in the legal sense) as well as one or more counties. Baltimore, St Louis, and Carson City are also independent of county structures. Finally, in the six New England states, MSAS are based on cities and 'towns' instead of counties. The New England 'town' is not an urban place, but an administrative subdivision of a county, similar to townships elsewhere. In New England, 'town' government is generally influential; elsewhere, the township is a relatively insignificant unit.

Ana, Riverside–San Bernardino, and Oxnard-Ventura. At the time of the 1980 census, a total of twenty-two CMSAS had been defined.

The delimitation of some CMSAS has been carried out rather generously, so that the resulting CMSA includes metropolitan centers that still retain their own distinctive identities, especially in the minds of the local inhabitants. An example is provided by the CMSA of Cleveland, which consists of the three PMSAS of Cleveland, Akron, and Lorain-Elyria. Although there are certainly close economic ties among these centers, each possesses a strong sense of its own individuality. In such cases, it seems justifiable to break the CMSA into its constituent elements. In the analysis below, most CMSAS have been wholly or partly disaggregated in this way, their components being treated as independent MSA units. However, a few of the more compact and closely integrated CMSAS, such as those of St Louis and Kansas City, have been left intact. The net result is a list of 320 metropolitan units (MSAS and MSA-equivalents) with 1980 populations that range from a high of 8,275,000 (New York) to a low of 63,000 (Enid, Oklahoma).

Using counties as the basic building blocks of MSAS is helpful because county boundaries are almost completely stable over long periods. County boundaries are much more stable than those of incorporated villages, towns, and cities, which often change through annexation of additional territory or amalgamation of adjacent municipalities. In the study of city growth it is desirable – some might say essential – that population totals for different dates should refer to fixed geographic areas. If this spatial constancy is not achieved, rates of growth become difficult to interpret. For example, a city whose population is static may appear to experience significant growth if it happens to expand its boundaries during the period covered by the study. The use of counties in constructing MSAS virtually eliminates problems of this type. Even when an entire MSA expands in size, leading to the addition of a new county to the official statistical unit, the population counts for earlier years can easily be adjusted by retroactively including the newly added county in the earlier totals. In fact, the census bureau makes these adjustments routinely, and thus time-series data in any individual report on MSAS are automatically provided on a constant-area basis.

Patterns of MSA growth 1970–80

Between 1970 and 1980, the total population living within the 320 MSAS that comprise the American metropolitan system increased from 154.6

TABLE 10.1
Cross-classification of 1970 size and 1970–80 percentage change in size

		Population size in 1970			
Category of change	All cities (%)	Under 250,000 (%)	250,000–500,000 (%)	500,000–1,000,000 (%)	Over 1,000,000 (%)
Absolute decline	10.6	7.1	6.1	14.3	34.3
Slow growth (0–20%)	57.2	57.6	65.2	60.0	37.1
Moderate growth (20–40%)	20.9	20.7	21.2	20.0	22.9
Rapid growth (over 40%)	11.3	14.7	7.6	5.7	5.7
All cities	100.0	100.0	100.0	100.0	100.0

SOURCE: Calculated from census data reported in US Bureau of the Census, *Statistical Abstract of the United States, 1984* (Washington, DC, 1983)

million to 170.8 million, the latter figure representing 75.4 per cent of the total 1980 US population. The absolute gain of 16.2 million persons is equivalent to a weighted average increase of 10.5 per cent for the metropolitan population as a whole. However, if rates of change are calculated separately for each of the 320 individual MSAS, the *unweighted* average increase is 17.0 per cent. The higher unweighted average rate means that in the decade 1970–80 the smaller MSAS tended to grow more quickly than the larger ones. This tendency is documented in Table 10.1, which cross-classifies MSAS by size in 1970 and percentage change from 1970 to 1980.

Consider, first of all, the category of most rapid growth, defined in Table 10.1 as a population increase of more than 40.0 per cent during the decade. Cities falling into this category accounted for 11.3 per cent of all 320 MSAS covered by the analysis. Compared with this general figure, however, rapidly growing centers accounted for 14.7 per cent of those MSAS containing fewer than 250,000 inhabitants, and for only 5.7 per cent of MSAS with one million inhabitants or more. Rapid growth, in other words, is proportionately overrepresented among small centers and underrepresented among large ones. The category of absolute decline, in contrast, shows precisely the opposite pattern. Centers that experienced actual losses of population accounted for 10.6 per cent of all the cities considered, but for only 7.1 per cent of the MSAS belonging to the smallest

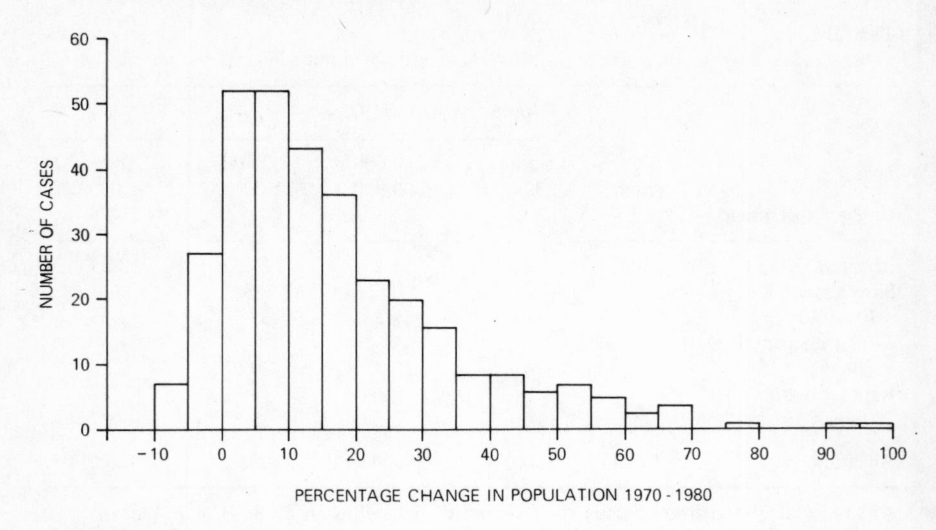

Figure 10.1
Frequency distribution of the rates of population growth in American metropolitan areas (MSAs), 1970–80 (N = 320). Note the positive skewness. SOURCE: 1980 US Census

size-category, and for a remarkable 34.3 per cent of those belonging to the largest size-category (Table 10.1). Thus the relatively small magnitude of the weighted average growth rate is attributable primarily to the fact that more than one-third of all MSAs with more than one million inhabitants actually lost population during the decade. The reasons for these striking losses are discussed in the final section of this chapter.

The frequency distribution of the 320 individual MSA growth rates is shown in Figure 10.1. Positive skewness is evident – the usual situation whenever a large number of city growth rates is considered. The positive skewness reflects the fact that, in general, substantial rates of increase are moderately common, whereas large rates of decline – implying rapid depopulation – are comparatively rare (Simmons and Flanagan 1981, 19). The average value of the present frequency distribution, as noted above, is 17.0 per cent. The median value, however, is only 12.5 per cent.

No MSA managed to double its population between 1970 and 1980, though the two Florida centers of Fort Myers and Fort Pierce came close, with increases of 95.1 and 91.7 per cent, respectively. In fact, Florida accounted for six of the ten fastest-growing centers: Fort Myers, Fort Pierce, Fort Lauderdale, West Palm Beach, Sarasota, and Ocala. The remaining members of the top ten were: Fort Collins, Colorado; Las Vegas, Nevada; Bryan–College Station, Texas; and Olympia, Washington. At the other end of the scale, seven MSAS achieved the rather remarkable feat of losing more than 5 per cent of their population during the ten-year period. These cities were, in descending order of size: New York, Pittsburgh, Cleveland, Buffalo, Jersey City, Utica-Rome (New York), and Pittsfield (Massachusetts). In 1970 the median size of these seven MSAS was 1,113,000; but the median size of the ten fastest-growing centers was only 98,000.

The foregoing comments on the upper and lower tails of Figure 10.1 suggest a geographical contrast between the 'Old Northeast' and the remainder of the country. None of the seven most rapidly declining MSAS lies south of Pittsburgh or west of Cleveland, whereas all of the ten most rapidly growing centers are located within the great arc of territory that extends westward from Florida through the sunbelt states and then northward to the Pacific Northwest. A more comprehensive portrayal of this contrast is provided by Figures 10.2 and 10.3. Figure 10.2 shows the locations of all MSAS that grew by less than 5 per cent during the decade. This map includes thirty-four centers that experienced absolute decline and fifty-two centers in which growth was positive but failed to attain even the modest level of 5 per cent. Figure 10.3, in contrast, shows all MSAS in which the ten-year increase was 25 per cent or more. Altogether there are eighty-six MSAS on Figure 10.2 and eighty on Figure 10.3. In effect, therefore, these two maps isolate the upper and lower quartiles of the frequency distribution shown in Figure 10.1.

The concentration of slow-growth and declining MSAS in the northeast is striking (Figure 10.2). More than 75 per cent of these lagging MSAS lie within 200 miles of New York, Pittsburgh, or Chicago. The median center, or point of minimum aggregate travel, for the entire set of eighty-six centers lies at Youngstown, Ohio – remarkably close to the median center identified earlier for 'manufacturing centers' – see chapter 3, especially Figure 3.5. Indeed, Figure 10.2 leaves us in no doubt that slow growth and absolute decline among MSAS are associated spatially with the traditional manufacturing belt. Manufacturing as a whole is no longer a rapidly expanding sector of the US economy. But the very poor growth performance of northeastern MSAS also suggests that new growth within the manufacturing sector, limited though it may be, is now showing a preference for areas outside the traditional region of manufacturing concentration – or at least, outside MSAS within this region.

Figure 10.2 includes San Francisco, a city that might have been

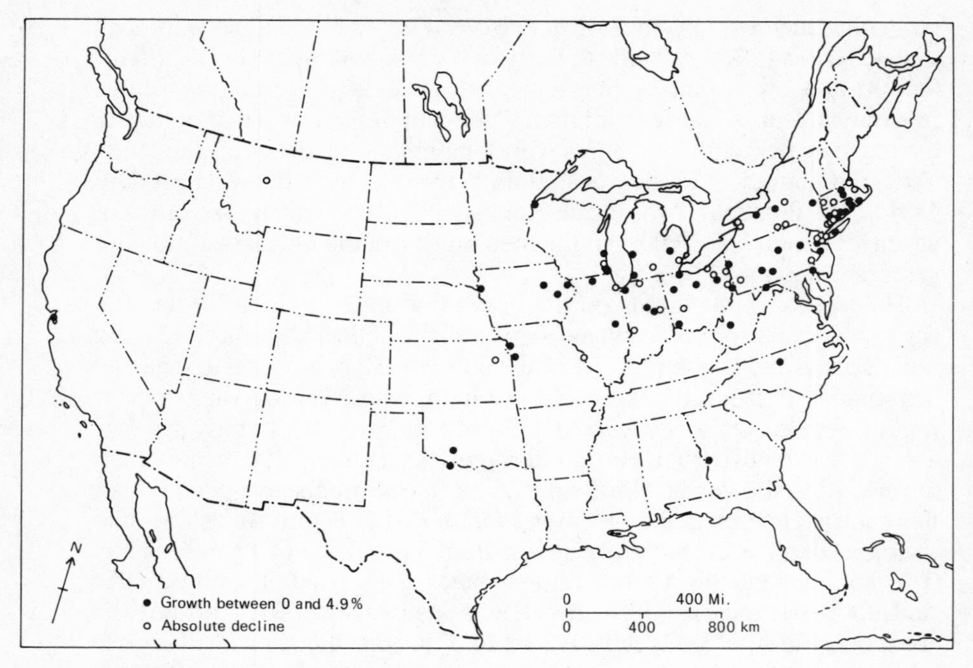

Figure 10.2
MSAs with growth rates of less than 5.0 per cent, 1970–80. SOURCE: 1980 US Census

expected to benefit from the continuing drift of population to the western states in general and to California in particular. However, the San Francisco that appears on Figure 10.2, and which grew by only 4.6 per cent between 1970 and 1980, is a combination of the two PMSAS of San Francisco and Oakland. It does not include the surrounding PMSAS of Santa Cruz, San Jose, Vallejo-Fairfield-Napa, and Santa Rosa–Petaluma, all of which grew relatively rapidly. (Three of these four appear on Figure 10.3.) Within the San Francisco Bay district, the San Francisco–Oakland combination performs the role of an 'old core,' like the 'inner city' portions of smaller and more self-contained metropolitan areas. Land suitable for urban expansion had been more or less fully used before 1970, and, accordingly, rapid growth had become virtually impossible. Surrounding satellite centers, such as Santa Cruz and Vallejo-Fairfield-Napa, now exhibit high rates of growth. A similar situation, in fact, exists in the area of Los Angeles. Between 1970 and 1980, the Los Angeles–Long Beach PMSA increased in population by only 6.2 per cent, while the adjacent

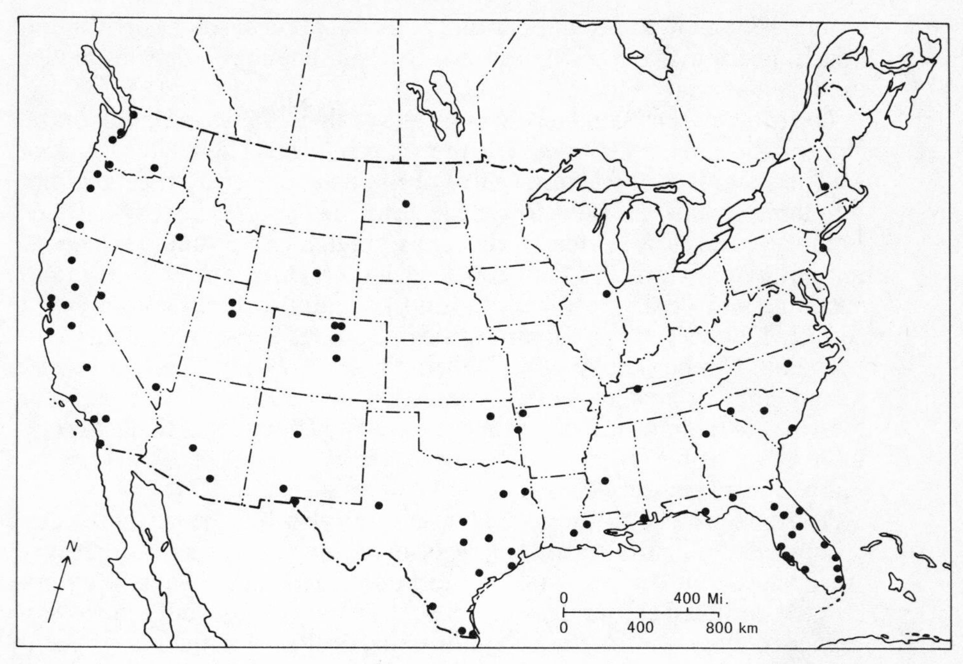

Figure 10.3
MSAs with growth rates of 25.0 per cent or higher, 1970–80. Compare with Figure 10.2. SOURCE: 1980 US Census

PMSAS of Anaheim–Santa Ana, Riverside–San Bernardino, and Oxnard-Ventura increased by percentages of 36.0, 36.8, and 39.8, respectively.

The overall pattern of fast-growth MSAs appearing in Figure 10.3 is essentially the inverse of the slow-growth pattern shown in Figure 10.2. The three archetypal sunbelt states of Florida, Texas, and California account for forty of the eighty fast-growth centers. All parts of Florida have participated in the boom. The Gulf Coast is represented by Tampa–St Petersburg, Bradenton, Sarasota, and Fort Myers; the Atlantic coast by a string of five MSAs extending northward from Miami-Hialeah through Fort Lauderdale, West Palm Beach, and Fort Pierce to Daytona Beach; the interior by Gainesville, Ocala, Orlando, and Lakeland–Winter Haven; and the Florida panhandle by Tallahassee and Panama City. The average 1970–80 population increase in these fifteen fast-growing Florida MSAs was an impressive 56.2 per cent, more than three times the national average of 17.0 per cent for all 320 MSAs. Even the four

Florida MSAs that do not appear on Figure 10.3 (Pensacola, Fort Walton Beach, Jacksonville, and Melbourne-Titusville) had growth rates higher than the national average.

In Texas, most of the fast-growth centers lie in the eastern half of the state – the only exceptions are the oil-producing High Plains city of Midland and the far western center of El Paso. The eastern centers fall into three groups: first, the lower Rio Grande cities of Laredo, McAllen-Edinburg, and Brownsville-Harlingen; second, a group situated close to, but not actually on, the Gulf coast, including Houston, Brazoria, and Victoria; and finally, a distinct group of inland centers consisting of Austin, Killeen-Temple, Bryan–College Station, Tyler, and Longview-Marshall. Neither Dallas–Fort Worth nor San Antonio, the two largest inland cities of eastern Texas, succeeded in attaining the 25 per cent growth rate required for inclusion in Figure 10.3, but each of them grew by more than 20 per cent. The northern Texas centers of Amarillo and Lubbock also performed well.

The fast-growth cities of California may also be divided into three groups. First, in the south, there is a group comprising San Diego, Anaheim–Santa Ana, Riverside–San Bernardino, and Oxnard-Ventura (but not, as noted earlier, Los Angeles–Long Beach). Second, around San Francisco Bay, we may note Santa Cruz, Vallejo-Fairfield-Napa, and Santa Rosa–Petaluma (but not San Francisco–Oakland). Third, a group of fast-growth MSAs is found within the Great Valley, extending from Redding in the north (a center that arguably belongs more to the Klamath Massif than to the Great Valley) through Chico, Sacramento, and Modesto to Visalia-Tulare in the south.

Outside the three leading states of Florida, Texas, and California, noteworthy clusters of fast-growing MSAs are found in South Carolina (Charleston, Columbia, and Anderson, with Florence and Greenville-Spartanburg also growing strongly), in Colorado (Denver-Boulder, Colorado Springs, Greeley, and Fort Collins–Loveland), and in the Pacific Northwest. Most of the fast-growing MSAs in the latter region are located in the Puget Sound–Willamette Valley Lowland, including Bellingham, Bremerton, Olympia, and Vancouver in Washington, and Salem and Eugene in Oregon (Figure 10.3). These cities are linked by way of Interstate Highway 5 through Grants Pass and the rapidly growing MSA of Medford in southern Oregon to the fast-growth centers of California's Great Valley. The result is a continuous chain of rapidly growing MSAs that extends all the way from Visalia-Tulare to the Canadian border.

Finally, mention must be made of the interior southwest – the desert states of Nevada, Utah, Arizona, and New Mexico. Figure 10.3 does not appear to show any notable concentration of fast-growth MSAS within this region, but these four states are very thinly populated and accordingly contain few towns of any size. As a matter of fact, all the MSAS located in these states appear on Figure 10.3: Las Vegas and Reno in Nevada; Salt Lake City–Ogden and Provo-Orem in Utah; Phoenix and Tucson in Arizona; and Albuquerque and Las Cruces in New Mexico. In other words, every MSA within the region qualifies as a fast-growth center. Not even Florida can boast such completeness of coverage.

The median center of the slow-growth MSAS is located at Youngstown, Ohio; that of the eighty fast-growth cities lies some ninety miles west of Tulsa, Oklahoma. Just as Youngstown lies close to the median center of manufacturing centers, so the point west of Tulsa calls to mind the median center of the 'nodal cities,' also in central Oklahoma – see Figure 3.5, in chapter 3. A city-by-city examination confirms that the set of fast-growth centers consists almost entirely of nodal cities. There is clearly an inverse relationship between rates of MSA growth and the relative importance of the manufacturing sector in each city's economy. The old manufacturing belt of the northeast is filled with MSAS whose populations are stagnant or declining, whereas the MSAS of the south and west are characterized both by rapid growth and by a relative lack of emphasis on manufacturing.

Additional insight concerning interregional differences in the growth performance of MSAS is provided by Figure 10.4 and Table 10.2. In Figure 10.4, the territory of the conterminous United States is divided into eleven regions, each consisting of one or more complete states. Table 10.2 shows the number of MSAS lying within each region, and, more important, the percentage of MSAS in each region classified as fast-growth centers during 1970–80. The percentage values fully confirm a sharp distinction between the 'Old Northeast' and the southern and western periphery. They also reveal, however, that both the southeast (region 7 – which excludes Florida) and the plains states (region 8) contain less than their 'fair share' of fast-growth MSAS when compared with the national average of 25.0 per cent. Of course, the southeast and the plains states are doing better than the northeast – perhaps thereby lending some credibility to the old Confederate promise that 'the South will rise again.' They are not doing nearly as well, however, as Florida and the southwest and west.

Natural and migrational components of change

Viewed in terms of immediate, as distinct from fundamental, causes, the population of a city can change in only three ways: by redefinition of the city's political boundary; by natural change – reflecting the difference

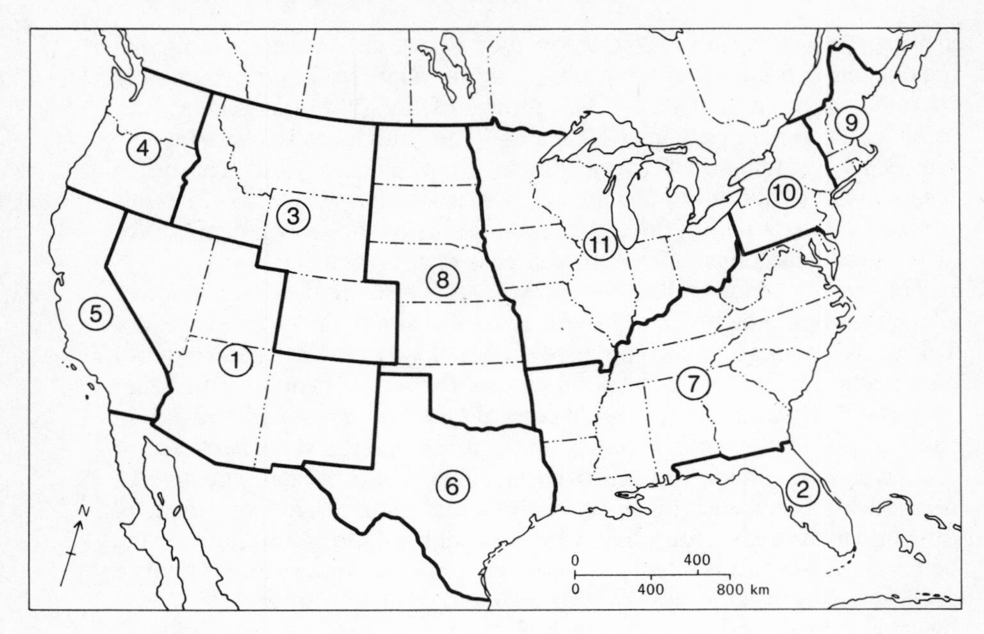

Figure 10.4
Division of the conterminous United States into eleven regions for the analysis of metropolitan growth rates. See also Tables 10.2 and 10.4.

between the number of births and the number of deaths during a particular period; and by migrational change – reflecting the net effect of the arrival of in-migrants and the departure of out-migrants. In the present context, because population data for MSAs are provided by the census bureau on a constant-area basis, the first type of change does not enter the picture. We are therefore left with natural change and migrational change as the immediate causes of gains and losses in MSA populations.

The laws of most countries require that all births and deaths be registered in writing with an appropriate government agency – usually a registrar of vital statistics at the state or provincial level. As a result, data pertaining to natural change are widely available and generally of good quality. However, direct monitoring of migratory movement is very rare, and it is therefore necessary to resort to indirect methods in order to estimate the direction and magnitude of migrational change at any particular place. The approach most commonly used is known as the

TABLE 10.2
Growth performance of metropolitan centers by region, 1970–80

Region[a]	Number of MSAS	Fast-growth MSAS as percentage of all MSAS in region[b]
1 Southwest	8	100.0
2 Florida	19	78.9
3 Mountain states	9	66.7
4 Pacific northwest	13	61.5
5 California	21	57.1
6 Texas	27	48.1
7 Southeast	69	18.8
8 Plains states	13	15.4
9 New England	31	3.2
10 Middle Atlantic	34	2.9
11 Midwest	76	1.3
All MSAS	320	25.0

SOURCE: See Table 10.1.
a Regions are ranked by growth performance (right-hand column). Rank numbers at left identify regions on Figure 10.4.
b Fast growth is defined as a 1970–80 population increase of 25 per cent or higher.

residual method: net migrational change is derived as a residual quantity after the impact of natural change has been taken into account.

The residual method may be illustrated by the following hypothetical example. Suppose that the total population of a particular MSA increased from 80,000 in 1970 to 86,000 in 1975; suppose, further, that the numbers of births and deaths occurring during the five-year period were 6,700 and 3,900, respectively. Since births exceeded deaths, natural change was positive: it produced growth rather than decline. (As we will see below, negative natural change is not an impossibility.) The magnitude of the natural change is simply the difference between the 6,700 births and the 3,900 deaths, or 2,800 persons. We now add this natural change to the 1970 population of the city in order to obtain the figure that would be achieved in 1975 provided that net migration is zero. In this example, the estimated 1975 population is therefore 82,800 persons. The actual 1975 population, however, is 86,000. Therefore this MSA must have experienced a net in-migration of 3,200 persons (86,000 minus 82,800).

Several words of caution are in order. First and foremost, both natural change and migrational change may turn out to be either positive or negative in sign. It is vital, therefore, to keep track of positive and negative signs while the calculations are being carried out. The various possible combinations of positive and negative signs are discussed in more detail below. Second, the entire approach is unworkable if the city experiences a change in its boundary during the period covered by the investigation. Suppose, for example, that the area of the city expands, and that the city's population is thereby increased, at some time after the beginning of the period of study. Some of the people living in the newly annexed territory may have been present before the study's starting-date; others may have been born there since that date; and still others may have arrived as migrants. Short of a person-by-person survey, there is no way to separate these categories of inhabitants. Hence the residual method cannot properly be applied. (As already noted, however, the problem of boundary changes does not arise in the present analysis.)

A further difficulty arises in cases where the timing of the counts of total population does not coincide precisely with the timing of the 'vital statistics year.' Most agencies concerned with vital statistics report births and deaths by calendar years, but censuses of total population are normally carried out in spring or early summer (to minimize difficulties caused by unpleasant weather and by people being away from home on vacation). A satisfactory solution is to prorate the birth and death totals for the end-years of the study according to the number of days required in order to achieve a match with the timing of the censuses. For example, if we are studying the five-year period from mid-1970 to mid-1975, we will need to assemble birth and death statistics for six calendar years – i.e. 1970 through 1975, inclusive. However, portions of the data for 1970 and 1975 will be discarded on a pro rata basis in order to yield an appropriate set of vital statistics for the required five-year intercensus time interval.

Two additional points concerning the character of migration estimates obtained by the residual method should be noted. First, these estimates indicate only the magnitude and direction of *net* migration; they do not tell us anything about the magnitudes of gross migratory movements. In the numerical example described above, the hypothetical city was found to have experienced a net in-migration of 3,200 persons. Clearly this particular net flow could be achieved in many different ways. It might, for example, represent a gross inflow of 10,000 combined with a gross outflow of 6,800. Alternatively, it might represent a combination of 3,500 in-migrants and only 300 out-migrants. On its own, the residual method can reveal nothing about the absolute sizes of the gross flows. Second, and closely related, the residual method is also silent on the subject of origins and destinations. It does not tell us where in-migrants have come from; nor does it identify the destinations of out-migrants. However, the residual method is a convenient technique for identifying one important aspect of the overall demographic 'personality' of each place.

Natural increase, as noted, is the difference between births and deaths. In the United States in recent years, the death rate has been stable and the birthrate declining, and so the rate of natural increase has also been undergoing a decline. In the early 1950s, at the height of the postwar 'baby boom,' the birthrate stood at twenty-five births per thousand population per year; by the mid-1970s, however, this rate had fallen to only fifteen per thousand per year. The death rate, in contrast, remained virtually constant throughout this period, hovering just below ten per thousand per year (Phillips and Brunn 1978, 280). It is generally agreed that the recent decline in the birthrate has been caused jointly by a growing preference for smaller families and by the introduction and widespread adoption of birth control pills. If the birthrate continues to fall, the United States could soon reach a state of 'zero population growth,' with the birthrate equal to or smaller than the death rate. Such a state is not necessarily a bad thing, but it is certainly without precedent in the past two centuries of US history.

In the following paragraphs, we will take as our focus the period 1970–5, the most recent period for which the relevant data at the level of MSAS are currently available (US Bureau of the Census 1978, Table 3). During this five-year period, the national rate of natural increase was 3.3 per cent, equivalent to an increment of 6.5 persons per thousand population per year, applied over five years. However, for the 258 MSAS that existed at that time, the average rate of natural increase was 4.2 per cent over the same period. This is equivalent to an increment of 8.3 persons per thousand population per year, as compared with 6.5 persons per thousand per year in the nation as a whole.

These figures are interesting for two reasons. First, contrary to a common 'anti-urban' belief, they show that the typical large city is not dependent on net in-migration to prevent its population from declining. The belief that large cities could not sustain their population numbers by natural increase alone seems to have originated in western Europe during the nineteenth century, when it was evidently easy to believe that unsanitary conditions in the rapidly expanding industrial cities would cause the death rate to exceed the birthrate. It is unclear whether this belief was true even then; but it is certainly not true today. Second, and again contrary to what is often believed, the figures reveal that the rate of natural increase is actually greater in cities of substantial (i.e. 'metropolitan size') size than in smaller towns and rural areas. This is implied by the fact that the rate for MSAS is greater than the corresponding rate for the nation as a whole during the same period.

Although an excess of births over deaths is the norm, five MSAS had negative rates of natural change during 1970–5: Scranton–Wilkes-Barre, in Pennsylvania, and Tampa–St Petersburg, Bradenton, Sarasota, and Daytona Beach, all in Florida. No obvious explanation exists for Scranton–Wilkes-Barre, but the negative rates in the four Florida cities are attributable to their role as retirement centers. The presence of elderly persons in large numbers both raises the death rate and also tends to lower the birthrate (other things being equal, a greater percentage of elderly persons implies a smaller percentage of persons in the child-bearing years). If these two mutually reinforcing tendencies become sufficiently pronounced, the number of births per year falls below the number of deaths, producing a natural decrease (or negative natural increase) in the city's population.

As the MSA data reveal, an excess of deaths over births is a rarity among cities of 50,000 inhabitants or more; Tampa–St Petersburg, a center containing 1.1 million inhabitants in 1970, is really quite exceptional. However, natural decrease is relatively common among small towns and villages, especially those not located within easy commuting distance of major centers. In many such small, isolated places, most of the young adults have departed in search of better economic opportunities in larger cities, leaving behind a population in which the older age-groups are proportionately overrepresented. This bias toward an 'old' age-structure is augmented by the common tendency for retired farm folk to settle 'in town' (i.e. in the local village) once their days on the land are at an end. In effect, the typical rural village is no less a retirement center than Tampa or Daytona Beach. Its clientele, however, is radically different in occupational background; and it cannot hope – and does not wish – to rival the Florida cities in terms of size. Demographically, nevertheless, the net result is the same in both cases: an 'elderly' population that requires the services of a funeral director more often than those of an obstetrician.

Just as the rate of natural increase tends to be low, and even negative, where the age-structure is biased toward the elderly, so a 'youthful' age-structure tends to promote a rate that lies above the national MSA average. Thus high rates of natural increase are found in Houston, Denver, Anaheim, and many other centers in the southwest and west that have become well established as popular destinations for young migrants from the 'Old Northeast.' Age-structure, however, is not the only factor influencing the level of natural change. Extremely high rates of natural increase are found in cities that contain large percentages of Spanish-speaking inhabitants, notably the Rio Grande cities of McAllen-Edinburg, Brownsville-Harlingen, Laredo, and El Paso. Exceptionally high rates are also found in Utah, apparently reflecting the Mormon preference for large families: Provo-Orem had the highest rate of natural increase (16.0 per cent) of any MSA during 1970–5. It is clear that rates of natural increase are affected by cultural factors as well as by the overall age-structure of a city's population.[2]

With net migration, the rates of migrational change are much more variable from place to place than the rates of natural change. For the 258 cities included in this portion of the analysis, the rates of natural change during 1970–5 displayed a mean value of 4.2 per cent and a standard deviation of 2.2 per cent. The rates of migrational change, in contrast, exhibited a mean value of 3.0 per cent and a standard deviation of 8.3 per cent. Comparison of the two standard deviations makes it clear that the migrational rates are significantly more variable from city to city than the rates of natural change. Natural change ranged from a low of −3.9 per cent (Sarasota) to a high of 16.0 per cent (Provo-Orem). Migrational change, however, ranged from a low of −14.1 per cent (Lawton, Oklahoma) to a high of 46.3 per cent (Fort Myers, Florida).

The foregoing results might suggest that the total change in any particular city's population is likely to depend more heavily on net migration than on the rate of natural change. However, mean values, standard deviations, and extremes, taken in isolation, can be deceptive. The absolute value – the numerical value regardless of sign – of the rate of natural change is greater than the absolute value of the rate of net migration in more than half of all cities (153 out of the total of 258). Therefore, despite the fact that migrational change varies more from place to place than natural change, the overall change in population for individual MSAS is attributable, in six cases out of ten, more to natural change than to net migration. In view of the fact that the national rate of natural increase has been declining, and also in view of the fact that

2 Although Mormonism evidently continues to encourage large families, Roman Catholicism seems not to – at least the faithful do not appear to be responding. In New England, where Catholicism tends to be strong, rates of natural increase are among the lowest in the United States. And Quebec, the most Catholic region of Canada, has the lowest birthrate of any Canadian province. On the whole, it seems likely that the high rates of natural increase along the Rio Grande are attributable more to poverty and ignorance than to religious affiliation as such.

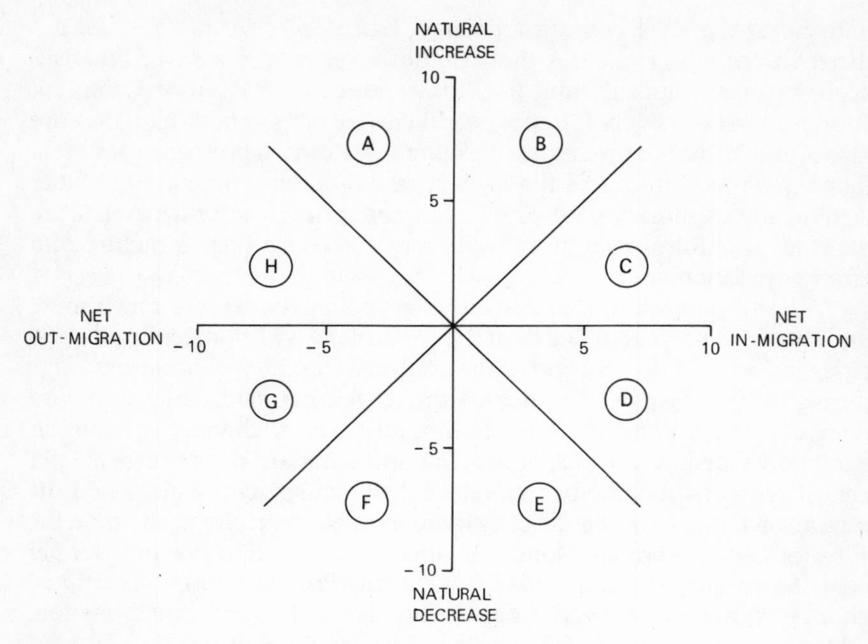

Figure 10.5
A conceptual framework for classification of cities or regions according to the directions and magnitudes of the natural and migrational components of population change. The numbers along each axis represent percentage change during a given period of time. The circled letters identify eight distinctive demographic types. See text for further explanation. After Webb (1963)

migration – especially to the sunbelt – has been the subject of much recent discussion, this finding is of considerable interest.

Negative values occur much more frequently in migrational change than in natural change. Only five MSAS (1.9 per cent of the total) were characterized by negative natural change during 1970–5, but 113 MSAS, or 43.8 per cent of the total, experienced net out-migration. Losses attributable to net out-migration affected cities of all sizes, from small centers, such as Battle Creek, Michigan, and Muncie, Indiana, to metropolitan giants, such as Detroit, Philadelphia, and New York.

Figure 10.5 shows how places may be classified into eight demographic types on the basis of the direction and magnitude of the natural and migrational components of population change (Webb 1963; Marshall 1972, 70–3). In this figure, the vertical axis represents natural change, the horizontal, migrational change. At the point where the two axes intersect, both have the value of zero. Each of the four quadrants created by the main axes is divided into two sectors on the basis of the relative magnitudes of the two components of change. The resulting eight demographic types are described as follows:

A: overall population increase; natural gain exceeds net out-migration.
B: overall population increase; natural gain exceeds net in-migration.
C: overall population increase; net in-migration exceeds natural gain.
D: overall population increase; net in-migration exceeds natural loss.
E: overall population decline; natural loss exceeds net in-migration.
F: overall population decline; natural loss exceeds net out-migration.
G: overall population decline; net out-migration exceeds natural loss.
H: overall population decline; net out-migration exceeds natural gain.

As a general rule, the numbers of cities falling into types D, E, F, and G are small. Taken together, types D, E, F, and G accounted for only 12 per cent of all urban centers in Webb's classic study of England and Wales during the 1920s (Webb 1963, 133). Similarly, an analysis of eighty-eight cities in Ontario for the period 1961–6 produced no centers of types E, F, and G, and only one of type D: Newmarket, which at that time, before it became a dormitory satellite of Toronto, was a small town containing a very large home for the elderly (Marshall 1972, 72). Types E, F, and G are also completely unrepresented in the present study of 258 American MSAS during the early 1970s. Type D is present, but accounts for only five places: namely, the cities in Florida and Pennsylvania with negative rates of natural change.

Table 10.3 shows the percentage of all MSAS belonging to each demographic category during three successive time intervals between 1950 and 1975. The percentages can change quite dramatically with time. During the 1950s, for example, centers of type B formed an absolute majority of all MSAS, but their relative frequency fell to only 28.7 per cent by the early 1970s. In contrast, types C, A, and H became relatively more common. Taking types A and H together, the percentage of cities experiencing net out-migration (regardless of the role performed by natural change) increased from 22.9 per cent in the 1950s to 43.8 per cent during the early 1970s. Conversely, the percentage of cities experiencing net in-migration declined. In other words, net out-migration – normally regarded as undesirable by municipal tax-collectors – has come to afflict an increasing proportion of all cities. We are naturally led to inquire

TABLE 10.3
Relative frequency of metropolitan areas by demographic type, 1950–60, 1960–70, and 1970–5

Demographic type	1950–60 (*N* = 192) (%)	1960–70 (*N* = 257) (%)	1970–5 (*N* = 258) (%)
H	4.7	6.6	13.2
A	18.2	33.1	30.6
B	63.0	40.1	28.7
C	14.1	19.5	25.6
D	0.0	0.8	1.9
All cities	100.0	100.0	100.0

SOURCES: Calculated from census data reported in US Bureau of the Census, *County and City Data Book* (Washington, DC, 1967, 1978).
NOTE: See text and Figure 10.5 for explanation of demographic types.

whether the MSAS experiencing net out-migration are concentrated in any particular sections of the United States.

An answer is provided by Figure 10.6 and Table 10.4. Figure 10.6 makes it clear that MSAS experiencing net out-migration during 1970–5 were strongly concentrated in the 'Old Northeast' – the area extending from the northeastern seaboard to the corn belt of the Midwest. The three neighboring states of Indiana, Ohio, and Michigan account for 25 per cent of all cities appearing on this map; the Michigan centers, moreover, are situated entirely in the southern portion of the state. Outside the 'Old Northeast,' most of the remaining cities are scattered unevenly across the south. Only seven of the 113 MSAS in Figure 10.6 lie west of the one hundredth meridian.

Table 10.4 provides additional information on the relative frequency of net out-migration in each region. The regions on which Table 10.4 is based are identical to those used for Table 10.2; these regions are shown in Figure 10.4. Tables 10.4 and 10.2 may be compared directly, though Table 10.4 covers only five years (1970–5), whereas Table 10.2 covers ten (1970–80). Note that the rank order of the eleven regions according to the relative significance of fast-growth MSAS (Table 10.2) is very close to being the exact opposite of the rank order of the same regions according to the relative significance of net out-migration (Table 10.4). In other words, the higher the percentage of MSAS experiencing net out-migration, the lower the percentage of MSAS experiencing rapid growth, and vice versa. Thus, regardless of the role performed by natural change, the presence of net

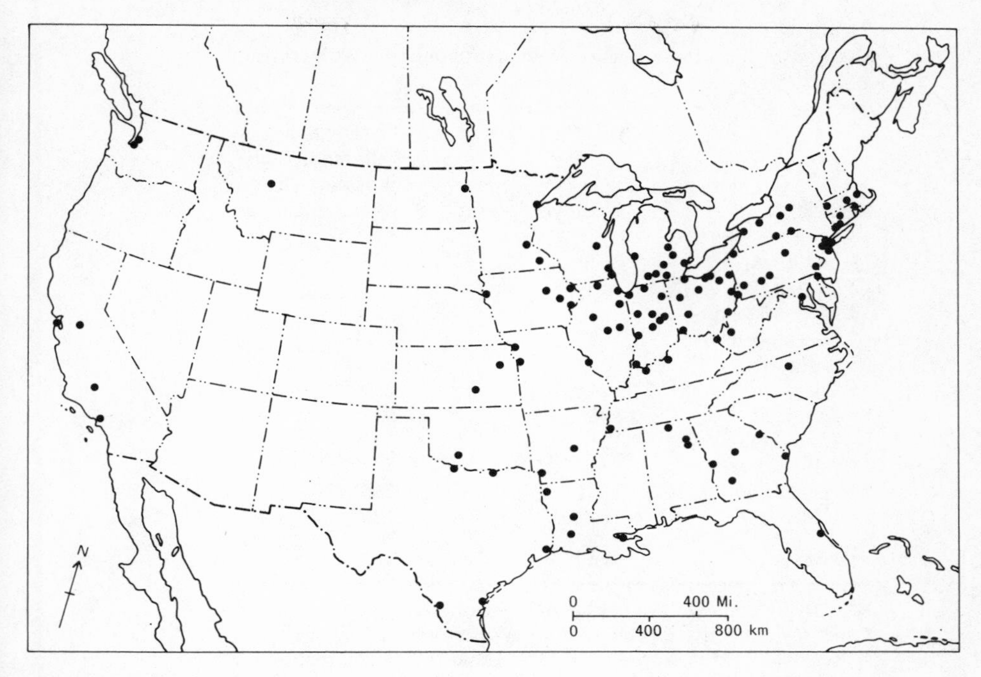

Figure 10.6
MSAs with net out-migration, 1970–5. Compare with Figure 10.2. SOURCE: US Bureau of the Census, *County and City Data Book, 1977* (Washington, DC: US Government Printing Office, 1978)

out-migration effectively relegates a city to the status of being a slow-growth center. This conclusion is corroborated by a comparison of Figures 10.6 and 10.2: cities of net out-migration and cities of slow growth (and decline) display essentially identical patterns of spatial distribution.

Some of the most striking demographic changes in the American urban system in recent years have occurred among the very largest cities. These changes are summarized in Tables 10.5 and 10.6, which deal exclusively with the thirty-five metropolitan areas that contained more than one million inhabitants in 1970. During the 1950s, none of these leading centers experienced an absolute decline in its total population, and only one member of the group – Pittsburgh – experienced net out-migration. But during the first half of the 1970s, in contrast, twenty-one (60 per cent) of these cities exhibited net out-migration, and eleven recorded absolute losses (Table 10.5).

Moreover, only four of the twenty-one large centers exhibiting net

TABLE 10.4
Occurrence of net out-migration from metropolitan areas by region, 1970–5

Region[a]	Number of MSAS[b]	MSAS with net out-migration as a percentage of all MSAS in region
1 Southwest	7	0.0
2 Florida	16	6.3
3 Mountain states	8	12.5
4 Pacific northwest	8	25.0
5 California	17	23.5
6 Texas	24	20.8
7 Southeast	58	39.7
8 Plains states	11	36.4
9 New England	13	53.8
10 Middle Atlantic	30	56.7
11 Midwest	66	74.2
All MSAS	258	43.8

SOURCE: Calculated from census data reported in US Bureau of the Census, *County and City Data Book* (Washington, DC, 1978).

a Regions listed in same order as in Table 10.2. Each region is identified by number on Figure 10.4.

b Numbers differ from Table 10.2 because of different time-periods. Moreover, for New England, Table 10.2 uses standard MSAS; the present table, for reasons of data availability, uses New England County Metropolitan Areas (NECMAS).

out-migration between 1970 and 1975 lay outside the 'Old Northeast': namely, Los Angeles–Long Beach, San Francisco–Oakland, Seattle-Everett, and New Orleans (Table 10.6). Los Angeles and San Francisco are examples of what was earlier termed the 'inner city effect.' To some extent, this effect may help to explain the net out-migration experienced by Seattle and New Orleans, but in these two cases it is possible that other factors were also at work. Seattle, which depends heavily on Boeing Corp., may have suffered losses as a result of a recession in the aircraft manufacturing industry. In New Orleans, the crucial factor appears to have been a decline (possibly only temporary) in the city's significance as a port.

In general, however, the salient feature is that net out-migration from large metropolitan centers is essentially a characteristic of the old manufacturing regions of the northeast and the Midwest.

TABLE 10.5
The changing demographic character of cities of more than one million population in 1970

Demographic type	1950–60 (no.)	1960–70 (no.)	1970–5[f] (no.)
H	0	1[c]	11
A	1[a]	6[d]	10
B	24	16	7
C	10[b]	12[e]	6
D	0	0	1
Total	35	35	35

SOURCES: See Table 10.3.
a Pittsburgh
b All in south and west except 'outer New York' (i.e. Bergen-Passaic and Nassau-Suffolk)
c Pittsburgh
d Buffalo, Cincinnati, Cleveland, Detroit, Milwaukee, and New York
e All in south and west except Nassau-Suffolk
f See Table 10.6 for complete listing of cities in this column.

Underlying causes

The foregoing description of recent demographic changes in American metropolitan areas confirms the existence of a substantial net movement of population from the northeastern section of the country to the south, southwest, and west. Several factors appear to be responsible for this trend.

In Boston, so runs an old saying, there are only two seasons: winter – and July! No doubt this claim overstates the case; but it makes an important point. Given adequate opportunities for employment – and, nowadays, given air conditioning – life in the sunbelt is clearly preferable to an annual confrontation with wind-chill, icy sidewalks, and snow-clogged driveways and streets. As one demographer has put it, the southwestward drift of population might well be viewed as 'merely a long-overdue correction of the original mistake made by the British settlers when they landed in the upper right-hand corner of the map instead of proceeding directly to Houston' (Morrison 1981, 1).

The first significant factor, then, is climate. However, the climate is by no means completely uniform throughout the south and west. Only the southern portion of this vast area can legitimately lay claim to the title

TABLE 10.6
Cities of more than one million population in 1970, showing demographic type for 1970–5

Net out-migration		Net in-migration		
Type H	Type A	Type B	Type C	Type D
Bergen-Passaic	Boston[a]	Baltimore	Anaheim–Santa Ana	Tampa–St Petersburg
Buffalo	Chicago	Columbus (Ohio)	Atlanta	
Cincinnati	Detroit	Dallas–Fort Worth	Denver	
Cleveland	Indianapolis	Nassau-Suffolk	Houston	
Los Angeles–Long Beach	Kansas City	Norfolk–Newport News	Miami-Hialeah	
New York	Milwaukee	Riverside–San Bernardino	San Diego	
Newark	Minneapolis–St Paul	San Jose		
Philadelphia	New Orleans			
Pittsburgh	San Francisco–Oakland			
St Louis	Washington, DC			
Seattle-Everett				

SOURCE: See Table 10.4.
a Demographic type estimated from county-level data

'sunbelt,' and even here the picture is not entirely rosy, as witness the frost hazard faced by Florida citrus-growers and the threat of damage from hurricanes along the Gulf coast. The mountain states and the Pacific northwest – regions 3 and 4 on Figure 10.4 – certainly cannot be described as winterless, though the Puget Sound–Willamette Valley Lowland, the most extensive zone of dense population in this section of the country, does not usually experience the extended periods of severe cold encountered in the northeast. What one really avoids by leaving the northeast is not the snow, as such, but the worst of the bitterly cold weather. Snow itself is not necessarily undesirable and may, under the right circumstances, be counted as an asset. (One wonders, however, if very many snowplow drivers are avid skiers.) The journalistic contrast between 'sunbelt' and 'snowbelt,' though not without point, is something of an oversimplification.

Another factor influencing migrants' decisions is the character of the physical landscape. The northeast is not lacking in local beauty spots – particularly in the Appalachians – but the region as a whole is rather drab and undistinguished. For all-round scenic grandeur, the northeast has nothing to equal the desert and mountain landscapes of the southwest and west. Given the current high levels of popular interest in access to nature and in landscape appreciation, it is reasonable to assume that at least some of the easterners who migrate westward are motivated by a love of western scenery as much as by other factors.

Taken together, climate and the quality of the physical landscape comprise the environmental amenities of a place. These factors are distinguished from other influences by their noneconomic manner of operation. In other words, they exert a type of influence on potential migrants that is essentially independent of the need to earn a living. But North American society, for better or worse, is materialistic, and hence the importance of these noneconomic factors for the distribution of population tends to be underestimated. Perhaps this situation will change as the sunbelt continues to emerge as the home of an increasing proportion of all Americans. Among scholars, the potential significance of environmental amenities for American regional development has been recognized for some time (Ullman 1954; Zelinsky 1978). The current ascendancy of southern and western cities seems to indicate that these factors are now coming into their own.

Other factors, explicitly economic in character, are also at work. In manufacturing, for example, it is becoming increasingly evident that the northeast no longer exercises undisputed leadership as the 'obvious' region in which to install new industrial capacity. There are several distinct reasons for this change (Sternlieb and Hughes 1975; Browning 1978). First, the cost of labor is generally higher in the northeast than elsewhere, largely as a result of the greater strength of labor unions in the traditional manufacturing belt. Industrial workers (and potential industrial workers) outside the northeast tend to be comparatively unskilled, but the growing use of automated production processes is reducing the need for highly skilled workers. Besides, skills can soon be learned. Second, and again related to unionization, relations between labor and management appear to be somewhat less harmonious in northeastern cities than in those of the south and west. Third, traffic congestion in and around the major cities of the northeast is increasingly perceived by manufacturers as a serious disadvantage. In the south and west, thanks to lower population densities both regionally and within the largest cities, congestion is not generally a significant problem. Fourth, many sections of the south and west have now achieved a 'critical mass' in terms of providing a market large enough to make manufacturing profitable. Such areas include eastern Texas, the Colorado Piedmont, the Puget Sound district, and the densely populated areas centered on San Francisco and Los Angeles. Other regions besides the northeast can now support manufacturing on a large scale. And finally, the building of the superb nationwide Interstate Highway System, begun during the 1950s and now almost complete, has effectively nullified the argument that a southern or western location would be too isolated to serve as a base of operations for any manufacturer wishing to reach more than a local market. Acting in combination, these influences account for the fact that new manufacturing jobs are being created outside the traditional manufacturing belt in greater numbers than ever before.[3]

Important though manufacturing undoubtedly is, it accounts for only one-fourth of all nonagricultural employment in the United States. It is the tertiary activities – retailing, education, medical care, personal services, and so forth – that dominate the employment profiles of American cities. These activities, moreover, possess two significant characteristics. First, they are relatively more prominent, in percentage terms, in cities of the south and west than in cities of the northeast. Second, their current rates of growth are higher than those of almost all activities in the manufacturing sector (see chapters 3 and 4). Because of these two characteristics alone, the growth rates of southern and western cities are higher, on average, than those of cities in the 'Old Northeast.'

This point may be clarified with the aid of a simple illustration. Consider two hypothetical cities, *A* and *B*. City *A* is located in the manufacturing belt, and city *B* somewhere in the south or west. In city *A*, manufacturing comprises 40 per cent of the total labor force, with the remaining 60 per cent engaged in tertiary activities. In city *B*, the corresponding values are 15 per cent and 85 per cent. (For simplicity, neither city has any employment in the primary sector – i.e. in mineral extraction.) We also make the following assumptions concerning growth rates: (a) the rates of growth of the manufacturing sectors are the same in both cities; (b) the rates of growth of the tertiary sectors are also the same; and (c) the growth rate of the tertiary sector is greater than that of manufacturing. Now, because the tertiary sector comprises a higher percentage of the labor force in city *B* than in city *A*, the rate of growth in total employment – and, by extension, in total population – will be greater during any given period in city *B* than in city *A*.

In practice, of course, assumptions (a), (b), and (c) will rarely be precisely fulfilled. However, they will be approximately true in most

3 At the same time that manufacturing employment is becoming decentralized nationally from the northeast to the south and west, it is also becoming decentralized at a more local scale from metropolitan counties (i.e. those included within MSAs) to non-metropolitan counties (Lonsdale and Seyler 1979). Both trends are geographically significant, but only the first is of immediate interest here.

cases. When these assumptions are not true, moreover, the truth will almost always favor city *B* rather than city *A*. To generalize, southern and western cities have high rates of growth, at least in part, simply because they contain disproportionately large shares of the fastest-growing types of employment. In short, some portion of the regional variation in city growth rates is purely a structural effect, reflecting differences in the composition of the urban labor force.[4]

Still another factor promoting the rapid growth of southern and western cities have high rates of growth, at least in part, simply because they contain disproportionately large shares of the fastest-growing types areas – including many from Canada as well as those from the northeastern United States – bring money into these regions from outside, thereby helping to sustain the host area's economy. The magnitude of the annual net income provided by the tourist dollar is not easy to determine, but there is no doubt about either its significance or its widespread areal impact. Second, tourism disseminates information concerning the benefits and opportunities available in southern and western states. This information may contribute to later decisions by some visitors to move to the south or west permanently.

The final factor that must be mentioned is the impact of retirement. First, owing to modern advances in nutrition and medical care, average life expectancy has been steadily increasing. Because of this, and because of the decline of the birthrate, people over fifty years of age now form the most rapidly growing age-group in the United States. Second, thanks to the development of financially realistic pension plans and to the tendency to take early retirement (that is, before the conventional age of sixty-five), retired persons are not only more numerous than ever before but also capable of moving to a region of their own choice instead of living out their final years in straitened circumstances in the communities where they spent their working lives. These developments have led to the current tendency for northerners to migrate to the sunbelt upon reaching retirement age. Moreover, the money spent by retired persons in the cities in which they choose to settle represents a net transfer into the host region

4 Readers familiar with the literature on regional economic development will recognize that the quantitative strength of this structural effect could be ascertained for individual MSAs by means of 'shift and share' analysis (Dunn 1980). For our present purposes, however, no formal quantitative analysis is needed. It is sufficient to note that differences in growth are partly attributable to differences in the structure of employment, and that this effect works to the advantage of cities in which the manufacturing sector is weak.

of funds earned elsewhere. In this respect the retirement dollar, like the tourist dollar, is 'new money,' creating new demands for goods and services and thus generating additional city growth.

Although retired people are found in increasing numbers throughout the sunbelt states, the role performed by Florida in this regard is worthy of special mention. Four Florida MSAS have such 'elderly' age-structures that they experience more deaths than births. Of the 320 MSAS existing in the United States at the time of the 1980 census, only sixteen (5 per cent) had more than 15 per cent of their total population aged sixty-five or older (US Bureau of the Census 1983, 20–5) – and ten of these sixteen were in Florida. Thus, Florida as a whole appears to have established itself as the retirement center par excellence in contemporary metropolitan America.

To recapitulate, six factors work to the advantage of southern and western MSAS: a warmer climate, a more scenically attractive physical landscape, increased ability to attract new investment in manufacturing, disproportionately large shares of employment in the fast-growing tertiary sector, the growth of tourism, and an increasing tendency on the part of retired people to move to southern and western regions, especially to Florida.

If the rates of population growth recorded by individual MSAS during the 1970s continue unchanged into the future, Dallas will be larger than Detroit by the end of the century and metropolitan Phoenix will overtake New York by 2015. Of course, there can be no guarantee that present rates of growth will be maintained. However, given current social and economic trends in the United States as a whole, it seems safe to predict that the cities of the south and west will continue to grow more rapidly than those of the northeast for many years to come.

Summary

This chapter has introduced a group of terms and concepts that are useful in the study of city growth rates. It has also provided some insight into the growth performance of American metropolises during the 1970–80 period.

Special attention was given to the concept of the Metropolitan Statistical Area, or MSA, as defined by the US Bureau of the Census. With a few exceptions, MSAS are delimited by using counties as the basic building blocks. The use of whole counties has two major advantages. First, it means that each MSA represents a much more complete and self-contained labor market than would be the case if the MSA were simply equated with the city's continuously builtup area. Second, because county boundaries are very stable through time, it facilitates the analysis of population growth on a constant-area basis.

Given a fixed boundary, any change in an urban center's total population can be fully accounted for as the sum of two distinct demographic elements: namely, natural change and net migration. Natural change is the net effect of additions through births and losses through deaths. Net migration is the difference between the number of in-migrants (arrivals) and the number of out-migrants (departures). If data on actual migrational movements are not available, net migration for a particular city can be estimated as the difference between the total change in the city's population and the natural change during the same period. This technique of estimation is known as the residual method.

In the analysis of rates of MSA growth in the conterminous United States between 1970 and 1980, we noted a marked contrast between the northeast and the remainder of the national territory. In the northeast, rates of city growth were generally low, even negative (i.e. some cities experienced absolute losses of population); in the south and west, rates were generally high. In terms of demographic mechanisms, the low growth rates of MSAS in the northeast were attributable primarily to the widespread occurrence of net out-migration. Conversely, the high growth rates of MSAS in the south and west reflected positive migrational change in these centers. Overall, migrational change during 1970–5 was negative in 113 centers, or 43.8 per cent of the total. Natural change, however, was negative in only five MSAS, four of them in Florida. In general, the rates of natural change were much less variable from city to city than the rates of net migration.

The relatively rapid growth of southern and western MSAS appears to be the result of six fundamental causes: (a) favorable climate; (b) attractive physical landscape; (c) increasing success in attracting new investment in manufacturing; (d) disproportionately large percentages of employment in the rapidly expanding tertiary sector of the economy; (e) the growth of tourism; and (f) the impact of the growing population of retired persons.

11

Rank-size relationships and population concentration

In this final chapter we will complete our survey of the various branches of urban systems analysis by examining a group of closely related topics that focus on the relative sizes of cities. There are four topics to consider: (1) the relationship between city size and city rank; (2) the problem of explaining observed rank-size regularities; (3) the concept of the primate city; and (4) the more general 'umbrella concept' of population concentration within an urban system.

The frequency distribution of city size

THE RANK-SIZE RULE: A CRITICAL INTRODUCTION

The existence of a regular statistical relationship between the populations of cities and their ranks within an urban system was briefly discussed in chapter 7, where it was demonstrated that the smooth continuum of city size revealed by most rank-size graphs is not incompatible with the concept of a hierarchically structured system of central places. In returning now to the theme of the 'smooth continuum,' we need to recall the manner in which a rank-size graph is constructed. First, the cities and towns of a particular area are ranked in descending order by population. Second, population size is plotted against rank on special graph paper on which both axes are scaled logarithmically. By general agreement, population size is plotted on the vertical axis and rank on the horizontal. An example is shown in Figure 11.1, which is based on the 200 largest metropolitan areas (MSAS) in the United States at the time of the 1980 census. Figure 11.1 may be compared with Figure 7.1, which shows the relationship between rank and size in a portion of Minnesota.

Rank-size graphs have been prepared by many researchers for many

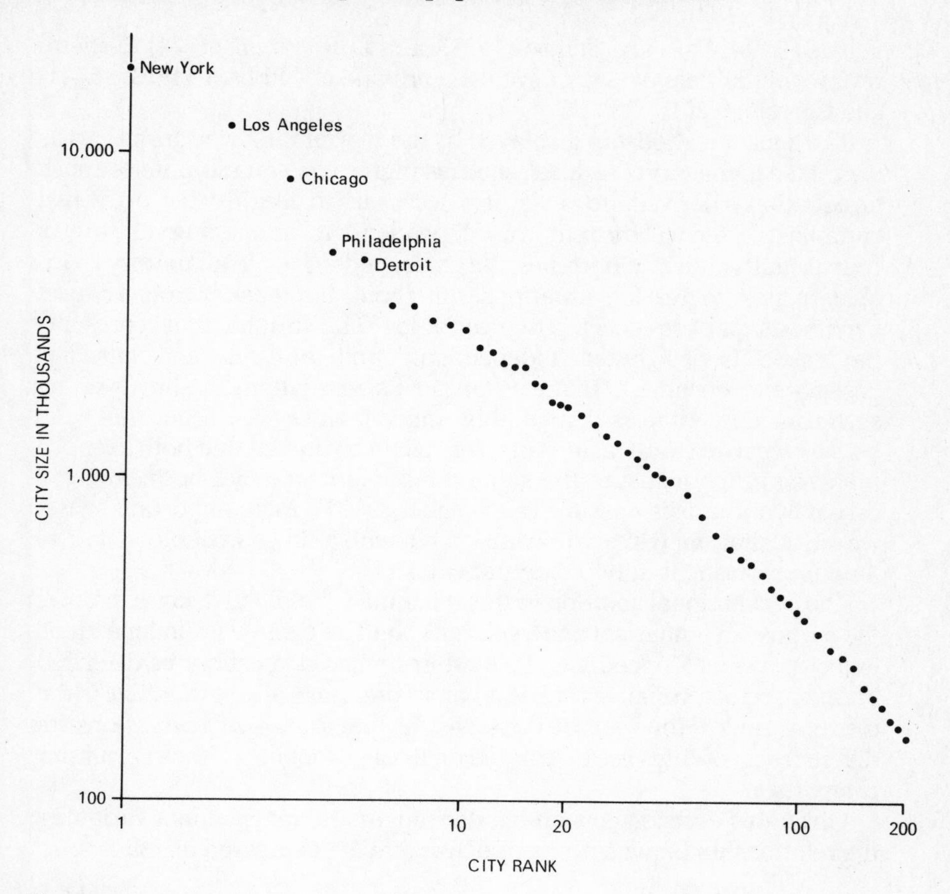

Figure 11.1
The rank-size continuum of city size in the United States in 1980 (logarithmic scales). Individual cities are plotted from number 1 to 20; then every second city from 20 to 40; then every fifth city from 40 to 100; then every tenth city down to 200. Compare with Figure 7.1. Population data from 1980 US Census

different areas, and the results are usually similar to those expressed by Figures 7.1 and 11.1: a straight-line relationship (in logarithms) with a negative slope of approximately forty-five degrees. The earliest writer to comment on this relationship appears to have been Auerbach (1913). Among the veritable flood of later studies, especially noteworthy contributions have been made by Singer (1936), Zipf (1949), Madden (1956a, 1956b, 1958), Berry (1961, 1971), Mandelbrot (1965), Rosing (1966),

Harris (1968), Parr and Suzuki (1973), and D.R. Vining (1974). General reviews of this field of study have been provided by Robson (1973, 16–41) and Carroll (1982).

The linear relationship displayed by the typical rank-size graph can be expressed in the form of a least-squares regression equation. Before such an equation is derived, however, it is necessary to decide which of the two variables plotted will be treated as 'dependent'; the remaining variable, of course, automatically becomes the 'independent' or 'explanatory' variable. In most regression situations, this choice flows easily from a causal hypothesis that connects the two variables. The variable that represents the 'cause' is designated 'independent,' and the presumed 'effect' is designated 'dependent.' In the urban rank-size relationship, however, no such causal hypothesis is available. Indeed, since 'size' and 'rank' can both be regarded as measures of 'size,' it can be argued that both axes of a rank-size graph measure the same thing and hence that no meaningful causal hypothesis is possible (Hakkenberg 1971). In an important sense, we are dealing not with two variables, but with a single variable – city size – and measuring it in two different ways.

The conventional solution to this dilemma is boldly to declare that, for the purpose of numerical analysis, rank shall be treated as 'independent' and city size as 'dependent.' This arbitrary decision cannot be defended in causal terms, because rank is no more the cause of size than size is the cause of rank. However, it does ensure that rank-size regressions for different sets of cities are calculated on a consistent basis, thus permitting comparison.[1]

Given this decision concerning dependent and independent variables, the relationship between rank and size may be expressed as follows:

$$\log P_i = a + b \log R_i + e_i, \tag{11.1}$$

where P_i is the population of city i, R_i is the rank of city i by population size, a is the intercept, b is the slope, and e_i is the residual or error term for city i. The parameters a and b are estimated from the empirical data

1 A bivariate linear regression can be done without specifying the dependent variable. The analysis makes use of the reduced major axis, which is the line bisecting the smaller of the two angles between the least-squares regression line with variable X dependent and the least-squares regression line with variable Y dependent. The reduced major axis is a 'neutral' line of regression, with no implications for the identities of 'cause' and 'effect' (Matthews 1981, 157). Despite its obvious attractions, the reduced major axis has apparently never been used in the study of urban rank-size relations. We will not attempt to break with tradition in the present discussion.

by standard least-squares linear regression. In practice, because the rank-size line necessarily slopes downward to the right, the value of the slope parameter, b, is invariably negative.[2]

The following equations, each based on the twenty largest cities (or metropolitan areas) of the country in question, may be taken as representative examples. Common logarithms (to the base ten) are used in all cases:

$$\text{India } (R^2 = 0.97)\text{: } \log \hat{P}_i = 7.021 - 1.061 \log R_i \qquad (11.2)$$

$$\text{France } (R^2 = 0.89)\text{: } \log \hat{P}_i = 6.596 - 1.005 \log R_i \qquad (11.3)$$

$$\text{Brazil } (R^2 = 0.96)\text{: } \log \hat{P}_i = 6.814 - 0.967 \log R_i \qquad (11.4)$$

$$\text{United States } (R^2 = 0.98)\text{: } \log \hat{P}_i = 7.230 - 0.803 \log R_i \qquad (11.5)$$

The value in parentheses is the coefficient of determination between the logarithms of size and of rank. Because rank is simply an alternative method of measuring size, all these coefficients of determination are statistically significant. The 'hat' notation – that is, $\hat{P}_i$ – is standard usage, signifying the estimated or 'expected' population of the city in question as distinct from the actual or 'observed' population, P_i.

Each of these equations may be regarded as a concise mathematical description of population size within the set of cities considered. Each equation, of course, describes only the set of data-points from which it is derived. (This qualification, in fact, applies to all regression equations, regardless of subject-matter.) Accordingly, a rank-size regression based, for example, on a country's top twenty cities does not necessarily describe accurately the sizes of cities ranked lower than number twenty. Putting this another way, the parameters a and b in equation 11.1 are not necessarily constant, even within a single country; they depend on the number of cities considered. In France, for example, the regression equation when only the top ten cities are considered is as follows:

$$\log \hat{P}_i = 6.715 - 1.250 \log R_i. \qquad (11.6)$$

Both the slope and the intercept differ in value from those of equation 11.3.

2 If, contrary to equation 11.1, we take size to be the 'cause' and rank to be the 'effect'; and if, in addition, we define the dependent variable as $R-1$, where R is rank; then the resulting regression equation is formally identical to the well-known Pareto equation for the frequency distribution of household income (Pareto 1964, 304–5; Champernowne 1953; Allen 1954). In general, therefore, it is acceptable to describe city size as a Pareto variable; 'the higher, the fewer,' obviously the case for both household incomes and city populations.

Because of this instability in regression parameters, international comparisons should be based on a fixed number of cities, as in the twenty-city examples given above.

In many cases, the slope parameter of a rank-size regression lies very close to -1 (that is, a slope of forty-five degrees). Some writers – notably Zipf (1949) – have regarded this tendency as implying that a slope of exactly -1 represents some kind of 'equilibrium' within the urban system. This claim, however, appears to be wishful thinking, since no one has ever identified either the precise nature of this 'equilibrium' or any plausible causal mechanism that might bring it into existence. Zipf, for example, speculated that equilibrium would be attained if the 'forces of unification' and the 'forces of diversification' in society were exactly equal in strength (Zipf 1949, 347–415). If the forces of unification were dominant, a country's urban population would become concentrated in a small number of large cities. Alternatively, if the forces of diversification got the upper hand, the population would live in a large number of small towns. A balance of forces would produce a perfectly regular rank-size series of urban populations. However, neither Zipf nor anyone else has been able to reveal how such opposing tendencies – assuming that their precise character could be specified – might give rise to a regression line with a slope of exactly forty-five degrees.

In short, there exists at present no convincing theoretical reason to associate a slope of -1 with 'equilibrium.' But let us observe what follows if such a slope is actually encountered. First, let us drop the error term, e_i, from equation 11.1. This implies that we must also replace the actual city population, P_i, by the estimated population, $\hat{P}_i$:

$$\log \hat{P}_i = a + b \log R_i. \tag{11.7}$$

Next, we write this equation in nonlogarithmic form:

$$\hat{P}_i = 10^a R_i^b. \tag{11.8}$$

Now, if the value of b is exactly -1, the parameter b can be removed, leaving us with

$$\hat{P}_i = 10^a / R_i. \tag{11.9}$$

For the system's largest city, $\hat{P}_1$, the term R_i takes the value of 1 by definition. For the largest city, therefore,

$$\hat{P}_1 = 10^a, \tag{11.10}$$

from which it follows that the parameter a is simply the (base-ten) logarithm of the estimated size of the country's largest city. Moreover, if all the cities fall exactly on the regression line, we obtain the relationship

$$P_i = P_1/R_i. \quad (11.11)$$

Equation 11.11, usually called the 'Zipf model' (Zipf 1949), states that the size of the city having rank R_i is equal to the size of the largest city divided by R_i. In such a system, in other words, the second-largest city is one-half the size of the largest; the third-largest city is one-third the size of the largest; the tenth-ranked city is one-tenth the size of the largest; and so forth. A structure this simple seems 'too good to be true,' yet equation 11.11 provides a surprisingly good fit to the population data in cases where (a) the slope of the true regression equation is close to -1 and (b) the largest city lies close to the true regression line.

When placed in proper perspective, the Zipf model (equation 11.11) is seen to be a special case of the general logarithmic model represented by equation 11.1. The latter formulation, because of its flexibility or 'openness,' should be used in detailed descriptive work. As we have mentioned, this general model is normally calibrated by transforming the data on size and rank into logarithmic form and then employing the standard statistical technique of least-squares linear regression. The use of logarithms, however, gives rise to two important technical problems.

1. *The problem of small-city bias*. The logarithmic transformation of a set of numbers gives greater weight to the smaller values. Consider, for example, two cities that contain 1,000,000 inhabitants and 100,000 inhabitants, respectively. In real terms the larger city is ten times as large as the smaller, but their logarithms are not so strikingly different, being 6.0 and 5.0, respectively. In a traditional logarithmic rank-size analysis, the values of the regression parameters are governed primarily by the closely packed mass of smaller centers. The less numerous larger centers do not receive a weight that is proportional to their size, and the rank-size structure of these larger places may therefore be misrepresented in the overall regression equation. Yet the investigator is usually more interested in the size relationships of the leading cities than in the characteristics of the smaller members of the system.

2. *The problem of mismatch in total urban population*. Suppose that the urban system under investigation contains N cities. When equation 11.1 is calibrated with the aid of a logarithmic transformation of the data, the sum of the N populations predicted by the regression equation may not equal the sum of the N populations actually observed. Although no comprehensive study of this lack of agreement has been undertaken, the

discrepancies are sometimes quite large. For example, the total urban population generated by equation 11.3, above, is 24 per cent below the actual total contained in the twenty French cities whose character the equation is supposed to describe. From the point of view of the esthetics of model-building, it does not seem unreasonable to demand that the hypothetical urban system described by the model should contain the same total number of inhabitants as the real urban system that is the subject of the analysis.

Nader (1984) has recently shown how both of the above difficulties can be overcome through the use of a method of calibration that does not involve a logarithmic transformation of the data. This method is not only theoretically well founded but also mathematically elegant, and it therefore seems likely to replace the logarithmic technique in future research in this field.

THE NADER NON-LOGARITHMIC CALIBRATION

The nonlogarithmic approach may conveniently be described in the form of a sequence of steps, as follows.

1. The initial step is the arbitrary selection of a 'possibly correct' value of the parameter b in equation 11.1. Because we know from experience that the value of b is generally close to -1, it makes sense to choose an initial value of b lying between -0.5 and -1.5. As we will see, this initial value may not be the best-fitting value for our purpose. But a selection must be made in order to allow us to get started.

2. Treating the chosen value of b as a positive number, we now find the numerical value of the following algebraic series:

$$\frac{1}{1^b} + \frac{1}{2^b} + \frac{1}{3^b} + \dots + \frac{1}{N^b}, \tag{11.12}$$

where N is the number of cities in the analysis. Let the sum of this series in any particular case be denoted by Z. That is:

$$Z = \frac{1}{1^b} + \frac{1}{2^b} + \frac{1}{3^b} + \dots + \frac{1}{N^b}, \tag{11.13}$$

3. Next, we obtain the estimated (that is, model-based) population of each of the N cities by multiplying each term in the above series by P/Z, where P is the actual total population of the N cities. Putting this more explicitly:

$$\hat{P}_1 = \frac{P}{1^b Z} = \frac{P}{Z}, \tag{11.14}$$

$$\hat{P}_2 = \frac{P}{2^b Z}, \tag{11.15}$$

$$\hat{P}_3 = \frac{P}{3^b Z}, \tag{11.16}$$

and so forth. The sum of the N estimates obtained in this way is necessarily equal to P, the observed total population of the urban system. In short:

$$\sum_{i=1}^{N} \hat{P}_i = \sum_{i=1}^{N} P_i = P. \tag{11.17}$$

4. However, equation 11.17 is true for any value of b. We therefore need to measure the degree of resemblance between the observed and the estimated city populations. Nader (1984) proposes that we should use the sum of the squares of the differences. That is, we find the value of S, where

$$S = \sum_{i=1}^{N} (P_i - \hat{P}_i)^2. \tag{11.18}$$

Naturally, the value of the sum of squares, S, will vary with different values of b.

5. Therefore, we repeat steps 1 through 4, using a different value of b each time, until we have determined the value of b that minimizes the value of S.

This nonlogarithmic approach to calibration is an example of an iterative search procedure. We start with an arbitrary (but plausible) value of b and employ this assumed value in order to derive a set of estimates of the urban populations. These estimates are then evaluated by means of a predetermined 'objective function,' or criterion of acceptability. The objective function here is the sum of squares, S (equation 11.18), and our goal is to find the value of b that minimizes the value of S. In practice we proceed by increasing or decreasing the assumed value of b very slightly at each iteration, noting how the value of S responds to each change. In this way, the search can be made to converge toward the required value of b. This would have been a tedious procedure in the days before electronic calculators and computers. The availability of

TABLE 11.1
Comparison of rank-size regression parameters obtained by logarithmic and non-logarithmic methods

Country	Number of cities	Logarithmic calibration		Nonlogarithmic calibration	
		Intercept	Slope	Intercept	Slope
India	20	7.021	−1.061	6.890	−0.854
France	20	6.596	−1.005	6.995	−1.703
Brazil	20	6.814	−0.967	6.870	−1.034
United States	20	7.230	−0.803	7.255	−0.832

these modern aids, however, makes iterative search procedures in general quite easy to apply.

The complete rank-size equation corresponding to the nonlogarithmic method of calibration is written as follows:

$$\log \hat{P}_i = \log (P/Z) + b \log R_i, \tag{11.19}$$

where $\hat{P}_i$ is the estimated population of city i, P is the total population of the urban system, Z is the sum of reciprocal powers defined by equation 11.13, b is the slope parameter, and R_i is the rank of city i. This equation, like equations 11.1 and 11.7, is written in logarithmic form. Indeed, equations 11.19 and 11.7 are structurally identical: the difference lies not in the form of the equation but in the method of calibration – the method by which the numerical values of the parameters are determined. As we have indicated, the nonlogarithmic method of calibration circumvents the problem of small-city bias and the problem of a mismatch in total urban population. It therefore provides a more satisfactory description of the rank-size structure of the urban system as a whole.

An impression of the effects of applying the nonlogarithmic method of calibration may be gained from Table 11.1. In preparing this table, the same data were used to produce both the logarithmic and the nonlogarithmic results for each country. As can be seen, the two methods produce different values for both the slope and the intercept in every case. The differences in some instances appear quite small, but small differences in the domain of logarithms translate into large differences in the numerical estimates of city size. Consider, for example, the estimated size of Paris, the highest-ranking city in France. Using the traditional logarithmic calibration, the estimate is 3,945,000 inhabitants (to the nearest thousand persons). Using the nonlogarithmic approach, the estimated number is

9,886,000, or more than twice as many as in the first estimate. The true figure is 9,863,000 inhabitants. In this example, as in many others that could be described, the superiority of the nonlogarithmic method of calibration is obvious.

It may be pointed out, as a small addition to Nader's contribution, that the sum of reciprocal powers denoted above by Z (see equation 11.13) is a form of the Riemann zeta function (N.L. Johnson and Kotz 1969, 240–4; Derman, Gleser, and Olkin 1973, 296–9). In the limiting case, the number of terms in the zeta function is infinite, but in the present context the number of terms must clearly be governed by the number of cities included in the study. In view of the important role played by this sum of reciprocal powers in describing the rank-size structure of urban populations, it is reasonable to say that the variable 'city size' conforms to a truncated Riemann zeta distribution, at least above a certain minimum size. Indeed, given the present state of our understanding of the frequency distribution of city size, this is probably the most appropriate descriptive label to employ.

The search for a model of growth

THE GIBRAT PROCESS

Regardless of the technique used to calibrate the regression equation relating size and rank, the relationship itself is log-linear in form, at least for each country's largest cities. (The term *log-linear* means simply that the relationship between the two variables appears as a straight line when the data are plotted on logarithmic graph paper.) Given such a relationship, we are led naturally to ask what kind of evolutionary process might have created it. Here we come face to face with the difficult problem of establishing a connection between the rank-size regularity and the character of city growth in a systemic context.

The Gibrat model (Gibrat 1931, 1957; Parr and Suzuki 1973) has been proposed as a possible explanation for the log-linearity of rank-size relationships. This model assumes that the members of a set of growing entities – in our case, cities – satisfy two basic conditions.

First, during any given time interval, the rates of city growth are statistically independent of city size. More specifically, if the cities are classified by size, the growth rates of cities within each size-class will vary randomly around an average growth rate that is identical for all size-classes. An equivalent statement is that the amount (not the rate) of growth experienced by each city is a randomly varying proportion of its absolute size. This latter statement is the 'law of proportionate effect.'

Second, during any given time interval, the variability of growth rates is identical in all size-classes. This means that the degree of dispersion of growth rates around the average value is the same for all size-classes. In statistical terminology, the growth rates are assumed to be homoscedastic with respect to city size.

Provided that both of these somewhat stringent conditions are fulfilled, and provided that growth continues for a sufficiently long time, the outcome of this process of growth will be a steady state in which the variable 'city size' follows a lognormal distribution. The mathematical argument by which this conclusion is reached lies beyond the scope of this book. However, a variable, X, conforms to a lognormal distribution if the values of the transformed variable, $\log X$, follow a normal distribution. A normal distribution, as is well known, is bell-shaped and perfectly symmetrical. A lognormal distribution is also bell-shaped after a fashion, but strongly skewed, with a very long upper tail (Aitchison and Brown 1957).

The relevance of the Gibrat model lies in the fact that the upper tail of a lognormal distribution takes the form of a straight line when the logarithm of size is plotted against the logarithm of rank. Therefore, if a system of cities grows in accordance with the Gibrat assumptions, its largest centers will ultimately exhibit a log-linear relationship between rank and size. In effect, then, an observed rank-size regularity can be accounted for as the outcome of a process of growth that leads to a steady-state lognormal distribution of city size (Berry 1971, 144–6; Parr and Suzuki 1973).

If the whole of a lognormal frequency distribution is plotted in rank-size form, the straight-line character of the uppermost section of the graph does not continue indefinitely. Beyond a certain point, the slope of the line becomes noticeably steeper, so that the graph as a whole is concave toward the origin. Such a change in slope does occur in some urban systems, provided that the plot extends downward to include towns of a sufficiently small size (Stewart 1958; Parr and Suzuki 1973). As yet, however, the evidence is fragmentary, and it would be premature to conclude that a downwardly concave shape is the norm. Nevertheless, the uniform slope displayed by most rank-size graphs may simply reflect the inclusion of too few cities. This is clearly a question that can be resolved by further empirical research.

If it should be found that steepening of the slope is normal, the hypothesis of lognormality would thereby receive significant support. We would then be in a position to refine our descriptive terminology. On the one hand, the frequency distribution of size within an urban system taken in its entirety could properly be described as lognormal. On the other hand, there would still be a straight-line relationship above a certain empirically identifiable level of size, and this uppermost section of the graph could still be described as a truncated zeta distribution. The zeta distribution, when plotted on logarithmic graph paper, maintains a constant angle of slope throughout the full extent of its range; but the lognormal distribution does not. Thus the applicability of each descriptive label depends on the level of size at which the ranked list of urban centers is truncated prior to construction of the graph.

Nevertheless, every rank-size graph, whether or not it includes a change in slope, is at least consistent with the hypothesis of lognormality. It is therefore tempting to conclude that the Gibrat model provides a satisfactory general explanation of the frequency distribution of city size. A serious difficulty arises, however, with the two assumptions on which the model is based: namely, that growth rates are uncorrelated with city size and that growth rates are homoscedastic. No urban system has yet been found in which these crucial assumptions hold true over an extended period of time. On the contrary, available evidence suggests that both assumptions are false. For example, in a major study of nineteenth-century urban growth in England and Wales, Robson has found that the growth rates in a majority of the decades were not independent of city size. Instead, there was a positive association – the larger the cities in a particular size-class, the higher their average rate of growth (Robson 1973, 71–89). A similar result has been reported for the period 1851–1971 in southern Ontario (Marshall and Smith 1978). Both studies also found that the growth rates were heteroscedastic with respect to city size, and not homoscedastic, as required by the Gibrat model. In nineteenth-century England and Wales, the variability of growth rates decreased significantly with an increase in city size: the larger the cities in a size-class, the less variable their rates of growth. This specific relationship was present in only four of the twelve decades covered by the southern Ontario study, but other forms of heteroscedasticity were present in other decades, and it is clear that the Gibrat assumption was not fulfilled. Nevertheless, though neither England and Wales nor southern Ontario satisfied either of the basic assumptions of the Gibrat model, the cities of both regions displayed statistically significant log-linear relationships between rank and size throughout the periods considered.

These findings compel us to conclude that the Gibrat model is unacceptable as a general explanation of urban rank-size relationships. It is conceivable, of course, that the lognormal frequency distribution may be derivable from assumptions that are less restrictive – and more realistic – than those of the Gibrat model; but no satisfactory alternative model has yet been proposed. If we maintain that a log-linear rank-size relationship represents the upper tail of a lognormal distribution, we must admit that the evolutionary dynamics underlying this statistical pattern are not adequately understood. Of course, we know equally little about the evolutionary dynamics of other relevant distributions, such as the zeta distribution. But the latter is intended to serve as a model only of structure; the Gibrat model has been proposed as a model of growth. Available evidence makes it clear that the process of growth occurring in the real world simply does not correspond to the process postulated in the Gibrat model. Accordingly, we currently lack a model of urban rank-size relations that is both mathematically rigorous and process-oriented.

We are not contending that the Gibrat assumptions do not lead to a lognormal distribution. On the contrary, we accept the validity of both the internal workings of the Gibrat model and the claim that observed frequency distributions of city size may be interpreted as being lognormal. However, in cases where the relevant historical information has been examined, the assumptions of the Gibrat model have been shown to be empirically false. Hence the frequency distributions in question, even if they are lognormal, cannot have been produced by the Gibrat process of growth. In short, based on present evidence, use of the Gibrat model to explain rank-size regularity appears to do no more than demonstrate that valid reasoning from false premises sometimes leads to a true conclusion.

It does not follow, however, that we can offer no explanatory suggestions whatsoever. We can take refuge – and perhaps not reluctantly – in the view that rank-size regularity arises as a distortion of the stratified character of the central place hierarchy. As noted in chapter 7, this approach asserts that a perfectly discrete stratification of cities by size will not normally occur over large areas under real-world conditions because the real world is not isotropic and cities do not function solely as central places. As a result of these two factors, sharp distinctions between hierarchical orders in terms of the population sizes of towns will be lost, thus leading to the smooth continuum that appears on the typical rank-size graph. The concepts of a hierarchy and of a rank-size continuum both call for an urban system in which places of successively larger size are progressively less numerous. As several writers have pointed out, the existence of a log-linear rank-size relationship does not preclude the simultaneous coexistence of an underlying hierarchical pattern of spatial organization (Beckmann 1958; Berry and Barnum 1962; Parr 1970). In fact, given the 'imperfections' in which reality abounds, the most satisfactory explanation of rank-size relationships may reside, after all, in the familiar concept of the central place hierarchy.

CITY SIZE AND VARIABILITY OF GROWTH RATES

As indicated above, Robson's detailed study of urban growth in nineteenth-century England and Wales revealed not only that growth rates were heteroscedastic with respect to city size but also that the heteroscedasticity exhibited a particular regularity of its own: namely, a decrease in the variability of growth rates with an increase in city size. Like any other persistent empirical regularity, this relationship cries out for explanation. Robson's proposed explanation links the behavior of city growth rates both to the concept of the urban hierarchy and to certain well-established ideas concerning the diffusion of innovations (Robson 1973, 186–213).

In brief, Robson posits that city growth may be regarded as a concomitant of the adoption of technical and entrepreneurial innovations in various sectors of the economy. Each successfully adopted innovation represents an investment that generates new employment opportunities and thereby creates growth, both by attracting migrants to the city from outside and by ensuring retention of residents who might otherwise move elsewhere. The rate of growth of a city during a given period is proportional to the number of innovations adopted in the recent past. The greater the number of adoptions, the higher the rate of growth.

The probability that an innovation will *originate* in a particular city is taken to be proportional to city size; it is assumed that the largest cities are the principal sources of new inventions that have the potential to generate growth in the urban system as a whole. Once an innovation has been 'born' into the system, the city of its origin begins to send out signals or 'stimuli' that inform other centers of the happy event, and in due course these other centers respond by adopting the innovation themselves. After they adopt, they begin to send out signals of their own, and in this way the innovation ultimately spreads to every corner of the urban system.

The likelihood that a particular city, C, receives a single stimulus from a transmitting center, T, is a function of the size of C and the distance between C and T. Following the logic of the standard gravity hypothesis of spatial interaction (Olsson 1965; Lowe and Moryadas 1975, 176–97), it is assumed that the probability of a 'hit' increases with an increase in the population size of *C* and decreases with an increase in the distance separating *C* from *T*. These assumptions are consistent with the accepted view that the diffusion of an innovation involves both a 'hierarchy effect' and a 'neighborhood effect' (Pedersen 1970; Lowe and Moryadas 1975, 226–47). The hierarchy effect is the tendency for a given innovation to be adopted in large cities before it is adopted in small centers. The neighborhood effect is the tendency for places close to the source of an innovation to become adopters earlier than places located farther afield.

These ideas are used by Robson as the basis of a Monte Carlo simulation model that describes the spread of a single 'composite innovation' that originates in the largest center of an imaginary urban system (Robson 1973, 187–92). Various adjustable rules are employed in order to specify such necessary details as (a) the relative strength of the hierarchy and neighborhood effects; (b) the number of stimuli emitted by each center during a given time-interval; (c) the number of stimuli that a town must receive before it adopts the composite growth-inducing innovation; and (d) the quantitative relationship between the receipt of stimuli and the growth of the town. The specific operational details need not concern us, but it should be noted that Robson's model, like all Monte Carlo simulation models, incorporates an element of chance as well as the systematic (nonrandom) components that represent, in this case, the hierarchy and neighborhood effects. The presence of the element of chance means that the hierarchy and neighborhood effects do not operate in a completely rigid, deterministic fashion. From time to time – just as in the real world – the unexpected occurs: for example, a small town, remote from all sources of stimuli, may become an adopter at an early stage of the process, or a large center close to a major source may nevertheless fail to adopt until comparatively late in the game.

As simulation models go, the Robson model is not especially complicated, yet it can replicate the phenomenon we wish to explain – the marked tendency for the variability of city growth rates to decline with an increase in city size. The model works because, at any given time, small centers differ among themselves to a greater degree than large centers with respect to the distance across which a growth-inducing stimulus must jump in order to reach any particular town. Because the distances from transmitters to potential adopters are thus more variable for smaller towns, the numbers of growth-inducing stimuli received, during a particular period, are also more variable for smaller centers. Accordingly, the growth rates for each period as a whole also vary more among smaller centers, just as we actually observe.

The specific mechanism of growth incorporated in the Robson model has not, of course, operated uniformly in every urban system at all times. We have seen, for example, that the variability of growth rates in southern Ontario over the period 1851–1971 was not always related to city size in the regular fashion encountered in nineteenth-century England and Wales (Marshall and Smith 1978). No doubt the historical experiences of different regions have been sufficiently unique that no single simulation model can fit all cases. However, we would be well advised, in future research, to pursue Robson's general approach, by assuming that the dynamics of growth in a systemic context can be effectively modeled as a stochastic process (or family of processes) involving the diffusion of growth-inducing impulses through a set of cities under the guidance of constraints that are, in part, locationally defined. The success of Robson's work in this direction suggests that a deeper understanding of the frequency distribution of city size is more likely to emerge from evolutionary studies – both theoretical and empirical – of the behavior of growth-inducing impulses than from any model in which growth rates and city sizes are assumed, a priori, to be uncorrelated. Could a simulation model of the Robson type, when permitted to run for enough time-intervals, produce the log-linear rank-size relationship as a steady-state result? This important question should be regarded as a major target of future research in this field.

Primate cities

Because population size seems a reasonable guide to the overall importance of an individual city within a national settlement system, scholars have shown an especially high level of interest in the population characteristics of each country's largest urban centers. Much of this interest has revolved around the concept of the primate city. As a preliminary definition, we may regard a city as being 'primate' if it is outstandingly dominant in terms of both size and economic influence within its particular urban system. However, the phrase 'outstandingly dominant' is notable chiefly for its vagueness. The alternative phrase, 'disproportionately large,' used by Jefferson when he introduced the concept of urban primacy in 1939, suffers from a similar lack of precision (Jefferson 1939, 231). Let us therefore begin by considering the actual extent to which the leading cities of various countries dominate their urban systems in terms of population size.

We must first decide on a method of measurement. The method used most commonly is to divide the population of the largest city by the population of the second-largest city. The greater the value of the quotient thus obtained, the higher the degree of primacy displayed by the leading city (Stewart 1958; Linsky 1965). In Brazil, to take one example, the populations of the two largest cities are as follows (using metropolitan area data): São Paulo, 7,199,000; Rio de Janeiro, 4,858,000. The index of primacy is therefore 1.5. In Argentina, by contrast, the relevant populations are as follows: Buenos Aires, 8,436,000; Rosario, 807,000. Argentina's index of primacy is therefore 10.5, seven times as large as the figure for Brazil.

An alternative measure of primacy may be obtained by expressing the size of the largest city as a percentage of the combined population of the top *n* cities, where *n* is a small number such as four or five. Taking *n* to be five, the values for Brazil and Argentina are 44.7 per cent and 76.8 per cent, respectively. Although this method is perfectly valid, experience has shown that it does not offer any significant advantage over the two-city index. In fact, as might be anticipated, the results produced by the two approaches are strongly correlated with each other. Mehta (1969, 308) reported a rank correlation of +0.86 in a study that included eighty-one countries. Given this high degree of association, we will opt for the first method – this being, by a small margin, the simpler of the two.

Using population data published in a recent edition of *The Geographical Digest* (Willett et al 1981), I calculated primacy indexes for ninety-one countries: namely, every country that contained at least two cities or urban agglomerations of 100,000 inhabitants or more. The city-states of Hong Kong and Singapore were excluded. The results are shown in Table 11.2.

The lowest index value occurs in Canada, where the first and second cities – Toronto and Montreal, respectively – are virtually identical in size. This situation yields a primacy index of 1.0 – by definition, the lowest value that can possibly occur. The highest calculable value occurs in Peru, at 10.9. Higher values occur in Uruguay, Cambodia, and Thailand, but my data-source provided information for only a single city in each case. Nevertheless, among all the countries included in Table 11.2, Thailand undoubtedly holds the record for high primacy, since metropolitan Krung Thep (Bangkok) contains 4.7 million inhabitants and no other Thai city appears to exceed 100,000. These figures imply a primacy index of at least 47.0.

The overall frequency distribution of the ninety-one primacy indexes is one in which small values are abundant and larger values are progressive-

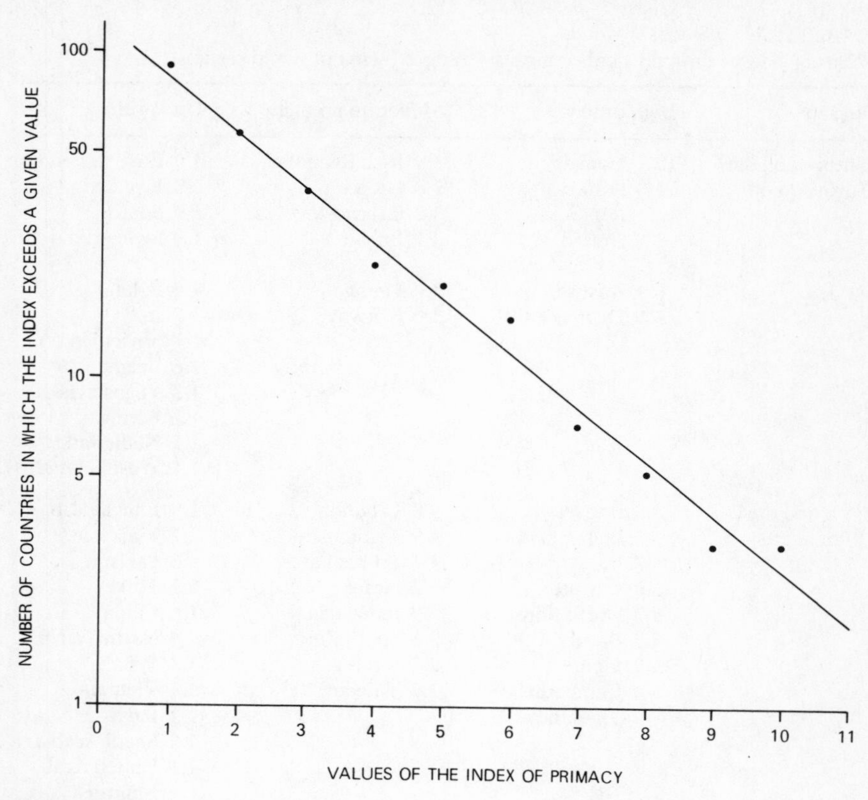

Figure 11.2
The relationship, at the world scale, between the index of urban primacy and the number of countries in which this index exceeds a designated value

ly less common. More specifically, there is a linear relationship between the value of the primacy index, X, and the logarithm of the number of countries having index values greater than X. This relationship is shown in Figure 11.2. For the ten data-points used in the construction of this graph, the best-fitting regression equation, using base-ten logarithms, is as follows:

$$\log Y_i = 2.096 - 0.171\, X_i, \qquad (11.20)$$

where X_i is an integer value of the primacy index (that is, $X_i = 1, 2, 3, \ldots, 10$), and Y_i is the number of countries having index values greater than X_i. The coefficient of determination, R^2, is 0.98.

TABLE 11.2
Primacy indexes for individual countries arranged by major world regions

Region[a]	High primacy	Moderate primacy	Low primacy
Europe and the Soviet Union	10.3 Hungary	3.9 Irish Republic	2.0 East Germany
	8.6 France	3.8 Greece	2.0 Sweden
	7.2 Romania	3.2 Czechoslovakia	1.9 Spain
	7.1 United Kingdom	3.1 Bulgaria	1.9 Switzerland
	6.4 Austria	3.0 Finland	1.8 Poland
	5.1 Denmark	3.0 Norway	1.7 Italy
			1.7 Soviet Union
			1.6 Belgium
			1.3 Yugoslavia
			1.2 Portugal
			1.1 Netherlands
			1.1 West Germany
Asia and the Middle East	8.0 Iraq	4.0 Lebanon	1.9 Bangladesh
	6.8 Philippines	3.9 Afghanistan	1.7 Malaysia
	6.7 Iran	3.7 Sri Lanka	1.6 Pakistan
	6.0 Kuwait	3.2 Israel	1.5 Turkey
	5.7 North Korea	2.9 Indonesia	1.4 China
	5.0 Burma	2.8 South Korea	1.4 North Yemen
	4.3 Japan	2.7 Jordan	1.3 Syria
	Cambodia[b]	2.7 Taiwan	1.3 Vietnam
	Thailand[b]		1.2 India
			1.2 Saudi Arabia
			1.1 United Arab Emirates
Africa	6.8 Senegal	3.3 Zaire	2.0 Congo
	6.8 Tanzania	3.1 Algeria	2.0 Libya
	Angola[b]	3.0 Ethopia	2.0 Tunisia
	Guinea[b]	2.9 Morocco	1.8 Zambia
	Madagascar[b]	2.7 Ivory Coast	1.7 Benin
	Mali[b]	2.7 Kenya	1.7 Nigeria
		2.2 Egypt	1.7 Zimbabwe
		2.2 Malawi	1.5 Burkina Faso
		2.1 Ghana	1.5 Cameroon
			1.3 South Africa
			1.1 Sudan
Latin America	10.9 Peru	3.5 El Salvador	1.8 Honduras
	10.5 Argentina	3.3 Venezuela	1.5 Brazil
	6.0 Chile	3.0 Panama	1.4 Ecuador
	6.0 Mexico	2.8 Bolivia	

TABLE 11.2 (continued)

Region[a]	High primacy	Moderate primacy	Low primacy
	5.9 Cuba	2.5 Colombia	
	4.6 Dominican Republic	2.5 Puerto Rico	
	Guatemala[b]		
	Haiti[b]		
	Jamaica[b]		
	Nicaragua[b]		
	Paraguay[b]		
	Uruguay[b]		
North America and Oceania		2.3 New Zealand	1.5 United States
			1.2 Australia
			1.0 Canada

SOURCE: *The Geographical Digest 1981* (Willett et al 1981)

a In each section of the table, each country is ranked by its primacy index, which appears in front of its name. The index is the quotient obtained by dividing the population of the largest city by the population of the second-largest city.

b Owing to data limitations, the value of the primacy index for certain countries could not be calculated. However, these countries are included if it is clear that the index is greater than 4.0.

At what point along the scale of primacy values does the first-ranked city become 'outstandingly dominant' or 'disproportionately large'? One reasonable answer is that a leading city should not be described as 'primate' unless the index is greater than 2.0, the value that would occur if the cities adhered to Zipf's simplified model of the relationship between rank and size (see equation 11.11, above). Countries in which the primacy index is greater than 2.0 may be divided into two groups: 'moderate primacy,' with index values greater than 2.0 but not greater than 4.0; and 'high primacy,' with index values greater than 4.0. These suggested threshold values of 2.0 and 4.0, in fact, have been used for the classification of countries in Table 11.2. The group totals are as follows: 33 countries with high primacy; 30 countries with moderate primacy; and 40 countries with no primate city.

When Jefferson discussed urban primacy in his seminal paper of 1939, he proposed a 'law' of primate cities: 'A country's leading city is *always* disproportionately large and exceptionally expressive of national capacity and feeling' (Jefferson 1939, 231; emphasis added). If we accept the view that the leading city is not 'disproportionately large' unless it is more than twice the size of the second-ranked city, the use of the term *law* in this context is inappropriate. A more satisfactory approach is to regard urban primacy as a continuous variable. When quantified by means of the simple two-city index employed in Table 11.2, this variable has a minimum possible value of 1.0 and, in principle, no upper limit. If the value attained by a particular country exceeds 2.0, the term *primate city* may properly be applied. If the value is 2.0 or less, no primate city is present.[3]

Most writers who have discussed the concept of the primate city have also attempted to identify the factors that cause the degree of urban primacy to vary from country to country (Stewart 1958; Berry 1961; Linsky 1965; Mehta 1969). On the whole, however, these attempts have met with frustration. The current consensus is that the worldwide pattern of variability in the degree of primacy cannot be accomodated within a single explanatory hypothesis. Different countries, in other words, seem to call for different explanations.

The most plausible general hypothesis is that the degree of primacy is negatively correlated with a country's level of economic development. This hypothesis presupposes that the less-developed countries (LDCs) are too poor and too agrarian to support, or to require, substantial numbers of large cities. Because manufacturing is only weakly represented, and because the general level of demand for consumer goods and services is low, most towns in the typical LDC remain small. One center, however, manages to attain metropolitan rank. In the ideal case, this center is a seaport rather than an inland town; and it is the principal, and perhaps the only, point through which economic links with the rest of the world are maintained. It also contains the lion's share of the country's limited manufacturing employment, and it is the sole provider of high-order services such as specialized medical care and post-secondary education.

Most LDCs were at one time colonial possessions of European powers. In the normal course of events, the mother country would select a single center in each colony to serve as the point of attachment for trade and as headquarters of the colonial administration. Although the center selected might already have been important in pre-colonial days, the European presence facilitated the emergence of the colonial capital as a strongly primate city. More often than not, the seat of government remained in the former colonial capital after independence, with the expanding indige-

3 If there is a 'law' of primate cities, it is expressed in equation 11.20: there is a mathematical relationship between the index of primacy and the number of countries in which each value of the index is exceeded. Figure 11.2 expresses this law as a graph.

nous bureaucracy contributing further to the continued growth of the metropolis.

In industrialized nations – so runs the theory – the above scenario does not apply. Developed countries require large cities in comparatively great numbers, partly to serve as manufacturing centers and partly to function as points of distribution for the many consumer goods and services demanded by a relatively affluent population. Because a developed country supports many large cities instead of only one, the urban system is unlikely to exhibit a high index of primacy. Growth impulses in a developed country are widely dispersed throughout the urban system, whereas in the typical LDC they tend to be strongly focused on a single city. The high concentration of growth impulses at a single location inevitably gives rise to a high degree of urban primacy.

Although the foregoing hypothesis has a certain degree of plausibility, it falls to the ground the moment it is confronted with the facts. It is true, of course, that some LDCs display high primacy – for example, Burma, Thailand, and Tanzania – and that some industrialized countries have low index values – for example, Belgium, Poland, and the United States. However, as Table 11.2 shows, some LDCs exhibit low primacy indexes – for example, Honduras, Burkina Faso (formerly Upper Volta), and Bangladesh – whereas some industrialized countries have values that are distinctly high – for example, the United Kingdom, France, and Japan. A detailed investigation shows that any appropriate statistical test involving the full array of countries listed in Table 11.2 leads to the conclusion that primacy and economic development are simply not related in any systematic way. The same conclusion, in fact, has been reached by other writers (Berry 1961; Mehta 1969). Accordingly, we must reject the hypothesis of a general negative correlation. The assumptions that lie behind this hypothesis may be valid for some countries, but they do not provide a comprehensive explanation that is acceptable in all cases.

The only other general hypothesis worth considering is that the level of primacy is related to the territorial extent of each country's ecumene, or effectively settled area. If the ecumene is small, a single city can perform all the metropolitan functions required by the entire country. The smallness of the ecumene should inhibit the development of competing centers, thereby permitting the lone metropolis to become preeminent in terms of size. However, no such monopoly of metropolitan functions is likely if the ecumene covers an extensive area. A large ecumene will require the services of several cities of metropolitan rank, each one serving as the focus of a different region. Under these circumstances, a high degree of primacy is unlikely for the country's urban system as a whole. In short, the hypothesis states that primacy should be high where the ecumene is small and low where the ecumene is large.

On the whole, this second hypothesis is slightly more successful in accounting for the facts. In particular, primacy is uniformly low in the world's seven giant-sized countries: namely, the Soviet Union, China, India, Australia, Canada, the United States, and Brazil. Several of these countries have an ecumene that is small in comparison to the total area of the national territory; but nevertheless, these ecumenes are still very large in relation to the inhabited areas of other countries. The average value of the primacy indexes for these seven countries is a mere 1.4, and none of the seven values is higher than 1.7 (Table 11.2). It is therefore reasonable to conclude that a large ecumene does inhibit the emergence of a national primate city.

However, if the above seven countries are excluded, the relationship between primacy and area breaks down. Ecumenes are small where primacy is high (except perhaps in Mexico and Argentina), but primacy is not generally high where ecumenes are small. Countries with decidedly small ecumenes, yet no primate city, include Portugal, Switzerland, Syria, Benin, and Ecuador (Table 11.2). At the world scale, therefore, the relationship between primacy and the size of the ecumene is asymmetrical. Primacy is low, as predicted, if the ecumene is notably large; but primacy is not necessarily high if the ecumene is small.

Additional insight into the relationship between primacy and area can be obtained by considering the characteristics of city size at the regional scale within individual countries. A particularly interesting example is Australia (Stewart 1958; Rose 1966; McCarty 1974), which has no primate city at the national scale (Table 11.2). However, each of the country's five largest cities – Sydney, Melbourne, Brisbane, Adelaide, and Perth – is strongly primate within its own state. During the nineteenth century, the British colonies that later evolved into the Australian states were, to a large extent, separate economic units, each having closer relations with London, halfway around the world, than with its neighbors on the Australian continent. In each case links with the mother country were channeled almost exclusively through a single point of entry, which developed in due course into a primate metropolis. Although high levels of primacy are not found in Tasmania or in the thinly populated Northern Territory, Australia as a whole demonstrates clearly that a country with no primate city at the national scale may nevertheless exhibit high levels of primacy within individual regions.

A somewhat similar result is obtained in Canada, another large country with no primate city at the national scale. In six of the ten Canadian provinces, the index of primacy is 3.0 or higher, with the average value for these six provinces being an impressive 7.4. (Primacy is also high in the Yukon at 13.0, but the index for the Northwest Territories is only 2.3.) The six provincial primate cities are: St John's (Newfoundland), Halifax (Nova Scotia), Montreal (Quebec), Toronto (Ontario), Winnipeg (Manitoba), and Vancouver (British Columbia). Montreal, Toronto, and Vancouver closely resemble their Australian counterparts in that they functioned historically as major points of entry for European traders and settlers. Moreover, St John's, Halifax, Toronto, and Winnipeg, like the Australian primates, are political capitals. The primacy indexes of Montreal and Vancouver would probably be even higher but for the fact that the political function in Quebec and British Columbia is housed in the second-ranked city – Quebec and Victoria, respectively.[4]

Clearly a city's degree of primacy depends on the extent of the area to which the analysis is applied. Sydney is not primate within Australia, but it is strongly primate within New South Wales. Winnipeg is not primate within Canada – indeed, it is not even the country's largest city – yet it is highly primate within Manitoba. Is there any 'proper' or 'natural' area to which a city ought to be related before its degree of primacy can be fairly assessed? This question has not been addressed in the literature, but it is nevertheless important. We therefore propose, as a point for discussion, that the most appropriate area to employ is the umland of the city in question. As explained in chapter 6, the umland is the region within which the city's role as a provider of goods and services is quantitatively dominant, exceeding the contributions of other cities of comparable size and functional complexity. Perhaps the comparative analysis of primacy should be pursued within a spatial framework defined by urban spheres of influence, not by administrative areas such as states or provinces.

Now sometimes, owing to the accidents of history, the boundaries of an administrative area coincide approximately with the natural economic divides that mark the limits of a particular city's umland. This is strikingly

4 High primacy at the national scale tends to be inhibited if a country's political capital is not its largest city. Among the 103 countries listed in Table 11.2, only twenty have a capital that is not the largest city. The average primacy index for these twenty countries is a mere 1.6, and only four have an index of 2.0 or higher (Morocco, Malawi, New Zealand, and Israel). Perhaps the absence of the capital function retards the largest city's development in relation to that of other cities, thereby fostering a low index of primacy.

true for the five leading cities of Australia – although the extreme northeastern portion of New South Wales looks to Brisbane rather than to Sydney. It is also substantially true in the six Canadian provinces where the largest city is notably primate. In these instances, therefore, the approximate correspondence of umlands and political areas is associated with high levels of urban primacy. Might it in fact be normal for cities to display high levels of primacy within the confines of their own umlands?

We cannot investigate this intriguing question in depth within the present discussion; it must essentially be left as a hypothesis to be tested in future research. However, a few pertinent observations may be offered. Consider, for example, the Canadian province of Alberta. Two basic points are of interest. First, the political boundary of Alberta encompasses an area considerably more extensive than the umland of the province's largest city. Second, Alberta has a low primacy index, Edmonton being only 1.2 times as large as Calgary. But now suppose that we divide the province into two parts by means of an east-west line positioned halfway between the two leading centers. As Davies and Gyuse have shown, the two resulting areas correspond closely to the Calgary and Edmonton umlands (Davies and Gyuse 1975, 126–7). If we now calculate primacy indexes for Calgary and Edmonton within their own umlands, the resulting values are 10.1 and 17.2, respectively. (Calgary is here compared with Lethbridge; Edmonton with Red Deer.) Similar – though not quite as dramatic – results are obtained if we partition Saskatchewan between Regina and Saskatoon, or New Brunswick between Saint John and Moncton. In each of these cases, the primacy indexes within the individual umlands (however crudely the latter are defined) are distinctly higher than the index for the political unit as an undifferentiated whole.

Moving now to Spain, we note that the two leading cities are Madrid and Barcelona, with the national index of primacy being 1.9 (Table 11.2). However, viewed in the context of its own sphere of influence – here defined as the historic province of Catalonia – Barcelona has a primacy index of 11.4; and the 'removal' of this territory raises the index in the remainder of Spain, even without further territorial subdivision, to 4.9. The example of Barcelona is of special interest because Catalonia has strong separatist sentiment. If Catalonia were to become an independent country with Barcelona as its capital, this city's index of primacy within its own sphere of influence would be likely to rise even higher.

As a final illustration, consider India, another country with a low index of primacy at the national scale. The four largest cities of this diamond-shaped country are so positioned that they serve as natural foci for four distinct sections of the national territory: Calcutta in the east, Bombay in the west, Delhi in the north, and Madras in the south. Each of these major centers has a higher index of primacy within its own section of the country than Calcutta exhibits for India as a whole. Further subdivision of the national territory into smaller areas, each focused on a large city and delimited by means of the gravity formula described in chapter 6, produces similar results, with a majority of the regional metropolises displaying primacy indexes well above the arbitrary critical value of 2.0.

Although the above examples are limited both in number and in scope, they do suggest that large cities typically exhibit high levels of primacy within their own spheres of influence. It would not be inappropriate to treat this suggestion as a hypothesis worthy of detailed investigation. Moreover, this hypothesis connects the concept of the primate city firmly to that of the central place hierarchy. The basic proposal is that the appropriate spatial framework for the study of primacy is the network of mutually exclusive cells formed by the umlands of cities of comparable hierarchical rank. The very act of subdividing an area into urban spheres of influence involves the assumption that the cities included as the centers of these tributary regions represent one or more complete orders in an implicit urban hierarchy. In addition, because each city possesses a nested series of umlands, one for each hierarchical level on which it functions, each city also possesses a series of distinct indexes of primacy, with one index corresponding to each hierarchical level. The detailed study of relationships between primacy and hierarchical structuring could well emerge as an important dimension of future research on the size and spacing of cities.

Finally, the existence of a strongly primate city does not preclude the existence of a statistically significant relationship between rank and size within the same urban system. A strongly primate city may be expected to appear as a positive residual in relation to the best-fitting regression line on a rank-size graph, but this can occur without reducing the coefficient of determination to an insignificantly low level. In the case of France, for example, the index of primacy is an impressive 8.6, but the coefficient of determination between the logarithms of rank and size is still 0.89, a statistically significant value. Rank-size relationships, of course, are normally examined for a relatively large number of cities, whereas the analysis of primacy usually involves no more than the two largest cities in the system. This being the case, it is probably wise to keep the study of urban primacy and the study of rank-size relationships in separate conceptual compartments. Taken by itself, an aggregate log-linear relationship between rank and size tells us nothing about the presence or absence of a primate city at the head of the urban system.

The concentration of urban population

The index of primacy employed in the preceding section may be viewed as a measure of the extent to which the total population of a two-city mini-system is concentrated in the larger of the two centers. The higher the index of primacy, the greater the degree of concentration. But what if we want to measure the degree of population concentration more comprehensively, taking more cities into account? One possible answer is provided by the slope parameter of the rank-size regression line. The steeper the slope, the greater the concentration of the population in the system's largest cities. However, two regression lines with identical slopes may be accompanied by quite different patterns of residuals, and thus the slope parameter may not be a sufficiently sensitive indicator of the variable that we wish to measure. In measuring concentration, we are interested in the population size of each individual city as a unique numerical quantity. If we consider only the general relationship between size and rank within the system as a whole, specific information on the sizes of individual cities is automatically suppressed.

It is helpful to approach the problem of measuring concentration by considering two extreme situations. At one extreme, the population of a set of cities may be said to be 'completely dispersed' if all the cities are identical in size. At the other extreme, 'complete concentration' is achieved if the entire population lives in a single city. Both of these situations, of course, are hypothetical; real urban systems occupy various points along the continuous scale that connects the two extremes. What we require is an index that measures the position of a particular urban system on this continuous scale. In fact, we have already met the index that most effectively performs this task: the Gini coefficient of concentration, which was employed in chapter 4 to measure industrial diversification.

However, there exists one crucial distinction between the measurement of industrial diversification and the measurement of population concentration within an urban system. In the former case, the standard of reference that is used to define 'complete diversification' involves employment percentages that are unequal in size (see chapter 4). In the latter case, the standard of reference that defines 'complete dispersal' (that is, a complete absence of concentration) involves percentages that are equal in size, representing equal shares of the total population within the urban system. From a purely mathematical viewpoint, the Gini coefficient can be employed just as easily with equal shares as with unequal shares; the only requirement is that the shares must add up to 100 per cent. The choice between equal and unequal shares depends entirely on the characteristics of the problem in hand; it is a conceptual decision, not one dictated by the technique of measurement. In the present context, unequal shares would make no sense as a standard of reference, whereas equal shares represent the limiting case of a uniform distribution in which all cities are identical in size and in which concentration is therefore completely absent.

When the standard of reference for a Gini coefficient is taken, as here, to involve equal shares of the relevant total, the calculations can be simplified by making use of a relationship between the Gini coefficient and a statistic known as the Gini mean difference (Kendall and Stuart 1969, 46–51). If we have N values of a variate X, the Gini mean difference, D, is defined as follows:

$$D = N^{-2} \sum_{i=1}^{N} \sum_{j=1}^{N} \left| X_i - X_j \right|. \tag{11.21}$$

The coefficient of concentration, C, is then given by

$$C = D/2M, \tag{11.22}$$

where M is the mean value of the variate. The value of C lies between zero, or complete dispersal (all cities the same size) and 1, or complete concentration. The 'short cut' provided by equations 11.21 and 11.22 can be used only if the standard of reference involves equal shares of the phenomenon under investigation (Marshall and Smith 1978, 33; W.R. Smith, Huh, and Demko 1983).

Table 11.3 illustrates the application of equation 11.21 to urban population data. The number of cities in this example has deliberately been kept small for the purpose of clarity. Each cell in the table shows the difference in population size (expressed in thousands) between two particular cities. Since the same set of cities forms both the rows and the columns, the matrix of differences is symmetrical. The sum of the differences is 18,920. The number of differences is 25, including the five

TABLE 11.3
Matrix of population differences for the five largest cities of Italy

	Rome (2,898)	Milan (1,706)	Naples (1,225)	Turin (1,182)	Genoa (795)
Rome (2,898)	0	1,192	1,673	1,716	2,103
Milan (1,706)	1,192	0	481	524	911
Naples (1,225)	1,673	481	0	43	430
Turin (1,182)	1,716	524	43	0	387
Genoa (795)	2,103	911	430	387	0

SOURCE: *The Geographical Digest 1981* (Willett et al 1981)
NOTE: Population figures are given in thousands. The Gini index of concentration for this table of differences is 0.242. See text for explanation of method.

self-differences – each of which, by definition, has the value zero. The resulting mean difference, therefore, is 18,920/25, or 756.8 thousand persons.

In order to complete the calculations, we find the mean value of the actual populations of the five cities: 1,561.2 thousand persons. Finally:

$$C = D/2M = (18{,}920/25)/(2)(1{,}561.2) = 0.242.$$

Thus the Gini index of population concentration for the five largest cities in Italy is 0.242 – a relatively modest level of concentration.

The numerical value of this index is sensitive to the number of cities included in the calculations. For example, if we consider the top ten cities of Italy instead of the top five, the index rises from 0.242 to 0.370. It can be shown, in fact, that the value of the index for any given urban system usually increases as more and more cities are included. In comparative studies, therefore, data must be assembled for a fixed number of cities in each urban system that is examined. If the number of cities, or 'level of truncation,' is allowed to vary from system to system, comparison of the results is unwarranted.

The Gini index of concentration can be used either to compare different urban systems at a single point in time or to analyze temporal changes within a single system. An example of the latter type of application is provided by Figure 11.3, which shows the behavior of the Gini index for the forty largest cities and towns in southern Ontario during the period from 1861 to 1971 (Marshall and Smith 1978, 33–4). Between 1861 and 1871, the value of the Gini index declined, indicating that the top forty

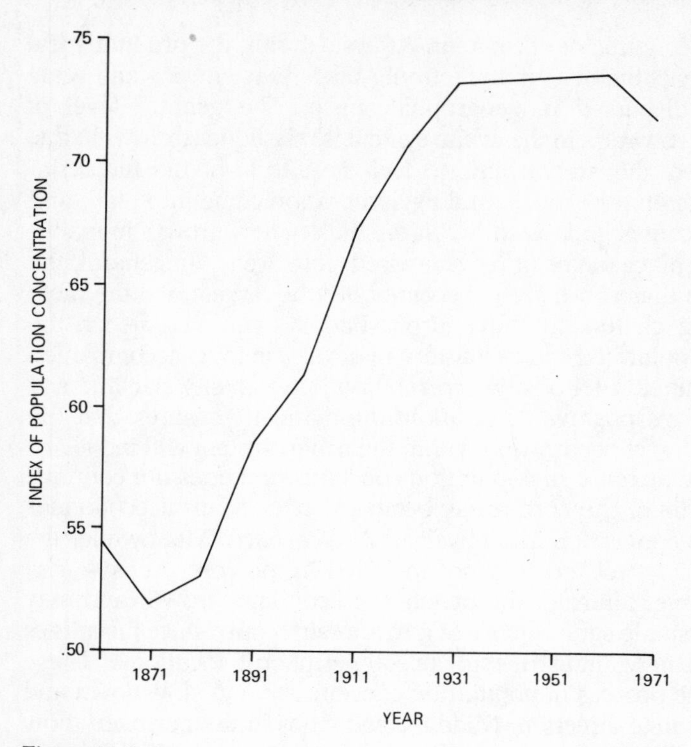

Figure 11.3
Change (1861–1971) in the Gini index of population concentration for the forty largest cities of southern Ontario. SOURCE: Marshall and Smith (1978, 34)

cities became more equal in size, rather than less equal, during that decade. Between 1871 and 1931, however, the index of concentration increased dramatically, indicating that the largest centers were capturing an increasing proportion of the total population in the forty-city set. From 1931 to 1961, the level of concentration remained virtually constant. The value of the index then experienced a slight decline, suggesting that the era of increasing population concentration may have come to an end.

The changes depicted in Figure 11.3 are consistent with the view that the level of population concentration within a set of cities should first rise and later fall as the area in question experiences a period of rapid industrialization. This view – the 'Williamson hypothesis' (Williamson 1965) – rests on the following theory concerning the spatial expression of the process of economic development. At first, during the pre-industrial period of an area's history, manufacturing takes place on a small scale and is widely dispersed in geographic space. The general level of population concentration in the urban system is consequently low. In due course, however, the area begins to feel the effects of the Industrial Revolution. Economies of scale and agglomeration come into play, and the territory becomes 'polarized,' with the bulk of new growth in manufacturing taking place in one or more favored 'core areas.' In general, the cities located in these core areas become both the largest and the most rapidly growing centers in the entire urban system, and hence the emergence of a polarized spatial pattern of development is accompanied by the establishment of a positive correlation between city size and city growth rates. This positive correlation automatically ensures that the level of population concentration within the urban system will increase.

However, the increase in population concentration does not continue indefinitely. In time, the core areas begin to suffer from diseconomies associated with congestion and physical obsolescence. Meanwhile, the periphery – that is, all territory not included in the core areas – may clamor for a larger share of the benefits of economic growth and may successfully persuade senior levels of government to introduce incentives that encourage new industries to select peripheral locations. These factors cause the process of population concentration to slow down and ultimately to change direction. Middle-sized cities in the periphery now grow more rapidly than the large cities of the core areas. The level of population concentration in the urban system as a whole begins to decline.[5]

As Figure 11.3 indicates, the experience of southern Ontario conforms quite closely to this hypothesized process of systemic development. The level of population concentration was low until late in the nineteenth century. It began to rise conspicuously just at the time when the manufacturing sector entered its period of most rapid expansion, creating two noteworthy industrial core areas: a large one centered on Toronto, and a small one centered on Windsor in the far southwest. The period of

5 The Williamson hypothesis was originally formulated for economic welfare rather than population size, and for regions rather than urban centers (Williamson 1965). It is generally agreed, however, that the central idea of a period of divergence followed by a period of convergence should apply equally to the relative sizes of cities. For commentary, both 'regional' and 'urban,' see Berry (1971), Johnston (1971), Friedmann (1972), El-Shakhs (1972), Brookfield (1975, 85–123), and Gore (1984, 55–67).

rapid increase in concentration came to an end in 1931, possibly reflecting the impact of the Depression. The highest recorded value, 0.736, was reached in 1961. After that date, the index showed a small decline, possibly signaling the onset of an era of postindustrial deconcentration (Vining and Kontuly 1977; Robert and Randolph 1983; J.F. Hart 1984).

In general, therefore, the historical behavior of the urban system of southern Ontario accords well with current theoretical ideas concerning the relationship between population concentration and the process of economic development. However, until additional longitudinal studies are completed, we cannot say whether or not the evolutionary sequence observed in southern Ontario is generally true of urban systems throughout the world. In short, extensive testing of the Williamson hypothesis stands as an important task for future research.

Summary

This chapter has examined a group of ideas that are interconnected by a common concern with systemic structure as reflected in city size. Historically, the earliest manifestation of this concern was the rank-size rule, which posits a log-linear relationship between city size and city rank. The discovery of this regularity provided a basis for the objective comparison of the overall size-structures of different urban systems through an examination of the parameters (slope and intercept) of the rank-size regression equations. However, the logarithmic method by which these parameters have traditionally been estimated suffers from two weaknesses: small-city bias and mismatch in total urban population. Nader has recently shown that both difficulties can be overcome by adopting a method of parameter estimation that does not involve a logarithmic transformation of the data.

Although the empirical validity of the rank-size rule is not in doubt, we currently lack a satisfactory quantitative model of the process of growth by which this regularity is created. The main line of attack on this problem has been to regard the rank-size rule as an expression of the right-hand tail of a lognormal distribution and to call on the Gibrat model of growth as the basis for an explanation of this structure. However, the Gibrat model rests on assumptions that have been shown to be empirically false. Accordingly, even though the notion of lognormality may be satisfactory on the descriptive level, the Gibrat model is unacceptable as an explanation of this pattern's evolution.

The two main assumptions of the Gibrat model are that the rates of growth of cities are statistically independent of city size and that the rates of growth are homoscedastic with respect to city size. Studies completed in Britain and Canada have shown, however, that rates of city growth, more often than not, are positively correlated with city size and are usually heteroscedastic. Research on the nature of these relationships in other urban systems is urgently needed. At present, however, the evidence indicates that systemic growth in real urban networks does not take place in accordance with the requirements of the Gibrat model.

An alternative approach to explanation lies in the idea that the rank-size regularity arises as a modification, under real-world conditions, of the discretely stratified character of the central place hierarchy. This approach has considerable appeal, but it currently lacks the explicit concern with process that a growth model of the Gibrat type would provide. However, the simulation of growth by means of Monte Carlo methods, as exemplified in the work of Robson, may prove capable of transforming a hypothetical hierarchically structured urban system from its initial hierarchical state into a stable terminal state that conforms to the rank-size regularity. This line of research should be actively pursued.

An important type of deviation from the strict rank-size regularity is the phenomenon of the primate city. The largest city in an urban system may be described as being primate if it is a positive residual with respect to the corresponding rank-size regression line. More specifically, primacy is present if the largest city is more than twice the size of the second-largest city in the system. The chief theoretical problem encountered in the analysis of primacy concerns the proper definition of the area within which the leading city's degree of primacy should be measured. National territories have traditionally been used, but some large cities that do not display primacy at the national scale are strongly primate within regions of more limited extent. The appropriate spatial unit may be not the national territory as such, but the umland of a particular city. The analysis of primacy in terms of umlands could provide a degree of comparability that has been absent from previous work, and might lead to the formulation of interesting hypotheses relating levels of primacy to other aspects of the structure of urban systems.

The rank-size regularity and the concept of the primate city are both encompassed within the more general concept of spatial concentration. Spatial concentration of the urban population may be said to be completely absent if all the cities in a system are equal in size. The actual degree of concentration present in any particular case can be measured by means of the Gini coefficient. Comparisons may be drawn both across space and through time. Longitudinal analyses are especially valuable because they permit the testing of the Williamson hypothesis, which postulates that the level of concentration first rises and later falls as an urban system experiences the effects of the Industrial Revolution. Extensive testing of this hypothesis, at both national and regional scales, should be a high priority in future research on the structure and functioning of urban systems.

References

Ackerman, W.V. 1978. 'Testing Central Place Concepts in Western Argentina.' *Professional Geographer*, 30:377–88

Adams, T. 1917. *Rural Planning and Development: A Study of Rural Conditions and Problems in Canada*. Ottawa: Commission of Conservation, Government of Canada

Aitchison, J., and Brown, J.A.C. 1957. *The Lognormal Distribution*. London: Cambridge University Press

Alder, H.L., and Roessler, E.B. 1964. *Introduction to Probability and Statistics*. Third edition. San Francisco: W.H. Freeman

Alexander, J.W. 1954. 'The Basic-Nonbasic Concept of Urban Economic Functions.' *Economic Geography*, 30:246–61

– 1958. 'Location of Manufacturing: Methods of Measurement.' *Annals, Association of American Geographers*, 48:20–6

Alker, H.R., Jr, and Russett, B.M. 1964. 'On Measuring Inequality.' *Behavioral Science*, 9:207–18

Allen, G.R. 1954. 'The *courbe des populations*: A Further Analysis.' *Bulletin of the Oxford University Institute of Statistics*, 16:179–89

Archer, J.C., and White, E.R. 1985. 'A Service Classification of American Metropolitan Areas.' *Urban Geography*, 6:122–51

Ashton, T.S. 1948. *The Industrial Revolution, 1760–1830*. London: Oxford University Press

Auerbach, F. 1913. 'Das Gesetz der Bevölkerungskonzentration.' *Petermanns Geographische Mitteilungen*, 59, no. 1:74–6

Aurousseau, M. 1921. 'The Distribution of Population: A Constructive Problem.' *Geographical Review*, 11:563–92

Barnum, H.G. 1966. *Market Centers and Hinterlands in Baden-Württemberg*. Research Paper No. 103. Chicago: Department of Geography, University of Chicago

Beavon, K.S.O. 1977. *Central Place Theory: A Reinterpretation.* London: Longman

Beckmann, M.J. 1958. 'City Hierarchies and the Distribution of City Size.' *Economic Development and Cultural Change*, 6:243–8

Beckmann, M.J., and McPherson, J.C. 1970. 'City Size Distribution in a Central Place Hierarchy: An Alternative Approach.' *Journal of Regional Science*, 10:25–33

Beresford, M. 1967. *New Towns of the Middle Ages: Town Plantation in England, Wales and Gascony.* London: Lutterworth Press

Berry, B.J.L. 1961. 'City Size Distributions and Economic Development.' *Economic Development and Cultural Change*, 9:573–88

– 1967. *Geography of Market Centers and Retail Distribution.* Englewood Cliffs: Prentice-Hall

– 1971. 'City Size and Economic Development: Conceptual Synthesis and Policy Problems, with Special Reference to South and Southeast Asia.' In L. Jakobson and V. Prakash, eds, *Urbanization and National Development.* Beverly Hills: Sage Publications

– 1973. *Growth Centers in the American Urban System.* Two volumes. Cambridge, Mass.: Ballinger

Berry, B.J.L., and Barnum, H.G. 1962. 'Aggregate Relations and Elemental Components of Central Place Systems.' *Journal of Regional Science*, 4, no. 1:35–68

Berry, B.J.L., Barnum, H.G., and Tennant, R.J. 1962. 'Retail Location and Consumer Behavior.' *Papers and Proceedings, Regional Science Association*, 9:65–106

Berry, B.J.L., and Garrison, W.L. 1958a. 'Recent Developments of Central Place Theory.' *Papers and Proceedings, Regional Science Association* 4:107–20

– 1958b. 'The Functional Bases of the Central Place Hierarchy.' *Economic Geography*, 34:145–54

Berry, B.J.L., and Smith, K.B., eds (1972). *City Classification Handbook: Methods and Applications.* New York: Wiley-Interscience

Berry, B.J.L., et al. (1965). *Central Place Studies: A Bibliography of Theory and Applications.* Bibliography Series No. 1, with supplement. Philadelphia: Regional Science Research Institute

Betjeman, J. 1943. *English Cities and Small Towns.* London: Collins

Blumenfeld, H. 1955. 'The Economic Base of the Metropolis: Critical Remarks on the "Basic-Nonbasic" Concept.' *Journal of the American Institute of Planners*, 21:114–32

Bogue, D.J. 1950. *The Structure of the Metropolitan Community: A Study of Dominance and Subdominance.* Ann Arbor: Horace H. Rackham School of Graduate Studies, University of Michigan

Borchert, J.R. 1967. 'American Metropolitan Evolution.' *Geographical Review*, 57:301–32
– 1978. 'Major Control Points in American Economic Geography.' *Annals, Association of American Geographers*, 68:214–32
Borchert, J.R., and Adams, R.B. 1963. *Trade Centers and Trade Areas of the Upper Midwest*. Upper Midwest Economic Study, Urban Report No. 3. Minneapolis: Upper Midwest Research and Development Council and University of Minnesota
Bracey, H.E. 1952. *Social Provision in Rural Wiltshire*. London: Methuen
– 1953. 'Towns as Rural Service Centres.' *Transactions and Papers, Institute of British Geographers*, no. 19 (old series): 95–105
– 1956. 'A Rural Component of Centrality Applied to Six Southern Counties in the United Kingdom.' *Economic Geography*, 32:38–50
– 1962. 'English Central Villages: Identification, Distribution and Functions.' In K. Norborg, ed, *Proceedings of the IGU Symposium in Urban Geography, Lund 1960*. Lund Studies in Geography, Series B, No. 24. Lund: C.W.K. Gleerup Publishers
Breese, G. 1966. *Urbanization in Newly Developing Countries*. Englewood Cliffs: Prentice-Hall
– ed. 1969. *The City in Newly Developing Countries: Readings on Urbanism and Urbanization*. Englewood Cliffs: Prentice-Hall
Bridenbaugh, C. 1938. *Cities in the Wilderness: The First Century of Urban Life in America, 1625–1742*. New York: Ronald Press
Britton, J.N.H. 1973. 'The Classification of Cities: Evaluation of Q-Mode Factor Analysis.' *Regional and Urban Economics*, 2:333–55
Brookfield, H.C. 1975. *Interdependent Development*. London: Methuen
Brookfield, H.C., and Hart, D. 1971. *Melanesia: A Geographical Interpretation of an Island World*. London: Methuen
Browning, C.E. 1978. 'Core and Fringe in Regional Development.' In J.D. Eyre, ed, *A Man for All Regions: The Contributions of Edward L. Ullman to Geography*. Studies in Geography No. 11. Chapel Hill: Department of Geography, University of North Carolina at Chapel Hill
Brush, J.E. 1953. 'The Hierarchy of Central Places in Southwestern Wisconsin.' *Geographical Review*, 43:380–402
Brush, J.E., and Bracey, H.E. 1955. 'Rural Service Centers in Southwestern Wisconsin and Southern England.' *Geographical Review*, 45:559–69
Bücher, C. 1901. *Industrial Evolution*. Translated by S.M. Wickett. New York: Henry Holt
Burghardt, A.F. 1971. 'A Hypothesis about Gateway Cities.' *Annals, Association of American Geographers*, 61:269–85

– 1979. 'The Origin of the Road and City Network of Roman Pannonia.' *Journal of Historical Geography*, 5: 1–20

Canada. 1986. *Air Passenger Origin and Destination: Domestic Report*. Catalogue No. 51-204. Ottawa: Transportation Division, Statistics Canada

Careless, J.M.S. 1954. 'Frontierism, Metropolitanism, and Canadian History.' *Canadian Historical Review*, 35:1–21

– 1970. 'The Development of the Winnipeg Business Community, 1870–1890.' *Transactions, Royal Society of Canada*, Fourth Series, 8:239–54

– 1974. 'Urban Development in Canada.' *Urban History Review*, no. 7 (no. 1-74): 9–14

Carol, H. 1970. 'Walter Christaller: A Personal Memoir.' *Canadian Geographer*, 14:67–9

Carroll, G.R. 1982. 'National City-Size Distributions: What Do We Know after 67 Years of Research?' *Progress in Human Geography*, 6:1–43

Carter, H. 1955. 'Urban Grades and Spheres of Influence in South West Wales: An Historical Consideration.' *Scottish Geographical Magazine*, 71:43–58

– 1956. 'The Urban Hierarchy and Historical Geography: A Consideration with Reference to North-East Wales.' *Geographical Studies*, 3:85–101

Carter, H., and Davies, M.E.L. 1963. 'The Hierarchy of Urban Fields in Cardiganshire, Wales.' *Tijdschrift voor Economische en Sociale Geografie*, 54:181–6

Carus-Wilson, E.M. 1958. 'Towns and Trade.' In A.L. Poole, ed, *Medieval England*, Volume 1. London: Oxford University Press

Cazalis, P. 1964. 'Sherbrooke: sa place dans la vie de relations des Cantons de l'Est.' *Cahiers de géographie de Québec*, no. 16:165–97

Champernowne, D.G. 1953. 'A Model of Income Distribution.' *Economic Journal*, 63:318–51

Chandler, T., and Fox, G. 1974. *3000 Years of Urban Growth*. New York: Academic Press

Chang, S. 1963. 'The Historical Trend of Chinese Urbanization.' *Annals, Association of American Geographers*, 53:109–43

Chatelain, A. 1957. 'Géographie sociologique de la presse et régions françaises.' *Revue de géographie de Lyon*, 32:127–34

Christaller, W. 1933. *Die zentralen Orte in Süddeutschland: Eine ökonomisch-geographische Untersuchung über die Gesetzmässigkeit der Verbreitung und Entwicklung der Siedlungen mit städtischen Funktionen*. Jena: Gustav Fischer

– 1966. *Central Places in Southern Germany*. Translated by C.W. Baskin. Englewood Cliffs: Prentice-Hall

– 1968. 'Wie ich zu der Theorie der zentralen Orte gekommen bin.' *Geographische Zeitschrift*, 56:88–101

Chung, R. 1970. 'Space-Time Diffusion of the Transition Model: The Twentieth Century Patterns.' In G.J. Demko, H.M. Rose, and G.A. Schnell, eds, *Population Geography: A Reader*. New York: McGraw-Hill
Cloke, P.J. 1979. *Key Settlements in Rural Areas*. London: Methuen
Collis, J. 1981. 'A Theoretical Study of Hill-Forts.' In G. Guilbert, ed, *Hill-Fort Studies: Essays for A.H.A. Hogg*. Leicester: Leicester University Press
Commoner, B. 1971. *The Closing Circle: Nature, Man, and Technology*. New York: Alfred A. Knopf
Conkling, E.C. 1963. 'South Wales: A Case Study in Industrial Diversification.' *Economic Geography*, 39:258–72
Constandse, A.K. 1963. 'Reclamation and Colonisation of New Areas.' *Tijdschrift voor Economische en Sociale Geografie*, 54:41–5
Converse, P.D. 1949. 'New Laws of Retail Gravitation.' *Journal of Marketing*, 14:379–84
– 1953. 'Comment on Movement of Retail Trade in Iowa.' *Journal of Marketing*, 18:170–1
Conzen, M.P. 1975. 'A Transport Interpretation of the Growth of Urban Regions: An American Example.' *Journal of Historical Geography*, 1:361–82
– 1977. 'The Maturing Urban System in the United States, 1840–1910.' *Annals, Association of American Geographers*, 67:88–108
Cook, S.F., and Simpson, L.B. 1948. *The Population of Central Mexico in the Sixteenth Century*. Ibero-Americana No. 31. Berkeley: University of California Press
Coppolani, J. 1959. *Le Réseau urbain de la France: sa structure et son aménagement*. Paris: Les Editions Ouvrières
Craig, G.M. 1963. *Upper Canada: The Formative Years, 1784–1841*. Canadian Centenary Series No. 7. Toronto: McClelland and Stewart
Crissman, L.W. 1976. 'Specific Central-Place Models for an Evolving System of Market Towns on the Changhua Plain, Taiwan.' In C.A. Smith, ed, *Regional Analysis, Volume 1: Economic Systems*. New York: Academic Press
Dacey, M.F. 1964. 'A Note on Some Number Properties of a Hexagonal Hierarchical Plane Lattice.' *Journal of Regional Science*, 5, no. 2:63–7
– 1965. 'An Interesting Number Property in Central Place Theory.' *Professional Geographer*, 17, no. 5:32–3
– 1966. 'Population of Places in a Central Place Hierarchy.' *Journal of Regional Science*, 6, no. 2:27–33
Dahms, F.A. 1975. 'Some Quantitative Approaches to the Study of Central Places in the Guelph Area, 1851–1970.' *Urban History Review*, no. 11 (no. 2-75):9–30
– 1981. 'The Evolution of Settlement Systems: A Canadian Example, 1851–1970.' *Journal of Urban History*, 7:169–204

Davies, W.K.D. 1967. 'Centrality and the Central Place Hierarchy.' *Urban Studies*, 4:61–79

– 1970. 'Towards an Integrated Study of Central Places: A South Wales Case Study.' In H. Carter and W.K.D. Davies, eds, *Urban Essays: Studies in the Geography of Wales*. London: Longman

Davies, W.K.D., and Gyuse, T.T. 1975. 'Changes in the Central Place System around Calgary: 1951–1971.' In B.M. Barr, ed, *Calgary: Metropolitan Structure and Influence*. Western Geographical Series, Vol. 11. Victoria: Department of Geography, University of Victoria

Deane, P. 1969. *The Industrial Revolution in England, 1700–1914*. Fontana Economic History of Europe, Volume 4, Section 2. London: Collins

Derman, C., Gleser, L.J., and Olkin, I. 1973. *A Guide to Probability Theory and Application*. New York: Holt, Rinehart and Winston

Dickinson, R.E. 1930. 'The Regional Functions and Zones of Influence of Leeds and Bradford.' *Geography*, 15:548–57

– 1932. 'The Distribution and Functions of the Smaller Urban Settlements of East Anglia.' *Geography*, 17:19–31

– 1934. 'The Metropolitan Regions of the United States.' *Geographical Review*, 24:278–91

– 1947. *City Region and Regionalism*. London: Routledge and Kegan Paul

– 1961. *The West European City: A Geographical Interpretation*. Second edition. London: Routledge and Kegan Paul

– 1964. *City and Region: A Geographical Interpretation*. London: Routledge and Kegan Paul

Duncan, B., and Lieberson, S. 1970. *Metropolis and Region in Transition*. Beverly Hills: Sage Publications

Duncan, J.S. 1955. 'New Zealand Towns as Service Centres.' *New Zealand Geographer*, 11:119–38

Duncan, O.D., and Duncan, B. 1955. 'A Methodological Analysis of Segregation Indexes.' *American Sociological Review*, 20:210–17

Duncan, O.D., et al. 1960. *Metropolis and Region*. Baltimore: Johns Hopkins Press for Resources for the Future

Dunn, E.S., Jr. 1980. *The Development of the U.S. Urban System*. Two volumes. Baltimore: Johns Hopkins University Press for Resources for the Future

Dupeux, G. 1981. *Atlas historique de l'urbanisation de la France (1811–1975)*. Paris: Centre national de la recherche scientifique

Earle, C.V. 1977. 'The First English Towns of North America.' *Geographical Review*, 67:34–50

El-Shakhs, S. 1972. 'Development, Primacy, and Systems of Cities.' *Journal of Developing Areas*, 7:11–36

Enggass, P.M. 1968. 'Land Reclamation and Resettlement in the Guadalquivir Delta – Las Marismas.' *Economic Geography*, 44:125–43

Ennen, E. 1979. *The Medieval Town*. Translated by N. Fryde. Amsterdam: North-Holland Publishing

Evenden, L.J. 1980. 'Regionalism and the Cognitive Hierarchy of Towns.' *Regional Studies*, 14:473–90

Florence, P.S. 1929. *The Statistical Method in Economics and Political Science*. London: Kegan Paul, Trench, Trubner

– 1937. 'Economic Research and Industrial Policy.' *Economic Journal*, 47:621–41

– 1955. 'Economic Efficiency in the Metropolis.' In R.M. Fisher, ed, *The Metropolis in Modern Life*. New York: Doubleday

Friedmann, J. 1972. 'A General Theory of Polarized Development.' In N.M. Hansen, ed, *Growth Centers in Regional Economic Development*. New York: Free Press

Friedmann, J., and Miller, J. 1965. 'The Urban Field.' *Journal of the American Institute of Planners*, 31:312–20

Galpin, C.J. 1915. *The Social Anatomy of an Agricultural Community*. Research Bulletin 34. Madison: Agricultural Experiment Station, University of Wisconsin

Garner, B.J. 1975. 'Post-War Changes in the Central Place Pattern of Wiltshire County, England, 1940–1970.' In B. Sárfalvi, ed, *Urbanization in Europe*. Budapest: Akadémiai Kiadó

Gauchy, M. 1955. 'Le Rayonnement des journaux Toulousains.' *Revue géographique des Pyrénées et du Sud-Ouest*, 26:100–12

Getis, A., and Boots, B. 1978. *Models of Spatial Processes: An Approach to the Study of Point, Line and Area Patterns*. Cambridge: Cambridge University Press

Gibrat, R. 1931. *Les Inégalités économiques*. Paris: Sirey

– 1957. 'On Economic Inequalities.' Translated by E. Henderson. *International Economic Papers*, 7:53–70

Gibson, L.J., and Worden, M.A. 1981. 'Estimating the Economic Base Multiplier: A Test of Alternative Procedures.' *Economic Geography*, 57:146–59

Gilbert, E.W. 1961. *The University Town in England and West Germany: Marburg, Göttingen, Heidelberg and Tübingen, Viewed Comparatively with Oxford and Cambridge*. Research Paper No. 71. Chicago: Department of Geography, University of Chicago

Gini, C. 1914. 'Sulla misura della concentrazione e della variabilità dei caratteri.' *Atti del Reale Istituto Veneto di Scienze, Lettere ed Arti*, 53:1203–48

– 1921. 'Measurement of Inequality of Incomes.' *Economic Journal*, 31:124–6
Golachowski, S. 1964. 'Rola teorii Christallera w planowaniu Hitlerowskim na Slasku.' *Studia Slaskie*, 10:167–77
Goldberg, M.A., and Mercer, J. 1986. *The Myth of the North American City: Continentalism Challenged*. Vancouver: University of British Columbia Press
Golledge, R.G., Rushton, G., and Clark, W.A.V. 1966. 'Some Spatial Characteristics of Iowa's Dispersed Farm Population and Their Implications for the Grouping of Central Place Functions.' *Economic Geography*, 42:261–72
Gore, C. 1984. *Regions in Question: Space, Development Theory and Regional Policy*. London: Methuen
Gould, P. 1985. *The Geographer at Work*. London: Routledge and Kegan Paul
Gourlay, R. 1822. *Statistical Account of Upper Canada, Compiled with a View to a Grand System of Emigration, in Connexion with a Reform of the Poor Laws*. Two volumes. London: Simpkin and Marshall
Gras, N.S.B. 1922. *An Introduction to Economic History*. New York: Harper and Brothers
– 1926. 'The Rise of the Metropolitan Community.' *Papers and Proceedings, American Sociological Society*, 20:155–63
Green, C.M. 1965. *The Rise of Urban America*. New York: Harper and Row
Green, F.H.W. 1950. 'Urban Hinterlands in England and Wales: An Analysis of Bus Services.' *Geographical Journal*, 116:64–81
– 1953. 'Community of Interest Areas in Western Europe: Some Geographical Aspects of Local Passenger Traffic.' *Economic Geography*, 29:283–98
– 1958. 'Community of Interest Areas: Notes on the Hierarchy of Central Places and Their Hinterlands.' *Economic Geography*, 34:210–26
– 1966. 'Urban Hinterlands: Fifteen Years On.' *Geographical Journal*, 132:263–6
Hainsworth, G.B. 1964. 'The Lorenz Curve as a General Tool of Economic Analysis.' *Economic Record*, 40:426–41
Hakkenberg, A. 1971. 'Who Will Be Surprised When a Regularity of Values of Different Entities Is Found, If the Second Entity Is Defined by the Quantity of the First Entity?' *Tijdschrift voor Economische en Sociale Geografie*, 62·61
Hall, P. 1970. *The Theory and Practice of Regional Planning*. London: Pemberton Books
Hammond, M. 1972. *The City in the Ancient World*. Cambridge, Mass.: Harvard University Press
Hansen, N.M., ed. 1972. *Growth Centers in Regional Economic Development*. New York: Free Press
Hardy, G.H. 1967. *A Mathematician's Apology*. London: Cambridge University Press
Harris, C.D. 1943. 'A Functional Classification of Cities in the United States.' *Geographical Review*, 33:86–99

– 1968. 'City and Region in the Soviet Union.' In R.P. Beckinsale and J.M. Houston, eds, *Urbanization and Its Problems: Essays in Honour of E.W. Gilbert*. Oxford: Basil Blackwell

– 1970. *Cities of the Soviet Union: Studies in Their Functions, Size, Density, and Growth*. AAG Monograph Series No. 5. Washington, DC: Association of American Geographers

Harris, C.D., and Ullman, E.L. 1945. 'The Nature of Cities.' *Annals, American Academy of Political and Social Science*, 242:7–17

Hart, J.F. 1984. 'Population Change in the Upper Lake States.' *Annals, Association of American Geographers*, 74:221–43

Hart, P.E. 1971. 'Entropy and Other Measures of Concentration.' *Journal of the Royal Statistical Society*, Series A, 134:73–85

Hassinger, E. 1957. 'The Relationship of Retail-Service Patterns to Trade-Center Population Change.' *Rural Sociology*, 22:235–40

Haughton, J.P. 1950. 'Irish Local Newspapers: A Geographical Study.' *Irish Geography*, 2, no. 2:52–7

Hettner, A. 1902. 'Die wirtschaftlichen Typen der Ansiedlungen.' *Geographische Zeitschrift*, 8:92–100

Hodge, G. 1965. 'The Prediction of Trade Center Viability in the Great Plains.' *Papers, Regional Science Association*, 15:87–115

Hoover, E.M., Jr. 1936. 'The Measurement of Industrial Localization.' *Review of Economic Statistics*, 18:162–71

Houston, J.M. 1963. *A Social Geography of Europe*. Revised edition. London: Duckworth

– 1968. 'The Foundation of Colonial Towns in Hispanic America.' In R.P. Beckinsale and J.M. Houston, eds, *Urbanization and Its Problems: Essays in Honour of E.W. Gilbert*. Oxford: Basil Blackwell

Howard, J. 1982. *Darwin*. Past Masters Series. Oxford: Oxford University Press

Hudson, J.C. 1979. 'The Plains Country Town.' In B.W. Blouet and F.C. Luebke, eds, *The Great Plains: Environment and Culture*. Lincoln: University of Nebraska Press

– 1985. *Plains Country Towns*. Minneapolis: University of Minnesota Press

Innis, H.A. 1954. *The Cod Fisheries: The History of an International Economy*. Revised edition. Toronto: University of Toronto Press

– 1956a. *Essays in Canadian Economic History*. Edited by M.Q. Innis. Toronto: University of Toronto Press

– 1956b. *The Fur Trade in Canada: An Introduction to Canadian Economic History*. Revised edition. Toronto: University of Toronto Press

Isard, W. 1956. *Location and Space-Economy: A General Theory Relating to Industrial Location, Market Areas, Land Use, Trade, and Urban Structure*. Cambridge, Mass.: M.I.T. Press

Isserman, A.M. 1980. 'Estimating Export Activity in a Regional Economy: A Theoretical and Empirical Analysis of Alternative Methods.' *International Regional Science Review*, 5:155–84

Jefferson, M. 1931. 'Distribution of the World's City Folks.' *Geographical Review*, 21:446–65

– 1939. 'The Law of the Primate City.' *Geographical Review*, 29:226–32

Johnson, G.A. 1972. 'A Test of the Utility of Central Place Theory in Archaeology.' In P.J. Ucko, R. Tringham, and G.W. Dimbleby, eds, *Man, Settlement and Urbanism*. London: Duckworth

Johnson, H.B. 1976. *Order upon the Land: The U.S. Rectangular Land Survey and the Upper Mississippi Country*. New York: Oxford University Press

Johnson, H.T. 1967. 'Cathedral Building and the Medieval Economy.' *Explorations in Entrepreneurial History*, 4:191–210

Johnson, N.L., and Kotz, S. 1969. *Distributions in Statistics: Discrete Distributions*. Boston: Houghton Mifflin

Johnston, R.J. 1971. 'On the Progression from Primacy to Rank-Size in an Urban System: The Deviant Case of New Zealand.' *Area*, 3:180–4

Jones, C.F. 1938. 'Areal Distribution of Manufacturing in the United States.' *Economic Geography*, 14:217–22

Jones, E. 1950. 'Tregaron, a Welsh Market Town.' *Geography*, 35:20–31

Jung, A.F. 1959. 'Is Reilly's Law of Retail Gravitation Always True?' *Journal of Marketing*, 24:62–3

Kaniowna, C. 1963. 'Problem osiedli centralnych w bylej rejencji Opolskiej.' *Materialy i Studia Opolskie*, 4:95–9

Kelley, K.B. 1976. 'Dendritic Central-Place Systems and the Regional Organization of Navajo Trading Posts.' In C.A. Smith, ed, *Regional Analysis, Volume 1: Economic Systems*. New York: Academic Press

Kendall, M.G. 1962. *Rank Correlation Methods*. London: Charles Griffin

Kendall, M.G., and Stuart, A. 1969. *The Advanced Theory of Statistics*. Third edition. London: Charles Griffin

Kerr, D.P. 1968. 'Metropolitan Dominance in Canada.' In J. Warkentin, ed, *Canada: A Geographical Interpretation*. Toronto: Methuen

– 1977. 'Wholesale Trade on the Canadian Plains in the Late Nineteenth Century: Winnipeg and Its Competition.' In H. Palmer, ed, *The Settlement of the West*. Calgary: University of Calgary and Comprint Publishing

King, L.J. 1966. 'Cross-Sectional Analysis of Canadian Urban Dimensions: 1951 and 1961.' *Canadian Geographer*, 10:205–24

Kohl, J.G. 1841. *Der Verkehr und die Ansiedlungen der Menschen in ihrer Abhängigkeit von der Gestaltung der Erdoberfläche*. Dresden

Kopec, R.J. 1963. 'An Alternative Method for the Construction of Thiessen Polygons.' *Professional Geographer*, 15, no. 5:24–6

Kuhn, T.S. 1970. *The Structure of Scientific Revolutions*. Second edition. Chicago: University of Chicago Press

Leftwich, R.H. 1960. *The Price System and Resource Allocation*. Revised edition. New York: Holt, Rinehart and Winston

Leigh, R. 1970. 'The Use of Location Quotients in Urban Economic Base Studies.' *Land Economics*, 46:202–5

Lewis, C.R. 1970. 'The Central Place Pattern of Mid-Wales and the Middle Welsh Borderland.' In H. Carter and W.K.D. Davies, eds, *Urban Essays: Studies in the Geography of Wales*. London: Longman

– 1975. 'The Analysis of Changes in Urban Status: A Case Study in Mid-Wales and the Middle Welsh Borderland.' *Transactions, Institute of British Geographers*, no. 64 (old series): 49–65

Linsky, A.S. 1965. 'Some Generalizations Concerning Primate Cities.' *Annals, Association of American Geographers*, 55:506--13

Lonsdale, R.E., and Seyler, H.L., eds. 1979. *Nonmetropolitan Industralization*. Washington, DC: V.H. Winston and Sons

Lopez, R.S. 1971. *The Commercial Revolution of the Middle Ages, 950–1350*. Englewood Cliffs: Prentice-Hall

Lorenz, M.O, 1905. 'Methods of Measuring the Concentration of Wealth.' *Publications of the American Statistical Association*, 9:209–19

Lösch, A. 1954. *The Economics of Location*. Translated by W.H. Woglom and W.F. Stolper. New Haven: Yale University Press

Lowe, J.C., and Moryadas, S. 1975. *The Geography of Movement*. Boston: Houghton Mifflin

Lower, A.R.M. 1938. *The North American Assault on the Canadian Forest*. Toronto: Ryerson Press

Lukermann, F. 1966. 'Empirical Expressions of Nodality and Hierarchy in a Circulation Manifold.' *East Lakes Geographer*, 2:17–44

Macaulay, J.U. 1954. 'The Oamaru Tributary Region.' *New Zealand Geographer*, 10:121–33

McCann, L.D. 1978. 'Urban Growth in a Staple Economy: The Emergence of Vancouver as a Regional Metropolis, 1886–1914.' In L.J. Evenden, ed, *Vancouver: Western Metropolis*. Western Geographical Series, Vol. 16. Victoria: Department of Geography, University of Victoria

McCarty, J.W. 1974. 'Australian Capital Cities in the Nineteenth Century.' In C.B. Schedvin and J.W. McCarty, eds, *Urbanization in Australia: The Nineteenth Century*. Sydney: Sydney University Press

McGee, T.G. 1967. *The Southeast Asian City: A Social Geography of the Primate Cities of Southeast Asia*. London: Bell

McKenzie, R.D., et al. 1933. *The Metropolitan Community*. New York: McGraw-Hill

Madden, C.H. 1956a. 'On Some Indications of Stability in the Growth of Cities in the United States.' *Economic Development and Cultural Change*, 4:236–52
– 1956b. 'Some Spatial Aspects of Urban Growth in the United States.' *Economic Development and Cultural Change*, 4:371–87
– 1958. 'Some Temporal Aspects of the Growth of Cities in the United States.' *Economic Development and Cultural Change*, 6:143–70
Magee, B. 1974. *Popper*. Modern Masters Series. London: Woburn Press
Mandelbrot, B. 1965. 'A Class of Long-Tailed Probability Distributions and the Empirical Distribution of City Sizes.' In F. Massarik and P. Ratoosh, eds, *Mathematical Explorations in Behavioral Science*. Homewood, Ill.: Irwin-Dorsey
Marcus, J. 1973. 'Territorial Organization of the Lowland Classic Maya.' *Science* 180 (Issue No. 4089):911–16
Marshall, J.U. 1969. *The Location of Service Towns: An Approach to the Analysis of Central Place Systems*. Research Publications No. 3. Toronto: Department of Geography, University of Toronto
– 1972. 'The Urban Network.' In L. Gentilcore, ed, *Ontario*. Studies in Canadian Geography Series. Toronto: University of Toronto Press
– 1975a. 'A Model of Size and Economic Structure in an Urban Hierarchy.' *Environment and Planning*, Series A, 7:637–49
– 1975b. 'City Size, Economic Diversity, and Functional Type: The Canadian Case.' *Economic Geography*, 51:37–49
– 1975c. 'The Löschian Numbers as a Problem in Number Theory.' *Geographical Analysis*, 7:421–6
– 1977a. 'Christallerian Networks in the Löschian Economic Landscape.' *Professional Geographer*, 29:153–9
– 1977b. 'The Construction of the Löschian Landscape.' *Geographical Analysis*, 9:1–13
– 1978. 'On the Structure of the Löschian Landscape.' *Journal of Regional Science*, 18:121–5
– 1981. 'Industrial Diversification in the Canadian Urban System.' *Canadian Geographer*, 25:316–32
– 1985. 'Geography as a Scientific Enterprise.' In R.J. Johnston, ed, *The Future of Geography*. London: Methuen
Marshall, J.U., and Smith, W.R. 1978. 'The Dynamics of Growth in a Regional Urban System: Southern Ontario, 1851–1971.' *Canadian Geographer*, 22:22–40
Marston, W.G. 1969. 'Social Class Segregation within Ethnic Groups in Toronto.' *Canadian Review of Sociology and Anthropology*, 6:65–79
Mather, E.C. 1972. 'The American Great Plains.' *Annals, Association of American Geographers*, 62:237–57

Mathieson, R.S. 1957. 'The Validity of Reilly's Law in Australia – Some Preliminary Considerations.' *Australian Geographer*, 7:27–32

– 1958. 'Socio-Economic Contact in the Melbourne-Sydney Penumbral Zone.' *Australian Geographer*, 7:97–102

Matthews, J.A. 1981. *Quantitative and Statistical Approaches to Geography: A Practical Manual*. Oxford: Pergamon Press

Maxwell, J.W. 1965. 'The Functional Structure of Canadian Cities: A Classification of Cities.' *Geographical Bulletin*, 7, no. 2:79–104

Medawar, P.B. 1967. *The Art of the Soluble*. London: Methuen

Mehta, S.K. 1969. 'Some Demographic and Economic Correlates of Primate Cities: A Case for Revaluation.' In G. Breese, ed, *The City in Newly Developing Countries: Readings on Urbanism and Urbanization*. Englewood Cliffs: Prentice-Hall

Menefee, S.C. 1936. 'Newspaper Circulation and Urban Regions.' *Sociology and Social Research*, 21:63–6

Meyer, D.R. 1980. 'A Dynamic Model of the Integration of Frontier Urban Places into the United States System of Cities.' *Economic Geography*, 56:120–40

Milani, L.D. 1971. *Robert Gourlay, Gadfly: Forerunner of the Rebellion in Upper Canada, 1837*. Toronto: Ampersand Press

Moisley, H.A. 1958. 'Glasgow's Spheres of Influence.' In R. Miller and J. Tivy, eds, *The Glasgow Region: A General Survey*. Edinburgh: T. and A. Constable

Mols, R. 1972. *Population in Europe 1500–1700*. Fontana Economic History of Europe, Volume 2, Section 1. London: Collins

Morrison, P.A. 1981. 'The Transition to Zero Population Growth in the Midwest.' In C.C. Roseman, A.J. Sofranko, and J.D. Williams, eds, *Population Redistribution in the Midwest*. Ames: North Central Regional Center for Rural Development, Iowa State University

Morrissett, I. 1958. 'The Economic Structure of American Cities.' *Papers and Proceedings, Regional Science Association*, 4:239–56

Moseley, M.J. 1974. *Growth Centres in Spatial Planning*. Oxford: Pergamon Press

Moser, C.A., and Scott, W. 1961. *British Towns: A Statistical Study of Their Social and Economic Differences*. Edinburgh: Oliver and Boyd

Muller, E.K. 1976. 'Selective Urban Growth in the Middle Ohio Valley, 1800–1860.' *Geographical Review*, 66:178–99

– 1977. 'Regional Urbanization and the Selective Growth of Towns in North American Regions.' *Journal of Historical Geography*, 3:21–39

Mumford, L. 1961. *The City in History: Its Origins, Its Transformations, and Its Prospects*. London: Secker and Warburg

Murdie, R.A. 1965. 'Cultural Differences in Consumer Travel.' *Economic Geography*, 41:211–33

Murphy, R.E. 1966. *The American City: An Urban Geography*. New York: McGraw-Hill

Nader, G.A. 1975. *Cities of Canada, Volume 1: Theoretical, Historical and Planning Perspectives*. Toronto: Macmillan

– 1984. 'The Rank-Size Model: A Non-Logarithmic Calibration.' *Professional Geographer*, 36:221–7

Neft, D.S. 1966. *Statistical Analysis for Areal Distributions*. Monograph Series No. 2. Philadelphia: Regional Science Research Institute

Nelson, H.J. 1955. 'A Service Classification of American Cities.' *Economic Geography*, 31:189–210

Nicholls, J.A.F. 1970. 'Transportation Development and Löschian Market Areas: An Historical Perspective.' *Land Economics*, 46:22–31

Norcliffe, G.B. 1982. *Inferential Statistics for Geographers: An Introduction*. Second edition. London: Hutchinson

Nordbeck, S. 1964. 'Computing Distances in Road Nets.' *Papers, Regional Science Association*, 12:207–20

Oakeshott, M. 1933. *Experience and Its Modes*. London: Cambridge University Press

O'Farrell, P.N. 1965. 'The Urban Hinterlands of New Ross and Enniscorthy.' *Irish Geography*, 5, no. 2:67–78

Olsson, G. 1965. *Distance and Human Interaction: A Review and Bibliography*. Bibliography Series No. 2. Philadelphia: Regional Science Research Institute

Pareto, V. 1964. *Cours d'économie politique. Oeuvres complètes, Tome 1*, ed. G. Busino. Geneva: Librairie Droz

Park, R.E. 1929. 'Urbanization As Measured by Newspaper Circulation.' *American Journal of Sociology*, 35:60–79

Park, R.E., and Newcomb, C. 1933. 'Newspaper Circulation and Metropolitan Regions.' In R.D. McKenzie et al, *The Metropolitan Community*. New York: McGraw-Hill

Parr, J.B. 1970. 'Models of City Size in an Urban System.' *Papers, Regional Science Association*, 25:221–53

– 1973. 'Structure and Size in the Urban System of Lösch.' *Economic Geography*, 49:185–212

– 1978. 'Models of the Central Place System: A More General Approach.' *Urban Studies*, 15:35–49

– 1980. 'Frequency Distributions of Central Places in Southern Germany: A Further Analysis.' *Economic Geography*, 56:141–54

Parr, J.B., and Suzuki, K. 1973. 'Settlement Populations and the Lognormal Distribution.' *Urban Studies*, 10: 335–52

Pattison, W.D. 1957. *Beginnings of the American Rectangular Land Survey System, 1784–1800*. Research Paper No. 50. Chicago: Department of Geography, University of Chicago

Pedersen, P.O. 1970. 'Innovation Diffusion within and between National Urban Systems.' *Geographical Analysis*, 2:203–54

Perry, D.C., and Watkins, A.J., eds. 1977. *The Rise of the Sunbelt Cities*. Urban Affairs Annual Reviews, Vol. 14. Beverly Hills: Sage Publications

Pfouts, R.W., ed. 1960. *The Techniques of Urban Economic Analysis*. West Trenton, NJ: Chandler-Davis Publishing

Phillips, P.D., and Brunn, S.D. 1978. 'Slow Growth: A New Epoch of American Metropolitan Evolution.' *Geographical Review*, 68:274–92

Pirenne, H. 1925. *Medieval Cities: Their Origins and the Revival of Trade*. Princeton: Princeton University Press

– 1937. *Economic and Social History of Medieval Europe*. New York: Harcourt, Brace

Popper, K.R. 1962. *Conjectures and Refutations: The Growth of Scientific Knowledge*. New York: Basic Books

Potter, R.B. 1980. 'Spatial and Structural Variations in the Quality Characteristics of Intra-Urban Retailing Centres.' *Transactions, Institute of British Geographers*, New Series, 5:207–28

Poulsen, T.M. 1959. 'Centrography in Russian Geography.' *Annals, Association of American Geographers*, 49:326–7

Pounds, N.J.G. 1969. 'The Urbanization of the Classical World.' *Annals, Association of American Geographers*, 59:135–57

Pownall, L.L. 1953. 'The Functions of New Zealand Towns.' *Annals, Association of American Geographers*, 43:332–50

Pratt, R.T. 1968. 'An Appraisal of the Minimum-Requirements Technique.' *Economic Geography*, 44:117–24

Pred, A.R. 1965. 'Industrialization, Initial Advantage, and American Metropolitan Growth.' *Geographical Review*, 55:158–85

– 1966. *The Spatial Dynamics of U.S. Urban-Industrial Growth, 1800–1914: Interpretive and Theoretical Essays*. Cambridge, Mass.: M.I.T. Press

Preston, R.E. 1971a. 'The Structure of Central Place Systems.' *Economic Geography*, 47:136–55

– 1971b. 'Toward Verification of a "Classical" Centrality Model.' *Tijdschrift voor Economische en Sociale Geografie*, 62:301–7

– 1975. 'A Comparison of Five Measures of Central Place Importance and of Settlement Size.' *Tijdschrift voor Economische en Sociale Geografie*, 66:178–87

– 1979. 'The Recent Evolution of Ontario Central Place Systems in the Light of Christaller's Concept of Centrality.' *Canadian Geographer*, 23:201–21

– 1983. 'The Dynamic Component of Christaller's Central Place Theory and the Theme of Change in His Research.' *Canadian Geographer*, 27:4–16

Prost, M.-A. 1965. *La Hiérarchie des villes en fonction de leurs activités de commerce et de service*. Paris: Gauthier-Villars

Quine, W.V. 1969. 'Natural Kinds.' In N. Rescher, ed, *Essays in Honor of Carl G. Hempel*. Dordrecht: D. Reidel

Raddall, T.H. 1971. *Halifax: Warden of the North*. Revised edition. Toronto: McClelland and Stewart

Rand McNally. 1984. *Commercial Atlas and Marketing Guide*. 115th edition. Chicago: Rand McNally

Reilly, W.J. 1929. *Methods for the Study of Retail Relationships*. Studies in Marketing No. 4. Austin: Bureau of Business Research, University of Texas

Reynolds, R.B. 1953. 'A Test of the Law of Retail Gravitation.' *Journal of Marketing*, 17:273–7

Reynolds, S. 1977. *An Introduction to the History of English Medieval Towns*. Oxford: Oxford University Press

Robert, S., and Randolph, W.G. 1983. 'Beyond Decentralization: The Evolution of Population Distribution in England and Wales, 1961–1981.' *Geoforum*, 14:75–102

Robson, B.T. 1973. *Urban Growth: An Approach*. London: Methuen

Rodgers, A. 1957. 'Some Aspects of Industrial Diversification in the United States.' *Economic Geography*, 33:16–30

Rose, A.J. 1966. 'Dissent from down under: Metropolitan Primacy as the Normal State.' *Pacific Viewpoint*, 7:1–27

Rosing, K.E. 1966. 'A Rejection of the Zipf Model (Rank Size Rule) in Relation to City Size.' *Professional Geographer*, 18:75–82

Rouget, B. 1972. 'Graph Theory and Hierarchisation Models.' *Regional and Urban Economics*, 2:263–95

Rowley, G. 1971. 'Central Places in Rural Wales.' *Annals, Association of American Geographers*, 61:537–50

Rubin, J. 1961. 'Canal or Railroad? Imitation and Innovation in the Response to the Erie Canal in Philadelphia, Baltimore, and Boston.' *Transactions, American Philosophical Society*, 51, part 7:1–106

Rushton, G., Golledge, R.G., and Clark, W.A.V. 1967. 'Formulation and Test of a Normative Model for the Spatial Allocation of Grocery Expenditures by a Dispersed Population.' *Annals, Association of American Geographers*, 57: 389–400

Russell, J.C. 1972. *Medieval Regions and Their Cities*. Newton Abbot: David and Charles

Sarbit, L.A., and Greer-Wootten, B. 1980. *Spatial Aspects of Structural Change in Central Place Systems: Southern Manitoba, 1961–1971*. Geographical Monographs No. 4. North York, Ont.: Department of Geography, Atkinson College, York University

Schlesinger, A.M. 1933. *The Rise of the City, 1878–1898*. New York: Macmillan

– 1940. 'The City in American History.' *Mississippi Valley Historical Review*, 27:43–66

Schwartz, G. 1962. 'Laws of Retail Gravitation: An Appraisal.' *University of Washington Business Review*, 22, no. 1:53–70

Scott, P. 1964. 'The Hierarchy of Central Places in Tasmania.' *Australian Geographer*, 9:134–47

Siegel, S. 1956. *Nonparametric Statistics for the Behavioral Sciences*. New York: McGraw-Hill

Simmons, J.W. 1986. *The Impact of Distribution Activities on the Canadian Urban System*. Research and Working Paper No. 17. Winnipeg: Institute of Urban Studies, University of Winnipeg

Simmons, J.W., and Flanagan, P.T. 1981. *The Movement of Growth Impulses through the Canadian Urban System*. Research Paper No. 120. Toronto: Centre for Urban and Community Studies, University of Toronto

Singer, H.W. 1936. 'The "courbe des populations": A Parallel to Pareto's Law.' *Economic Journal*, 46:254–63

Skinner, G.W. 1964. 'Marketing and Social Structure in Rural China: Part I.' *Journal of Asian Studies*, 24:3–43

– 1965a. 'Marketing and Social Structure in Rural China: Part II.' *Journal of Asian Studies*, 24:195–228

– 1965b. 'Marketing and Social Structure in Rural China: Part III.' *Journal of Asian Studies*, 24:363–99

– 1976. 'Mobility Strategies in Late Imperial China: A Regional Systems Analysis.' In C.A. Smith, ed, *Regional Analysis, Volume 1: Economic Systems*. New York: Academic Press

– 1977. 'Cities and the Hierarchy of Local Systems.' In G.W. Skinner, ed, *The City in Late Imperial China*. Stanford: Stanford University Press

Smailes, A.E. 1943. 'Ill-Balanced Communities – A Problem in Planning.' In E.A. Gutkind, ed, *Creative Demobilisation, Volume 2: Case Studies in National Planning*. London: Kegan Paul, Trench, Trubner

– 1944. 'The Urban Hierarchy in England and Wales.' *Geography*, 29:41–51

– 1947. 'The Analysis and Delimitation of Urban Fields.' *Geography*, 32:151–61

– 1966. *The Geography of Towns*. Fifth edition. London: Hutchinson

Smith, C.A. 1976. 'Causes and Consequences of Central-Place Types in Western Guatemala.' In C.A. Smith, ed, *Regional Analysis, Volume 1: Economic Systems*. New York: Academic Press

Smith, C.T. 1978. *An Historical Geography of Western Europe before 1800*. Revised edition. London: Longman

Smith, R.H.T. 1965a. 'Method and Purpose in Functional Town Classification.' *Annals, Association of American Geographers*, 55:539–48

– 1965b. 'The Functions of Australian Towns.' *Tijdschrift voor Economische en Sociale Geografie*, 56:81–92

Smith, W.R., Huh, W., and Demko, G.J. 1983. 'Population Concentration in an Urban System: Korea 1949–1980.' *Urban Geography*, 4:63–79

Sohns, R. 1978. 'Lösch and the Theory of Trade.' In R. Funck and J.B. Parr, eds, *The Analysis of Regional Structure: Essays in Honour of August Lösch*. London: Pion

Stabler, J.C., and Williams, P.R. 1973. 'The Changing Structure of the Central Place Hierarchy.' *Land Economics*, 49:454–8

Stanard, M.N. 1923. *Richmond: Its People and Its Story*. Philadelphia: J.B. Lippincott

Stanback, T.M., Jr, and Knight, R.V. 1970. *The Metropolitan Economy: The Process of Employment Expansion*. New York: Columbia University Press

Steigenga, W. 1955. 'A Comparative Analysis and a Classification of Netherlands Towns.' *Tijdschrift voor Economische en Sociale Geografie*, 46:105–19

Sternlieb, G., and Hughes, J.W., eds. 1975. *Post-Industrial America: Metropolitan Decline and Inter-Regional Job Shifts*. New Brunswick: Center for Urban Policy Research, Rutgers – State University of New Jersey

Stewart, C.T. 1958. 'The Size and Spacing of Cities.' *Geographical Review*, 48:222–45

Still, B. 1941. 'Patterns of Mid-Nineteenth Century Urbanization in the Middle West.' *Mississippi Valley Historical Review*, 28:187–206

Strong, H.M. 1937. 'Regions of Manufacturing Intensity in the United States.' *Annals, Association of American Geographers*, 27:23–43

Sviatlovsky, E.E., and Eells, W.C. 1937. 'The Centrographical Method and Regional Analysis.' *Geographical Review*, 27:240–54

Taeuber, K.E. and Taeuber, A.F. 1965. *Negroes in Cities: Residential Segregation and Neighborhood Change*. Chicago: Aldine Publishing

Takes, C.A.P. 1960. 'The Settlement Pattern in the Dutch Zuiderzee Reclamation Scheme.' *Tijdschrift van het Koninklijk Nederlandsch Aardrijkskundig Genootschap*, 77:347–53

Tarrant, J.R. 1967. *Retail Distribution in Eastern Yorkshire in Relation to Central Place Theory: A Methodological Study*. Occasional Papers in Geography No. 8. Kingston upon Hull: University of Hull Publications

– 1973. 'Comments on the Lösch Central Place System.' *Geographical Analysis*, 5:113–21

Taylor, G.R., ed. 1972. *The Turner Thesis Concerning the Role of the Frontier in American History*. Third edition. Lexington: D.C. Heath
Thiessen, A.H. 1911. 'Precipitation Averages for Large Areas.' *Monthly Weather Review*, 39:1082–4
Thompson, W.R. 1965. *A Preface to Urban Economics*. Baltimore: Johns Hopkins Press for Resources for the Future
Tiebout, C.M. 1962. *The Community Economic Base Study*. Supplementary Paper No. 16. New York: Committee for Economic Development
Toynbee, A. 1967. 'Cities in History.' In A. Toynbee, ed, *Cities of Destiny*. London: Thames and Hudson
Trabut-Cussac, J.-P. 1954. 'Bastides ou forteresses? Les bastides de l'Aquitaine anglaise et les intentions de leurs fondateurs.' *Le Moyen Age*, 60:81–135
Tress, R.C. 1938. 'Unemployment and the Diversification of Industry.' *Manchester School of Economic and Social Studies*, 9:140–52
Turner, F.J. 1920. *The Frontier in American History*. New York: Henry Holt
Ullman, E.L. 1941. 'A Theory of Location for Cities.' *American Journal of Sociology*, 46:853–64
– 1954. 'Amenities as a Factor in Regional Growth.' *Geographical Review*, 44: 119–32
– 1968. 'Minimum Requirements after a Decade: A Critique and an Appraisal.' *Economic Geography*, 44:364–9
Ullman, E.L., and Dacey, M.F. 1960. 'The Minimum Requirements Approach to the Urban Economic Base.' *Papers and Proceedings, Regional Science Association*, 6:175–94
US Bureau of the Census. 1967. *County and City Data Book, 1967*. Washington, DC: US Government Printing Office
– 1970. *Nineteenth Census of the United States. Population Characteristics*. Washington, DC: US Government Printing Office
– 1978. *County and City Data Book, 1977*. Washington, DC: US Government Printing Office
– 1983. *Statistical Abstract of the United States*. Washington, DC: US Government Printing Office
Van Cleef, E. 1937. *Trade Centers and Trade Routes*. New York: D. Appleton-Century
Van Hulten, M.H.M. 1969. 'Plan and Reality in the Ijsselmeerpolders.' *Tijdschrift voor Economische en Sociale Geografie*, 60:67–76
Vance, J.E., Jr. 1970. *The Merchant's World: The Geography of Wholesaling*. Englewood Cliffs: Prentice-Hall
Vining, D.R., Jr. 1974. 'On the Sources of Instability in the Rank-Size Rule: Some Simple Tests of Gibrat's Law.' *Geographical Analysis*, 6:313–29

Vining, D.R., and Kontuly, T. 1977. 'Increasing Returns to City Size in the Face of an Impending Decline in the Sizes of Large Cities: Which Is the Bogus Fact?' *Environment and Planning*, Series A, 9:59–62

Vining, R. 1955. 'A Description of Certain Spatial Aspects of an Economic System.' *Economic Development and Cultural Change*, 3:147–95

Wade, R.C. 1959. *The Urban Frontier: The Rise of Western Cities, 1790–1830*. Cambridge, Mass.: Harvard University Press

Wagner, W.B. 1974. 'An Empirical Test of Reilly's Law of Retail Gravitation.' *Growth and Change*, 5, no. 3:30–5

Walsh, M. 1981. *The American Frontier Revisited*. Atlantic Highlands: Humanities Press for the Economic History Society

Ward, D. 1971. *Cities and Immigrants: A Geography of Change in Nineteenth-Century America*. New York: Oxford University Press

Watkins, M.H. 1963. 'A Staple Theory of Economic Growth.' *Canadian Journal of Economics and Political Science*, 29:141–58

Webb, J.W. 1963. 'The Natural and Migrational Components of Population Changes in England and Wales, 1921–1931.' *Economic Geography*, 39:130–48

Webber, M.J. 1971. 'Empirical Verifiability of Classical Central Place Theory.' *Geographical Analysis*, 3:15–28

Weimer, A.M., and Hoyt, H. 1939. *Principles of Urban Real Estate*. New York: Ronald Press

Whebell, C.F.J. 1969. 'Corridors: A Theory of Urban Systems.' *Annals, Association of American Geographers*, 59:1–26

Wheeler, J.O. 1986. 'Corporate Spatial Links with Financial Institutions: The Role of the Metropolitan Hierarchy.' *Annals, Association of American Geographers*, 76:262–74

Wheeler, J.O., and Dillon, P.M. 1985. 'The Wealth of the Nation: Spatial Dimensions of US Metropolitan Commercial Banking, 1970–1980.' *Urban Geography*, 6:297–315

Whitelaw, J.S. 1962. 'The Measurement of Urban Influence in the Waikato.' *New Zealand Geographer*, 18:72–92

Whittlesey, D. 1954. 'The Regional Concept and the Regional Method.' In P.E. James and C.F. Jones, eds, *American Geography: Inventory and Prospect*. Syracuse: Syracuse University Press

Willett, B.M., et al. 1981. *The Geographical Digest 1981*. London: George Philip and Son

Williamson, J.G. 1965. 'Regional Inequality and the Process of National Development: A Description of the Patterns.' *Economic Development and Cultural Change*, 13:3–45

Woldenberg, M.J. 1968. 'Energy Flow and Spatial Order: Mixed Hexagonal Hierarchies of Central Places.' *Geographical Review*, 58:552–74
Wolfe, R.I. 1968. 'Economic Development.' In J. Warkentin, ed, *Canada: A Geographical Interpretation*. Toronto: Methuen
Wright, A.J. 1938. 'Manufacturing Districts of the United States.' *Economic Geography*, 14:195–200
Wright, A.L. 1956. 'The Genesis of the Multiplier Theory.' *Oxford Economic Papers*, 8:181–93
Wynn, G. 1981. *Timber Colony: A Historical Geography of Early Nineteenth Century New Brunswick*. Toronto: University of Toronto Press
Zelinsky, W. 1978. 'The Amenity Factor Revisited.' In J.D. Eyre, ed, *A Man for All Regions: The Contributions of Edward L. Ullman to Geography*. Studies in Geography No. 11. Chapel Hill: Department of Geography, University of North Carolina at Chapel Hill
Zipf, G.K. 1949. *Human Behavior and the Principle of Least Effort*. Cambridge, Mass.: Addison-Wesley Press

Index of cities

NOTE: Places with compound names (e.g. Miami-Hialeah, Kitchener-Waterloo) are indexed by full name only, even though a shorter form (e.g. Miami, Kitchener) may be used at some points in the text.

Subject index